EXS 92

Molecular Systematics and Evolution: Theory and Practice

Edited by R. DeSalle, G. Giribet and W. Wheeler

Birkhäuser Verlag
Basel · Boston · Berlin

Editors

Dr. Rob DeSalle
American Museum of Natural History
Division of Invertebrate Zoology
Central Park West at 79th Street
New York, NY 10024-5192
USA

Dr. Gonzalo Giribet
Harvard University
Department of Organismic and
Evolutionary Biology
16 Divinity Avenue
Cambridge, MA 02138
USA

Dr. Ward Wheeler
American Museum of Natural History
Division of Invertebrate Zoology
Central Park West at 79th Street
New York, NY 10024-5192
USA

Library of Congress Cataloging-in-Publication Data
Molecular systematics and evolution : theory and practice / edited by R. DeSalle, G. Giribet,
W. Wheeler.
    p. cm. -- (EXS ; 92)
    Includes bibliographical references and index.
    ISBN 3764365447 (alk. paper) -- ISBN 0-8176-6544-7 (alk. paper)
    1. Biology--Classification--Molecular aspects. 2. Molecular evolution. I. DeSalle, Rob.
II. Giribet, Gonzalo. III. Wheeler, Ward. VI. Series.
    QH83 M67 2001
    572.8'38--dc21   2001037878

Deutsche Bibliothek Cataloging-in-Publication Data
Molecular systematics and evolution: theory and practice / ed. by Rob DeSalle... - Basel;
Boston; Berlin: Birkhäuser, 2002
    (EXS ; 92)
    ISBN 3-7643-6544-7

ISBN 3-7643-6544-7 Birkhäuser Verlag, Basel - Boston - Berlin

© 2002 Birkhäuser Verlag, P.O. Box 133, CH-4010 Basel, Switzerland
Member of the BertelsmannSpringer Publishing Group
Printed on acid-free paper produced from chlorine-free pulp. TFC ∞
Printed in Germany
Cover illustration: Distribution of datasets in character-space. Nu DNA and amino acids sampled are
mapped onto the chromosomes of *Homo sapiens* (for details see p. 47)

ISBN 3-7643-6544-7

9 8 7 6 5 4 3 2 1

QH
83
M67
2002

http://www.birkhauser.ch

Contents

List of contributors

Richard H. Baker, The Galton Laboratory, University College London, 4 Stephenson Way, London NW1 2HE, UK; e-mail: richard.baker@ucl.ac.uk

Ranhy Bang, Graduate Training Program in Arthropod Systematics, Department of Entomology, Cornell University, Ithaca, NY 14853, USA; e-mail: hahaha@amnh.org

Andrew V.Z. Brower, Department of Entomology, Oregon State University, Corvallis, OR 97331, USA; e-mail: browera@bcc.orst.edu

Joel L. Cracraft, Department of Ornithology, American Museum of Natural History, Central Park West at 79th Street, New York, NY 10024-5192, USA; e-mail: jlc@amnh.org

Rob DeSalle, Division of Invertebrate Zoology, American Museum of Natural History, Central Park West at 79th Street, New York, NY 10024-5192, USA; e-mail: desalle@amnh.org

John Gatesy, Department of Molecular Biology, University of Wyoming, Laramie, WY 82071-3944, USA; e-mail: JohnGA@citrus.ucr.edu

Gonzalo Giribet, Department of Organismic and Evolutionary Biology, Museum of Comparative Zoology, Harvard University, 26 Oxford Street, Cambridge, MA 02138, USA; e-mail: ggiribet@oeb.harvard.edu

Paul Z. Goldstein, Division of Insects, Field Museum of Natural History, Roosevelt Road at Lake Shore Drive, Chicago, IL 60605, USA; e-mail: pgoldstein@fmnh.org

Cheryl Y. Hayashi, Department of Molecular Biology, University of Wyoming, Laramie, WY 82071-3944, USA. Current address: Department of Biology, University of California, Riverside, CA 92521, USA; e-mail: chayashi@citrus.ucr.edu

Daniel A. Janies, Division of Invertebrate Zoology, American Museum of Natural History, Central Park West at 79th Street, New York, NY 10024-5192, USA; e-mail: janies@amnh.org

Jyrki Muona, Zoological Museum, Division of Entomology, Finnish Museum of Natural History, P.O. Box 17, FIN-00014 Helsinki, University of Helsinki, Finland; e-mail: jyrki.muona@helsinki.fi

Michele K. Nishiguchi, Department of Biology, New Mexico State University, MSC 3AF, Box 30001, Las Cruces, NM 88003, USA; e-mail: nish@nmsu.edu

Patrick M. O'Grady, Division of Invertebrate Zoology, American Museum of Natural History, Central Park West at 79th Street, New York, NY 10024-5192, USA; e-mail: ogrady@amnh.org

Paul J. Planet, Department of Microbiology, Columbia University College of Physicians and Surgeons, 701 W 168th St., Rm 1518, New York, NY 10032, USA; e-mail: pjp23@columbia.edu

Andy Purvis, Department of Biology, Imperial College at Silwood Park, Ascot, Berkshire SL5 7PY, UK; e-mail: a.purvis@ic.ac.uk

Ted R. Schultz, Department of Entomology, MRC 188, National Museum of Natural History, Smithsonian Institution, Washington, DC 20560, USA; e-mail: schultz@onyx.si.edu

Scott E. Stanley, Department of Ornithology, American Museum of Natural History, Central Park West at 79th Street, New York, NY 10024-5192, USA; e-mail: scoob@amnh.org. Current address: Genaissance Pharmaceuticals, 5 Science Park, New Haven, CT 06511-1966, USA; e-mail: s.stanley@genaissance.com

Mike Steel, Biomathematics Research Centre, University of Canterbury, Private Bag 4800, Christchurch, New Zealand; e-mail: m.steel@math.canterbury.ac.nz

Joe Thornton, Columbia University Earth Institue, Columbia University #2430, New York NY 10027, USA; e-mail: jt121@columbia.edu

Alfried P. Vogler, Department of Entomology, The Natural History Museum, Cromwell Rd., London, SW7 5BD, UK; e-mail: apv@nhm.ac.uk

Ward C. Wheeler, Division of Invertebrate Zoology, American Museum of Natural History, Central Park West at 79th Street, New York, NY 10024-5192, USA; e-mail: wheeler@amnh.org

Michael F. Whiting, Department of Zoology, Brigham Young University, 574 WIDB, Provo, UT 84602, USA; e-mail: Michael_Whiting@byu.edu

General introduction

Systematics and evolution have seen a revolution of sorts over the last few decades because of the accessibility of the technology used to generate DNA sequence data. With the proliferation of genome-level sequencing, the field will continue to see advances and will more importantly, require continual overhaul with respect to the methods used to collect and analyze data. For instance, as questions are approached with unprecedented amounts of primary data, data analysis becomes more and more reliant on computer-based approaches. In addition techniques now mostly used in molecular biology, developmental biology and genomics will become important in how we observe the history of life.

We have divided the volume into three major sections that allow us to give examples of present progress in molecular systematics and evolution. The first section examines the utility of molecular information in systematics and evolution through several empirical examples. These chapters are meant to demonstrate how molecular data can be used in studies at different hierarchical levels. The second section looks at problems that have arisen as a result of the use of molecular data in evolutionary biology such as alignment problems, data combination, and the existence of multiple methods for analysis of data. The final section of the volume focuses on new problems and study systems that have been made available as a result of the availability of new techniques.

This volume is decidedly cladistic in approach and the reader will no doubt notice the large number of American Museum of Natural History (AMNH) associates involved in the author list. The editors of this volume used this author lineup for a number of reasons, the first being that we felt that this approach would allow us to present a cohesive and clear picture of our view of systematics. It is also more comfortable haranguing and annoying authors to meet their deadlines when they are associates and colleagues as opposed to complete strangers.

We hope that readers see the chapters of this volume as stimulating starting points for discussion of important subjects in systematics and evolution. While we do not expect all readers to agree completely or even in part with the ideas in this volume, we do hope that the chapters herein stimulate discussion and perhaps even focus research for some readers. We thank all AMNH associates in the Molecular Laboratories over the past 10 years for their hard work and for contributing to making the labs in the museum a stimulating atmosphere. In particular we thank Mike Novacek for his continued support of the work of the editors and authors of this volume.

Rob DeSalle, Gonzalo Giribet and Ward Wheeler

Part 1
Evolutionary analysis
at different levels

Introduction to part 1

Evolutionary analysis at different levels

The first part of this volume consists of six papers demonstrating the broad hierarchical expanse that molecular analyses have influenced. The section is organized so that each chapter leads to higher and higher taxonomic levels. In organizing the section this way we hope that the reader sees that different problems in data analysis and interpretation arise in comparison to the next level. The section is roughly organized so that the reader is exposed to population-level problems and species-boundary problems, then to problems of organizing species into genera, then to problems related to ordinal relationships and finally to higher-order hierarchies.

Brower leads off the section with a chapter on *Heliconius* butterflies and the problems relevant to understanding hierarchical relationships around the species boundary. O'Grady then examines the problem of species and species groups in a much-studied group, the Drosophilidae. Both of these chapters examine species-level phenomena and attempt to give the reader a framework for understanding hierarchy (or the lack of it) at and around this level. Stanley and Cracraft examine the higher-level hierarchical organization of bird taxa and present some interesting new inferences that can be made about this level of organization in birds. Gatesy takes this examination of phylogenetic organization a step further by examining levels of character support and assessments of character "quality" in establishing cladistic hierarchies in mammals. The final two chapters in this section examine higher-level relationships, one at the level of insect orders (Whiting) and the other at the level of metazoan phyla (Giribet). Both chapters indicate that great progress has been made on understanding the phylogeny of groups at these higher levels, and both point out shortcomings of the analyses and analytical methods that have been applied at these levels.

R. DeSalle, G. Giribet and W. Wheeler

Molecular Systematics and Evolution: Theory and Practice
ed. by R. DeSalle, G. Giribet and W. Wheeler
© 2002 Birkhäuser Verlag/Switzerland

Cladistics, populations and species in geographical space: the case of *Heliconius* butterflies

Andrew V.Z. Brower

Department of Entomology, Oregon State University, Corvallis, OR 97331, USA

Summary. The paradox of the species in evolutionary thought has promoted much debate and numerous incompatible definitions and concepts. This chapter argues that although the phylogenetic species concept (the author's version of it, at least) is no more accurate a description of "speciesness" than any other species concept (indeed, the notion of accuracy is irrelevant, as will be seen), it links species definition to species diagnosis via explicit criteria, which renders phylogenetic species more amenable to empirical testing than species defined by other concepts. The practical implications of cladistic species concepts for determining the boundaries between geographically differentiated sister taxa are explored using the example of *Heliconius*, based on my work and the recent studies of Mallet and others. The problem of circumscription is also addressed, with particular reference to the concepts of subspecies and geographical races.

What are species?

The conjecture of terminal super-organismal entities in systematics takes place prior to formation of a hypothesis of relationships among them. Every systematist who wishes to infer a cladogram either implicitly or explicitly invokes some notion of more inclusive groups when s/he selects representative exemplar organisms for study. Whatever additional criteria may be employed in their description/definition/diagnosis, these collective terminal entities represent hypotheses of species.

Some authors consider species to have a metaphysical existence independent of human knowledge, rendering our identifications of them correct or incorrect with respect to intrinsic criteria (e.g., the potential to interbreed or genealogical coalescence). The notion that there is an intrinsic defining attribute of speciesness is an essentialist view of the species problem. A good example of an essentialist species concept is the Biological Species Concept (BSC, Mayr, 1940), which relies on the criterion of potential to interbreed (whether or not interbreeding actually occurs), and makes untestable predictions about the trajectories of future evolution among differentiated but still potentially interbreeding populations. (Ironically, it is Mayr himself [e.g., 1982] who has most tirelessly campaigned against essentialism in species concepts, but he has done so merely to advance his own ontological assertions against what he deems to be the incorrect metaphysical claims of others.)

Empirical population biologists and systematists have moved away from essentialist views in recent years, rejecting that perspective as irrelevant to empirical research, and regarding species instead as hypotheses of grouping subject to corroboration or rejection by observation and data analysis. Both Mallet's Genealogical Cluster Species Concept (GCSC, Mallet, 1995; see below) and the Phylogenetic Species Concept (PSC, Eldredge and Cracraft, 1980; Nixon and Wheeler, 1990) are nominalist approaches, delimiting taxa on the basis of empirical evidence (including the observation of actual inter-breeding), rather than on unobservable criteria (for a table comparing various aspects of the GCSC, various versions of the PSC, and other concepts, see Goldstein and Brower, this volume). Species are thus groups hypothesized to exist, based on observed similarities, differences and interactions among individual organisms. Species are recognized because systematists explicitly identify material features that allow them to be distinguished. In this view, particular species do not have definitions separate from their diagnoses. The groups of organisms belonging to the category species exhibit a unique combination of character states that render them diagnosable as distinct (Nixon and Wheeler, 1990). To the extent that systematists discover and interpret characters, tabulate data matrices and infer cladograms, the resultant taxa are whatever the systematist (via rigorous analysis of the evidence) says they are.

As mentioned above and elsewhere (Brower, 2000a), the accusation of essentialism is a popular tactic to besmirch one's opponents' views in the species concept debate arena. Attacking the GCSC, Shaw (1996) argued that Mallet's (1995) "notion that 'good' species can be revealed to us by taxonomic authorities is steeped in the essentialistic outlook that Mallet (and others) seek to condemn." Mallet (1996) responded by accusing both the PSC and Baum and Shaw's (1995) Genealogical Species Concept of essentialist assumptions. But there is nothing intrinsically essentialistic about the judgement of systematic authorities—they are simply those researchers who have bothered to look more carefully at particular hypothetical taxa than other people, and are therefore able to muster better evidence to support their claims. If they are doing empirical science, then their evidence can be checked and the logic of their claims can be assessed by anyone who should care to do so. Essentialism only creeps in when belief-claims are made in the absence of, or in contradiction to the evidence. For additional discussion on the theory of species and comparisons among various species concepts, see the chapter by Goldstein and Brower (this volume).

Traditional species boundaries in *Heliconius*

The butterflies of the genus *Heliconius* represent a complex challenge to systematists for two reasons. First, they engage in interspecific Müllerian mimicry, so that members of separate species appear superficially very similar to one another, displaying virtually identical, brightly colored aposematic wing patterns. The older taxonomic literature of the group is replete with conflation of species that are quite distinct in features other than wing patterns, but additional mimetic relationships between more closely related taxa have only recently been revealed (Brower, 1996a).

Second, and more challenging from a theoretical perspective, is the fragmentation of "biological species" into parapatric or allopatric "geographical races." Current taxonomy recognizes aggregates of these geographically differentiated populations as single species under the BSC (Mayr, 1940). A single biological species may be composed of as many as 25 allopatric or parapatric, morphologically differentiated mimetic races (Sheppard et al., 1985). Some of the geographical races abut one another and freely hybridize in nature. For example, the "rayed" *H. erato emma* (yellow-banded FW, redrayed HW) and the "postman" patterned *H. erato favorinus* (red-banded FW, yellow striped HW) meet and cross freely in a narrow hybrid zone in the Huallaga Valley of eastern Peru (cf. Mallet and Barton, 1989). Phenotypes of both parental forms co-occur within the hybrid zone, as well as numerous F1, F2 and recombinant backcross forms (the genes responsible for specifying the wing patterns in *Heliconius* segregate into a dozen or more independent linkage groups, resulting in hundreds of phenotypic variants). The result of this uninhibited crossing between abutting races is that in the center of a hybrid zone, it is often difficult to find two specimens with the same wing pattern. Outside the zone, on the other hand, the "parental" races are monomorphic, apparently due to positive frequency-dependent selection by avian predators. Hybrid individuals and vagrant individuals from the other side of the contact zone are unable to invade the territory of a neighboring race because their novel wing patterns are not recognized as aposematic and thus suffer disproportionate mortality (Mallet and Barton, 1989).

Allopatric geographical races that each hybridize with an intervening race have been transitively linked in a Rassenkreis (polytypic species) by the BSC (if A = B and B = C then A = C), even though they never meet in nature. For example, *H. erato petiverana* from Central America and *H. erato cyrbia* from western Ecuador each hybridize with *H. erato venus* from western Colombia. And even races that are not connected by obvious transitive links of natural hybridization, such as the various *H. erato* representatives east *versus* west of the Andes, are considered conspecific under the BSC because they have been mated in captivity (Sheppard et al., 1985), thus exhibiting the "potential" to interbreed.

If two hypothetical species co-occur at a given place and time and are not observed to interbreed, the conclusion derived from the BSC that they are distinct is reasonable. But if allopatry is the major engine of speciation, as Mayr's writings (e.g., 1963) imply, then many sister taxa that do not encounter one another in nature will not exhibit the "non-dimensional" BSC condition of possession of intrinsic isolating mechanisms. The capacity to interbreed is a symplesiomorphy, a feature that may be retained long after lineages have developed other features that allow their differentiation and imply separate, divergent histories. Furthermore, the BSC's criterion of "actual or potential interbreeding" implies the existence of all-or-nothing barriers between species that is not supported by the patterns of natural diversity in many instances: for example, the "semispecies" *H. erato cyrbia* and *H. himera* meet and cross in nature, but the overlapping populations exhibit a significant deficit of hybrid individuals (Jiggins et al., 1996). Further, Mallet et al. (1998) report a number of museum records of what have been interpreted to be natural interspecific hybrids between ostensibly distinct *Heliconius* species (see Mallet's web site [http://abacus.gene.ucl.ac.uk/jim/Hyb/hybtab.html] for a growing list of such records with linked images). The observation of intrinsic isolating mechanisms (i.e., behaviors that discourage mating, or genetic/cytological incompatibilities that result in reduced fertility or viability of offspring) is thus a sufficient but not necessary criterion for recognition of species, at least if we are comfortable with species boundaries as they are currently circumscribed between such sympatric and normally non-hybridizing taxa as *H. melpomene* and *H. cydno*. If the parenthood of all of Mallet's hybrids is correctly determined, rigorously applying the BSC would reduce the number of *Heliconius* species from the currently-recognized 39 to 27.

Mallet's genealogical cluster species concept in *Heliconius*

Jim Mallet's extensive research on *Heliconius* hybrid zones has led him to endorse Darwin's opinion that species are arbitrarily recognized levels in a continually divergent evolutionary hierarchy (Mallet et al., 1998). He has promoted a new means to identify species boundaries, the GCSC, under which separate species are distinguished on the basis of statistically differentiable population genetic profiles: "we see two species rather than one if there are two identifiable genotypic clusters. These clusters are recognized by a deficit of intermediates, both at single loci (heterozygote deficits) and at multiple loci (strong correlations or disequilibria between loci that are divergent between clusters)" (Mallet, 1995:296).

Right away the discerning reader will see serious flaws with this approach. First of all, as Mallet has acknowledged, the GCSC is directly applicable only to species in sympatry or parapatry (a flaw it shares with the BSC). In spite of arguing that allopatric conspecific entities lack reality and cohesion, he nevertheless concluded, "if there is no evidence for separate species from sympatric overlap, closely related allopatric forms should mostly be considered conspecific" (Mallet, 1995:296). This is presumably because observed genetic differences that occur between allopatric populations might break down if the populations came into contact. It seems odd that Mallet is unwilling to recognize empirical evidence supporting the hypothesis of separate histories as a significant species-distinguishing factor, given his rejection of Mayrian essentialism (Mallet et al., 1998), but he apparently requires not only evidence of divergence, but the erection of irreversible barriers to future fusion between allopatric taxa that are to be recognized as distinct in his system.

The second problem lies in the vague and apparently arbitrary cutoff point at which genotypic clusters become identifiably discrete. Mallet suggested that alternate fixation of alleles between putative species is not necessary, but if not, how much genetic difference is necessary, and how are individuals with the "wrong" alleles to be interpreted? The specter of the notoriously subjective "75% rule" (Amadon, 1949) looms large. Third, Mallet's averaged hybrid indices imply uniform divergence at multiple loci between populations, but what if some loci exhibit strong disequilibria, but others do not? The limited evidence available suggest that this is a real concern: allozyme loci do not differ significantly in frequency among *Heliconius* races (Turner et al., 1979; Jiggins et al., 1997), while wing-pattern alleles obviously do. The failure of phenetic measures to account adequately for the complexities of mosaic evolution was pointed out long ago by Farris (1971), yet the GCSC founders on the same difficulty. See Brower (2000b) for further discussion of the GCSC in relation to *Heliconius* species delimitation.

The phylogenetic species concept in *Heliconius*

There are now a diversity of different concepts under the rubric of the PSC (see Goldstein and Brower, this volume), so it is important to specify which will be employed here. Nixon and Wheeler (1990:218) defined phylogenetic species as "the smallest aggregations of populations (sexual) or lineages (asexual) diagnosable by a unique combination of character states in comparable individuals (semaphoronts)." Nixon and Wheeler stated that constituent populations "must be historical entities and not defined as essential entities on the basis of potential interbreeding or what has been termed 'cohesiveness'," but as will be discussed further below, the historicity of hypothetical populations is not an *a priori* condition, but an inference based on observation of holomorphological similarity (or interbreeding, if it is observed within sexually reproducing taxa). By avoiding metaphysical claims and relying instead upon the aggregation of directly observed or parsimoniously inferred local populations, this version of the PSC is explicitly operational and empirical. Phylogenetic species are theoretical concepts based on well-corroborated initial hypotheses of unit populations and explicit analysis of evidence of their grouping into more inclusive entities.

Davis and Nixon (1992) developed a formal technique, Population Aggregation Analysis (PAA), to test the conspecificity of populations based on the criterion of differential fixation of inherited features. According to that method, features that are shared between populations are considered "traits," and do not constitute evidence of historical separation between the populations, while features that exhibit fixed alternate states represent "characters," which support the hypothesis of separateness. The discovery of one alternately fixed character is sufficient to support the hypothesis that the populations are specifically distinct. In population genetic terms, characters represent evidence that gene flow has been historically interrupted between the populations for a long enough period for alternate fixation of character states to be achieved via selection and/or genetic drift. Thus, even though alleles at some or even the majority of loci may be shared due to ancestral polymorphism or limited hybridization, a single fixed difference implies that for some period in the past, the populations have been separated. Of course, PAA does not guarantee the permanence of species into the future—patterns of gene flow may change and result in the erasure of formerly distinct taxa (e.g., *Anas superciliosa*—see Rhymer et al., 1994).

The procedure for PAA is as follows:

1. Identify individual organisms as representatives of specified local populations.
2. Identify attributes of individual organisms for comparison among populations.
3. Tabulate attributes of representative organisms within local populations to develop each population's attribute profile.
4. Compare attribute profiles among populations and divide individual attributes into characters (those with fixed differences between populations) and traits (those not so fixed).
5. Aggregate systems of populations that are not differentiated by characters into single phylogenetic species, and divide systems of populations that are differentiated by characters into separate phylogenetic species.

Brower (1999) noted some peculiarities of PAA in assessing molecular data, and proposed a complementary method, Cladistic Haplotype Analysis (CHA), for inferring the distinctness of phylogenetic species from DNA sequences. PAA can treat DNA sequences as single characters (PAA1), or as strings of linked characters (PAA2). PAA1 tends to over-resolve taxa in which sequences exhibit a lot of individual variation and is not particularly successful at rejecting hypotheses of specific distinctness. PAA2 assumes that characters segregate independently and dismisses homoplastic traits as irrelevant and of zero weight. However, it is evident that DNA sequences exist as strings of linked nucleotides, so that individual nucleotide "characters" can only be interpreted as supporting distinctness of populations if they are not contradicted by the weight of the other nucleotides in the sequence when interpreted in a parsimony framework. Brower (1999) presented the following stepwise procedure for CHA, modified from the PAA procedure outlined above.

1. Identify individual organisms as representatives of specified local populations.
2. Identify DNA sequences (haplotypes) of individual organisms for comparison.
3. Tabulate haplotypes of representative organisms within local populations to develop each population's haplotype profile.

4. Aggregate systems of populations that share identical haplotypes into single phylogenetic species (see also Doyle, 1995).
5. Conduct cladistic analysis of haplotypes to test for structure among remaining unaggregated populations.
6. Divide systems of populations that are topologically distinct into separate phylogenetic species.

CHA was described in the context of analyzing mitochondrial DNA sequences, but will perform equally well for any contiguous stretch of DNA as long as the gene region is not susceptible to frequent recombination, which would disrupt the hierarchical pattern of character change manifest in the sequences.

Brower (1999) performed PAA1, PAA2 and CHA on mtDNA sequence data for geographical races of *H. erato* as an empirical example, arriving at the same conclusions he reached based on traditional cladistic analysis of the same data (Brower, 1994). PAA1 supported some initial hypotheses of grouping but rejected others, implying no consistent patterns of grouping. PAA2 and CHA implied congruent patterns, both supporting two polytypic phylogenetic species composed of multiple parapatric races (*H. erato*, s.s., occupying the Amazon Basin, and *H. petiverana* [the oldest available name] in Central America + the Pacific slope of South America), and two monotypic phylogenetic species (*H. himera*, from western Ecuador, and *H. chestertonii*, from the upper Cauca Valley in Colombia).

Case study: geographical diversity in *Heliconius melpomene* and *H. cydno*

As an illustration of the PAA and CHA methods, an analysis is performed here for the *H. melpomene* group, using the mitochondrial DNA sequence data set published in Brower (1996b). The data set contains eight representatives of the *H. cydno* group[1], including a sequence from *H. timareta* (Brower and Egan, 1997). Three previously unpublished partial sequences of *H. melpomene mocoa* are also included. Figure 1 shows the variable positions in the sequences and the results of PAA1 and PAA2.

PAA1 (right column) identifies 32 individuals as unique, and five groups of multiple individuals. Only two of the groups hypothesized *a priori* (*H. melpomene nanna*, from southeastern Brazil and *H. m. thelxiopeia* from French Guiana) are corroborated actively by this approach. Seven hypothetical groups (races) are clustered into three polytypic "species" by shared identical sequences: *H. melpomene rosina* is not distinct from *H. melpomene melpomene* W[2], *H. melpomene amaryllis* is not distinct from *H. melpomene cognata* or *H. melpomene mocoa*, and *H. melpomene aglaope* is not distinct from *H.*

---

[1] Like *H. melpomene*, *H. cydno* is traditionally viewed as a polytypic biological species containing numerous geographical races. Some of these (*H. heurippa, H. pachinus, H. timareta, H. tristero*) have been elevated to specific status in the literature, but available evidence (Brower, 1996a) suggests that they are equivalent to races of *H. cydno* in their degree of divergence.

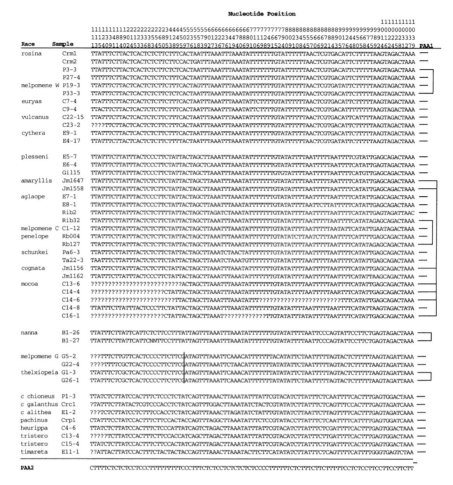

Figure 1. PAA1 and PAA2 analyses of mtDNA COI-COII sequences from geographical races of *Heliconius melpomene* and *H. cydno* and its relatives. See text for details.

melpomene penelope. The remaining groups (*H. melpomene euryas, H. m. vulcanus, H. m. cythera, H. m. plesseni, H. m. melpomene* C, *H. m. schunkei, H. m. melpomene* G, and the seven *H. cydno* groups) are trivially diagnosable by arbitrary lumping of individuals with unique haplotypes. Like the examples presented in Brower (1999), PAA1 in this case fails again to provide robust or consistent support for the hypothesized groups. The gene-as-a-single-attribute approach seems not to be particularly efficacious as a means for distinguishing species.

---

[2] Allopatric geographical races from Panama and northern Colombia, eastern Colombia and French Guiana bear a superficially similar phenotype and have been lumped under the nominate race, *H. melpomene melpomene*. The codes W ("western"), C ("Colombian") and G ("Guianan") differentiate representatives of these entities in Figure 1.

The PAA2 characters and traits are summarized on the bottom line of Figure 1, and characters that diagnose particular groups are boxed in the data matrix. Of 92 variable nucleotide sites in the original data, 21 represent characters uniting multiple individuals among groups hypothesized *a priori*. An additional 18 are autapomorphies that diagnose groups represented by a single sequence. Such characters represent only nominal corroboration of the distinctness of these "groups," given their minimal sample size. Seventeen other autapomorphies occur in one sequence of a group represented by more than a single individual, and represent traits. The remaining 36 sites are traits shared between multiple representatives of at least two hypothetical groups. The 12 groups diagnosed as phylogenetic species by PAA2 are *H. melpomene euryas*, (*H. melpomene rosina + melpomene* W + *vulcanus + cythera*), *H. melpomene melpomene* C, (*H. melpomene plesseni + amaryllis + aglaope + penelope + schunkei + cognata + bellula*), *H. melpomene nanna*, (*H. melpomene* G + *thelxiopeia*), *H. cydno galanthus*, *H. cydno alithea*, *H. pachinus*, *H. heurippa*, *H. timareta* and (*H. cydno chioneus + H. tristero*).

Figure 2 shows the results of CHA for the same data set. Groups identified as distinct are *H. melpomene euryas*, the other four Central American and western South American *melpomene* races comprising the "western group," the eight Amazonian *H. melpomene* races, *H. melpomene nanna*, the two Guianan *H. melpomene* races, the *H. cydno* group, and *H. tristero* (taxa represented by a single individual are also trivially distinct). Although PAA2 and CHA are generally in agreement, support for groups in CHA as implied by branch support (BS) values (Bremer, 1988, 1994) is generally greater than that implied by counting characters that support a given group in PAA2. For example, the western group (including *euryas*) received BS of five, but is supported by only two characters in Figure 1. *Heliconius tristero* is not supported by any characters in PAA2, but also received BS of five in CHA. The greater support for implied groups in CHA is due to its more inclusive treatment of the evidence, as discussed in Brower (1999).

Traditional interpretation of the CHA consensus tree implies the "paraphyly" of *H. cydno* with respect to *H. tristero* and the other splinter species, and the "paraphyly" of the *H. melpomene* group s. l. (the biological species, *H. melpomene*) with respect to the *H. cydno* group. Some phylogenetic species concepts demand "species monophyly," but Nixon and Wheeler's (1990) does not, and Davis and Nixon (1992) admonish their readers to avoid thinking of relationships within and between terminal sister taxa as represented by –phyly of any sort. In the first of these cases, the sampling is insufficient to draw a robust conclusion regarding relationships within the *H. cydno* group—most hypothetical terminals are represented by only a single sequence. The *melpomene-cydno* problem is clearer: the mtDNA data imply (in either PAA2 or CHA) that the traditional biological species *H. melpomene* represents at least four phylogenetic species, corresponding to discrete geographical regions, as identified by the black bars in Figure 2. *H. melpomene euryas* is also distinct, but suffers the minimal sample problem like the members of the *cydno* group.

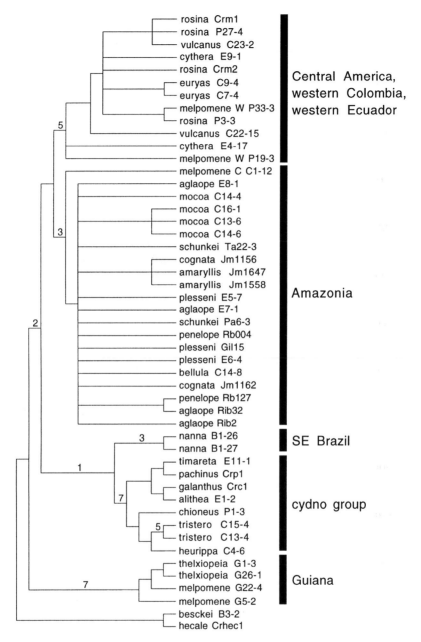

Figure 2. CHA analysis of mtDNA COI-COII sequences from geographical races of *Heliconius melpomene* and *H. cydno* and its relatives (rooted with two silvaniform *Heliconius* species). The tree is a strict consensus of 4654 equally parsimonious cladograms (L = 202, CIx = 0.580, RI = 0.856), based on an equal-weighted heuristic analysis of unique sequences (multiple individuals with identical sequences were excluded from the analysis and reattached to their sister groups after tree inference). Branch support for selected internal nodes is indicated. Calculations were performed in PAUP 3.1.1.

Taxa and names

Systematics, like most other branches of science, relies upon reification of abstract groups of similar phenomena about which predictive generalizations may be drawn. Thus, taxa are diagnosed by the shared possession of particular character states among their members. The names of taxa represent elements of a common language used by systematists and other biologists for groups about which they care to communicate. For animal taxa, the International Code of Zoological Nomenclature (ICZN, 1999) offers a formal structure that attempts to preserve the universality and continuity of this language, based on the conservative principle of priority and the acceptance of the organizational convention of the Linnean taxonomic hierarchy.

An increasingly popular belief among phylogenetic essentialists (e.g., de Queiroz and Gauthier, 1992) is that Linnean nomenclature is inadequate to provide labels for the various parts of the hierarchical pattern of groups nested within groups into which modern systematists organize the diversity of living and extinct organisms. It is clear, however, that if we adopt a phylogenetic model as the ontological foundation for our classifications, then the logical basis for recognizing discrete groups is undermined, and taxa become only arbitrary clusters of organisms rendered "discrete" by incomplete sampling of the evolutionary continuum, rather than coherent entities that are distinct from other such entities (Rieppel, 1988). If they are interested in producing predictive taxonomy, systematists should base their inferences on evidence rather than on metaphysics, and base their classifications on empirically supported clades, not on woolly, circular arguments about common ancestry and individuality.

However, even when metaphysical concerns are dismissed, the species-group represents a particularly difficult taxonomic problem, since species are not clades, and may be diagnosable only by the complementary absence of features that when present diagnose their sister species. The object of Davis and Nixon's (1992) and Brower's (1999) methods is to identify the least inclusive assemblages of individual organisms that exhibit fixed character differences, which imply historical separation. Under the PSC, these are species, the terminal taxa. Subspecies, geographical races, forms, varieties etc., have no formal taxonomic legitimacy as taxa separate from the species group, and such traditionally recognized entities should be tested for distinctness and raised, or sunk, accordingly.

Acknowledgements
Thanks to Darlene Judd and Gareth Nelson for discussion of concepts, even if we don't always agree. Thanks to Rob DeSalle and one anonymous reviewer for critiques that helped sharpen the manuscript. This work was supported by the H. and L. Rice Endowment for Systematic Entomology, Oregon State University.

Molecular Systematics and Evolution: Theory and Practice
ed. by R. DeSalle, G. Giribet and W. Wheeler
© 2002 Birkhäuser Verlag/Switzerland

Species to genera: phylogenetic inference in the Hawaiian Drosophilidae

Patrick M. O'Grady

Division of Invertebrate Zoology, American Museum of Natural History, New York, NY 10024, USA

Summary. Systematic studies at any taxonomic level require careful planning, even before genes are sequenced or morphological characters scored. Molecular systematists working at the level between species and genera must select gene regions which are suitable for the divergence within the group being examined and decide how many and which ingroup and outgroup taxa to sample in the analysis. This chapter will use studies of the Hawaiian Drosophilidae to illustrate strategies useful in (1) selecting nucleotide sequences for divergences between the levels of species and genera, (2) designing ingroup and outgroup taxon sampling schemes, and (3) performing phylogenetic analysis on data sets with large numbers of taxa and characters.

The Hawaiian Drosophilidae

The Hawaiian Drosophilidae have long been recognized as the premier example of adaptive radiation and rapid speciation in nature (Carson, 1987). Over 500 named species are currently known and an equal number remain to be described from this clade (Hardy and Kaneshiro, 1981; Kaneshiro et al., 1995). This high species diversity is accompanied by an impressive array of morphological variation. Most species have spectacular sexually dimorphic modifications of their mouthparts, wings and/or forelegs. Molecular clock estimates suggest that this group colonized the Hawaiian Archipelago approximately 25 million years ago (Thomas and Hunt, 1991; DeSalle, 1992). It is amazing to consider that one thousand species of *Drosophila*, approximately one-quarter of the world's diversity of this family, could have evolved on only 0.01% of the earth's land area in such a relatively short period of time (Carson et al., 1970). Extreme isolation, as a result of the rugged geography of this island chain, as well as the many habitats and microhabitats present within a small region, is no doubt partially responsible for this diversity. Within a few miles one can go from the Alakai Swamp, which receives over 460 inches of rain a year, to desert-like conditions where annual rainfall seldom rises above 10 inches. This impressive geographic variation is further segregated by the high degree of host plant specificity displayed by most Hawaiian drosophilids (Heed, 1968, 1971; Carson, 1971; Montgomery, 1975) which only serves to accentuate the isolation between species. The majority of species in this clade are not only restricted to a single host plant, but a single part of that plant (Craddock and Kambysellis, 1997). Ecological habitats within the Hawaiian Drosophilidae

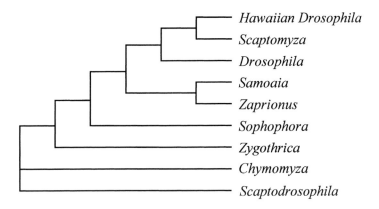

Figure 1. Phylogenetic hypothesis for the relationships of nine drosophilid taxa based on molecular data.

range from decaying flowers, leaves, stems and bark of host plants to fungi and spider egg sacs (Heed, 1968).

In spite of extensive study, the nearest continental relative of the endemic Hawaiian fauna also remains enigmatic (reviewed in DeSalle and Grimaldi, 1991, 1992; Powell and DeSalle, 1995; Powell, 1997). Morphological (Throckmorton, 1966, 1975; Carson and Kaneshiro, 1976), behavioral (Speith, 1982; Kaneshiro and Boake, 1987), and molecular (Beverly and Wislon, 1985; Thomas and Hunt, 1991; DeSalle, 1992; Russo et al., 1995; Baker and DeSalle, 1997; Remsen and DeSalle, 1998) studies suggest that a member of the subgenus *Drosophila* is closest to the Hawaiian Drosophilidae, although which clade is the exact sister group is unclear (Fig. 1). Stalker (1972), based on similarity in polytene chromosome banding patterns, proposed that the *Drosophila robusta* species group was the sister clade of the Hawaiian *Drosophila*. Grimaldi's (1990) morphological phylogeny was drastically different from all previous and subsequent work. It placed the Hawaiian *Drosophila* closest to a clade of mycophagous Neotropical genera, the *Zygothrica* genus group, and all *Scaptomyza* as sister taxa to the subgenus *Drosophila* (Fig. 2). Under this hypothesis, the Hawaiian Drosophilidae are not monophyletic and the genus *Scaptomyza* is actually more closely related to the subgenus *Drosophila* than to the Hawaiian *Drosophila* (Grimaldi, 1990). However, these studies are not without problems, all either did not use rigorous cladistic methodology (e.g., Throckmorton, 1966, 1975; Grimaldi, 1990) or did not include multiple representatives of the genus *Scaptomyza* (e.g., Thomas and Hunt, 1991; DeSalle, 1992). In order to fully test previously hypothesized relationships, both within the Hawaiian Drosophilidae and between this clade and its nearest relative(s), more taxa must be sampled.

Within the Hawaiian Drosophilidae, both systematic relationships and taxonomic ranks have been unsettled. Hardy (1965) described 400 species of Hawaiian Drosophilidae, placed in nine genera. Throckmorton (1966) recog-

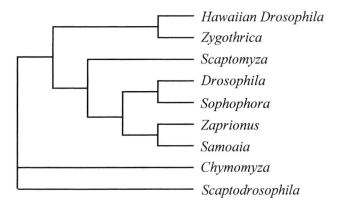

Figure 2. Phylogenetic hypothesis for the relationships of nine drosophilid taxa based on morphological data.

nized two main lineages, the "drosophiloids," which included the genera *Antopocerus*, *Ateledrosophila*, *Drosophila*, *Idiomyia*, and *Nudidrosophila*, and the "scaptoids," which were comprised of the genera *Celidosoma*, *Grimshawomyia*, *Scaptomyza*, and *Titanochaeta* (Fig. 3). In his revision of the Hawaiian Drosophilidae, Kaneshiro (1976) sank the "drosophiloid" genera into the genus *Drosophila* and referred to this clade as the Hawaiian *Drosophila*. Several poorly known "scaptoid" genera remain to be examined, although it is likely that these clades will also be sunk to the level of species groups within the Hawaiian *Scaptomyza* (Kaneshiro, personal communication). Grimaldi (1990) also examined relationships within the Hawaiian

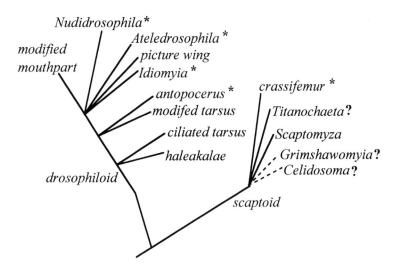

Figure 3. Relationships of Hawaiian scaptoids and drosophiloids from Throckmorton (1966).

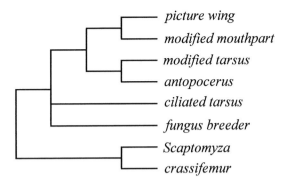

Figure 4. Relationships of the major species groups of Hawaiian Drosophilidae based on molecular work of Kambysellis et al., 1995; Baker and DeSalle, 1997.

Drosophilidae. In addition to finding that this clade was not monophyletic (Fig. 2), his study also suggests that many of the recognized species groups of Hawaiian *Drosophila* are also not monophyletic (Grimaldi, 1990). Such results have never been suggested with either morphological or molecular data. Two recent molecular studies (Kambysellis et al., 1995; Baker and DeSalle, 1997) have resolved phylogenetic relationships within some lineages of the Hawaiian *Drosophila,* but many groups recognized by taxonomists remain unsampled (Fig. 4).

The Hawaiian Drosophilidae present an extreme situation in phylogenetic reconstruction. This group has achieved great diversity, in terms of both species numbers and morphological characters, in a relatively short period of time. This implies that many of the internal branches in the Hawaiian Drosophilidae will be extremely short, with few synapomorphies present at each locus to support many clades. This paper considers some factors critical to the design and execution of a phylogenetic study, particularly one of a large, diverse and rapidly evolving group like the Hawaiian Drosophilidae.

Pilot studies in experimental design

Pilot studies with a reduced set of representative taxa are usually advocated to help select the appropriate sequence(s) for a study and to decide on a taxon sampling strategy prior to making a large-scale commitment of time and resources (e.g., Baverstock and Moritz, 1990, 1996). Such studies are invaluable and should be undertaken, especially when information about the taxonomy, morphology and evolutionary time-scale of the group in question is ambiguous or lacking. A pilot study must assess the effectiveness of a given gene sequence in estimating phylogenetic relationships and the influence of a certain taxon sampling scheme on ingroup structure and stability simultaneously. Often, one factor will have a reciprocal effect on the other. For example, if one uses a rapidly evolving mitochondrial locus to infer genus-level relationships, the taxon sampling scheme must, necessarily, be very dense in order to reduce long branches on the tree and offset the effect of convergent changes on the estimation of that tree's topology. Conversely, if one is using a slowly evolving locus to infer relationships among closely related taxa within a species subgroup, sampling can be much less dense. Some key issues involved in pilot study design, particularly sequence selection and taxonomic sampling, and data analysis are discussed below.

Sequence selection

Selecting which sequences to invest the time and resources in for molecular systematics is part of what Brower and DeSalle (1994) refer to as "stage 1 sieving." This is typically done by surveying the literature and selecting sequences which have been previously studied in clades of similar age and/or taxonomic diversity as the desired ingroup. The candidate sequences are then used in a pilot study to assess whether they are phylogenetically useful for the question being addressed. Although recent technological advances in genomics and automated nucleotide sequencing, particularly the use of high-throughput techniques and multiplex primers (Whiting, this volume), have made the generation of characters for molecular systematics much more efficient, information about a sequence and its performance in phylogenetic analysis, based on previous analyses and pilot studies, is still critical to the success of systematic studies. Several mitochondrial (Simon et al., 1994) and nuclear (Brower and DeSalle, 1994) loci used in insect systematics have been evaluated. Some additional characteristics to consider when selecting a locus include genomic copy number and ease of amplification, gene size and structure, nucleotide composition and pairwise sequence divergence.

   Genomic copy number can be both a benefit and a detriment to a systematic study. Mitochondrial and nuclear ribosomal sequences are often selected because their high genomic copy number makes them easier to amplify. These sequences are so effective because they are usually homogeneous from copy to copy within the same species, making them excellent markers for systematics. However, high within-species and within-individual variability, which may confound systematic studies, has been observed in both mitochondria (Moritz et al., 1987; Harrison, 1989) and ribosomal loci (Campbell et al., 1997) for some groups. Zhang and Hewitt (1996) also review several cases where mitochondrial sequences have been transferred to the nucleus. Such a "recent nuclear" sequence may be amplified and sequenced in a molecular systematic study, even though the target was the mitochondrial copy. This can lead to a systematic error because, in essence, paralogous sequences are being used to infer phylogeny. Nuclear loci belonging to multi-gene families can pose an analogous problem to phylogeny reconstruction. Care should be taken to identify all potential amplification artifacts when using a multicopy gene or a member of a gene family in molecular systematics, even if this was not a problem in previous studies.

The size, structure and composition of a target locus are also important. Longer sequences will have more characters for phylogenetic studies. When selecting a locus, it is important to consider the difference in time and resources between amplifying and sequencing a single 1000 base pair (bp) fragment *versus* four 250 bp ones. Major structural components of genes, such as introns and repeat regions, can also influence their utility as systematic markers. For example, a pair of primers designed to amplify across a large intron region may not be optimal if, because of high divergence, that intron is difficult to align. Even though some characters will be useful, many may be excluded from the analysis, reducing the efficiency of the experiment. Likewise, duplicated regions within a gene can also cause amplification and sequencing problems. Not only is the homology of these regions often difficult to determine, their amplification can also pose technical problems for amplification and sequencing. Finally, genes which display a major nucleotide bias may create problems for phylogeny reconstruction algorithms by increasing the degree of homoplasy per site (Lockhart et al., 1994).

The rate of evolution of a sequence is, arguably, the most important factor to consider when planning a study. Sequences selected for molecular systematic analyses would ideally be evolving rapidly enough to provide strong support at all nodes within the tree, but not so rapidly that alignment, and therefore homology statements, are uncertain. Those loci which evolve too slowly and lack variablilty, as well as those loci which are too variable to confidently align within a specified group of taxa, are useless to the systematist. Friedlander et al. (1994) suggested a conservative window between 0.2–0.3 percent pairwise divergence to reduce the chance that multiple substitutions at a single site would lead tree construction algorithms astray. Divergences below this level would yield very few characters, even in sequences hundreds of bases long. Above this point, saturation becomes evident when divergence is plotted against time (or ti/tv ratio) for several loci. However, since pairwise divergences are unable to reveal all substitutions in the manner of a phylogenetic tree, this range may be too conservative. In fact, Friedlander et al. (1994) found that, in several cases, third codon positions were able to recover the test phylogeny, even though they were saturated. Several other studies have noted that saturated sites also perform quite well (Bjorklund, 1999). This may be because, although the entire sequence is saturated, there are more characters which track the historical branching events than homoplastic characters.

Many authors have noted that a single locus is seldom able to resolve all nodes in a phylogeny with strong support (Hillis, 1987). Empirical and theoretical studies suggest that sampling multiple genes and using them in concert may be the most effective way to obtain highly resolved and strongly supported phylogenetic hypotheses. Cummings et al. (1995) used whole-genome sequences of mitchondrial DNA to demonstrate that single genes were poor estimators of the genome tree and that the data converged on the genome tree as more characters were added to the analysis. In fact, several authors have argued that increasing the number of characters per taxon in systematic studies will increase phylogenetic accuracy and support (e.g., Lecointre, 1993, 1994; Hillis et al., 1994). However, Graybeal (1998) points out that, under certain circumstances, indiscriminately adding characters can actually decrease phylogenetic accuracy.

Taxonomic sampling

Outgroup selection

Ingroup topology can be highly dependent upon the outgroup choice for an analysis. Outgroups are used to polarize ingroup characters by estimating the ancestral state of each character (Lundberg, 1972; Watrous and Wheeler, 1981; Maddison et al., 1984). When the outgroup chosen is the exclusive sister group (i.e., the closest relative to the ingroup without actually being inside the ingroup), rooting can be quite robust. However, when the outgroup is highly divergent it may actually represent a random collection of character states which have little or no value in polarizing characters or elucidating ingroup relationships (Wheeler, 1990). Random outgroups can be a particular problem with nucleotide sequences because there are only four character states possible at each position. Given a random sequence there is a 25% chance that any given character in the outgroup will match the corresponding character in the ingroup simply by chance alone (Wheeler, 1990; Maddison et al., 1992). The use of multiple outgroups is also critical to inferring relationships in the ingroup because they more rigorously test ingroup relationships and can yield a more robust estimation of ancestral states (Maddison et al., 1984).

It is clear that, in order to determine the sister clade of the Hawaiian Drosophilidae with any degree of reliability, outgroup sampling must be quite extensive. Several candidate sister taxa, including members of the subgenus *Drosophila* and the *Zygothrica* genus complex, can be selected on the basis of previous studies (Throckmorton, 1975; Grimaldi, 1990). The genus *Scaptomyza*, which may or may not be a clade within the Hawaiian Drosophilidae (Grimaldi, 1990), should also be adequately represented.

Ingroup sampling

It is seldom possible, or even necessary, to sample every single species within a given clade. For this reason, a small number of taxa are selected to represent the diversity within the ingroup. It is clear, however, that the taxa sampled may greatly influence the resultant tree topology. Furthermore, failing to sample critical ingroup taxa may introduce long branches which can lead to "positively misleading" phylogenetic results (Felsenstein, 1978).

The problem of which and how many taxa to sample in a study is especially difficult when you consider a group with the size and diversity of the Hawaiian Drosophilidae. Clearly, taxonomic rank and relative placement of the ingroup clades, as well as the number of species in a group, are all important factors to consider in the case of the Hawaiian Drosophilidae, even though these factors are somewhat subjective and not rigorous in any cladistic sense. While taxonomic level is not necessarily correlated with phylogenetic distance, it does serve as a useful heuristic tool in experimental design, especially when no other *a priori* knowledge of relationships is available. Any well-designed pilot study would consider this taxonomic information, as well as information from previous cladistic analyses (if available), and attempt to include as many representatives of the major lineages as possible.

For example, roughly nine major lineages (i.e., species group or higher in rank) are currently recognized in the Hawaiian *Drosophila*. In designing a pilot study, one would try to include all of these simply in an attempt to fully sample the ingroup. How many taxa to sample from each genus in a pilot study is a matter of convenience rather than methodology. Including at least two divergent members of each genus sampled tests the monophyly of the major ingroup clades and spreads the changes more evenly across the branches of the tree. Once an initial tree is constructed, additional taxa can be added to increase sampling at critical nodes (e.g., to reduce long branches). Sampling within large diverse groups, such as the *picture wing* species group, is more important than obtaining both representatives of a small group, like the *ateledrosophila* species group.

Two recent studies have examined relationships within the Hawaiian Drosophilidae. Kambysellis et al. (1995) used yolk protein genes (YP1 and YP2) to examine the phylogenetic relationships among 46 species of Drosophilidae. Thirty-six species were in the picture wing clade, six were placed in other endemic Hawaiian groups, and four were from continental taxa. This tree shows that the picture wing clade, which consists of the *adiastola, planitibia, primaeva,* and *grimshawi* species groups, is the monophyletic sister group of the *modified mouthpart* species group. The *antopocerus* species group is the most basal clade of Hawaiian *Drosophila* in this tree. The genus *Scaptomyza*, which is thought to have originated on Hawai'i (Throckmorton, 1966), is sister to the Hawaiian *Drosophila*. These relationships are all congruent with those proposed by Throckmorton (1966) based on morphology. However, they say very little about the relationships within the Hawaiian Drosophilidae as a whole. Many clades, both within the drosophiloid and scaptoid lineages were not sampled. Ideally, these would have to be examined to (1) place some of the enigmatic taxa and (2) test some of the statements of monophyly made in previous non-cladistic studies and brought into question by Grimaldi's recent work. The dense sampling within the picture wing group, while excellent for determining relationships within this clade and tracing the evolution of host plant usage, does not allow us to make any statements about evolution across the entire Hawaiian Drosophilidae.

Baker and DeSalle (1997) took a slightly different approach in their examination of Hawaiian Drosophilidae phylogeny. They used a variety of genes (hunchback, wingless, acetylcholinesterase, alcohol dehydrogenase, cytochrome oxidase II, 16S, NADH dehydrogenase 1 and cytochrome oxidase III) to reconstruct their phylogenetic trees. Their sampling strategy was quite different from that of Kambysellis et al. (1995), however. Baker and DeSalle (1997) included representatives from several major clades. They examined five *picture wing* species and only two taxa each from the *modified mouthpart*, *antopocerus*, *modified tarsus* and *haleakalae* species groups. Baker and DeSalle (1997) also included two Hawaiian outgroups from the *Scaptomyza* clade and two continental outgroups. This sampling strategy produced an "outline" of phylogenetic relationships among the major Hawaiian Drosophilidae clades but, since it did not look at any group in great detail, said little about relationships within those clades. These three indicate that the *picture wing* species are the sister group of the *modified mouthpart* species group. This clade is the sister group of a "leaf breeder" clade consisting of the *modified tarsus* and *antopocerus* species groups. The *haleakalae* species group is at the base of the Hawaiian *Drosophila* clade. As Throckmorton (1975) had suggested, the Hawaiian *Drosophila* and *Scaptomyza* lineages are sister groups. These results were congruent with those based on morphology (Throckmorton, 1975) and the yolk protein genes (Kambysellis et al., 1995).

Both of the above studies were able to reconstruct phylogenetic relationships within the Hawaiian Drosophilidae which were congruent with one another and previous hypotheses based on morphology, in spite of their being quite different with respect to sequence selection and sampling strategy. Each study suffers from the fact that several clades of Hawaiian Drosophilidae were not sampled. This can be corrected by either collecting and including these other taxa in future work or by using supertree building algorithms to combine previously published phylogenetic studies into one tree (see below).

Adding taxa *versus* adding characters

When designing a systematic study, one must consider the trade-off, in both time and resources, between increasing taxon sampling and increasing the number of characters for a smaller subset of ingroup taxa. It is not clear which approach, denser taxon sampling or greater character sampling, has a greater effect on the reconstruction of phylogenetic relationships. Several papers, many of which employ numerical simulations, have discussed the ramifications of adding characters *versus* adding taxa (Hillis, 1996; Kim, 1996; Graybeal, 1998; Hillis, 1998; Poe and Swofford, 1999). These studies typically measure the accuracy or inconsistency of phylogenetic reconstruction methods in light of increasing taxa or characters.

Notions of accuracy or consistency are, to some systematists, odd to consider with respect to phylogeny because the exact historical relationships can never be "known." However, within the framework of simulation studies, which model the performance of different phylogeny reconstruction algorithms given a set of variables, or experimental phylogenies, which have been manipulated by the investigator and have a set branching pattern, accuracy and consistency are measurable variables (Hillis et al., 1994, 1996). Methods are considered to be consistent when they converge on the correct phylogeny with the addition of infinite data. Efficient methods are able to converge on the correct tree with less data. Inconsistency becomes a problem when the assumptions of a given method are violated sufficiently such that they will never converge on the correct tree, even with the addition of infinite data. The so-called Felsenstein Zone, where long branches cause parsimony to be positively misleading (Felsenstein, 1978), is an example of such a case. Accuracy can be thought of as how many nodes of a known phylogeny are reconstructed under a given model.

Often, the results of such studies seem to be in opposition to one another. Kim (1996), for example, concluded that "it seems best to use fewer taxa because larger trees are more likely to contain inconsistent branches," while Hillis (1996) stated that "including large numbers of taxa in an analysis may be the best way to ensure phylogenetic accuracy." Hillis (1998) points out that such apparent contradiction is often the result of different investigators examining different variables.

Of those investigators who favor the addition of taxa over the addition of characters, all agree that taxa should be added judiciously, with a sampling scheme in mind. Hillis (1996) argued that adding taxa disperses noise in the data set. Increasing numbers of taxa reduce covarying patterns of homoplasy, giving a more accurate reconstruction of character states. Graybeal (1998) argued that the relative placement of each taxon was very important. She suggested that taxa be added to reduce long branches, easing problems that some phylogeny reconstruction methods have with multiple changes.

Several systematists have suggested that adding taxa can actually hinder phylogenetic reconstruction. Kim (1996) concluded that problems with inconsistency can become more acute with the addition of taxa, especially when the rate of change along branches is high. He also pointed out that trees with equal length branches may be just as susceptible to inconsistency issues as those with long branches. Kim (1996) concluded that fewer taxa with lower rates of change are better if one wants to avoid inconsistency in phylogenetic estimation and, if taxa must be added, they should be placed in such a way so they counter long branches, preferably close to the common ancestor of the clade. A recent study by Poe and Swofford (1999) agrees with Kim's (1996) results. While long branch subdivision did increase accuracy when simulating trees in the so-called Felsenstein Zone (see Felsenstein, 1978), this approach was not effective for other long branch situations. Their simulations indicate that, in some instances, adding taxa to break up two branches thought to be subject to long branch attraction, can actually increase problems of inconsistency. Long branch subdivision can actually work to the benefit and detriment of the systematist. According to Poe and Swofford (1999), "convergent changes may cause spurious attraction between lineages, but they may also help parsimony recover relationships correctly by preventing long branches from being drawn from their proper places by other long branches." They argue that subdividing such a long branch can create an imbalance of homoplasy that results in a misleading reconstruction of relationships.

The different conclusions and recommendations from the above simulations are dependent on where taxa are added to the tree. Graybeal (1998) added taxa specifically to subdivide long branches while Kim (1996) added taxa at random. Such differences can determine whether a given taxon sampling strategy can be effective or not. The prudent addition of taxa, particularly in those areas of the tree with long branches, seems to have a beneficial effect on phylogeny reconstruction. Poe and Swofford (1999) caution, however, that problems may actually be generated, rather than alleviated, by long branch subdivision and very dense taxon sampling may be required for accuracy.

Clearly, the exact relationship between adding taxa or characters remains somewhat unclear and further studies, in the form of additional simulations as well as empirical data, are needed to elucidate this relationship. Most authors agree, however, that when taxa are added to a phylogeny, they should be added judiciously, just as is the case when additional characters are selected.

Analysis of large data sets

The generation of multiple nucleotide sequences for a large number of taxa presents certain unique theoretical and empirical challenges in terms of data management and analysis. New techniques, such as high-throughput DNA sequencing (Whiting, this volume), will rapidly generate many nucleotide sequences for many taxa. One concern is whether it is appropriate to analyze such sequences in combined analysis or, instead, if it is better to keep different data sets separate. Aspects of combined *versus* individual analysis of nucleotide sequences are discussed elsewhere in the literature (Kluge, 1989; Bull et al., 1993; Miyamoto and Fitch, 1995; de Queiroz et al., 1995; Nixon and Carpenter, 1996), so they will not be addressed here. This section will review some of the analytical methods employed in large analyses, particularly those with complex parameter spaces.

Phylogenetic problems which either deal with large numbers of taxa or large numbers of equally parsimonious trees are difficult to solve. Three strategies have been proposed to deal with these situations: (1) reduce the fraction of search space which needs to be examined by using heuristic search algorithms, (2) reduce the computational complexity of the problem by employing a less combinatorily difficult algorithm (e.g., neighbor joining), or (3) reduce the size of the problem by using exemplar or placeholder taxa (Rice et al., 1997). Although the use of computationally simpler algorithms will not be addressed here, the use of heuristic search algorithms and placeholder taxa will be discussed below. Once the most parsimonious trees are found, systematists often will perform successive approximations character weighting to choose between multiple solutions. Some cautionary comments dealing with the use of this method are outlined below.

Heuristic search strategies

Any search strategy must strike a balance between thoroughness and speed, especially when examining a large data set which requires the use of heuristic search algorithms. Maddison (1990) pointed out that exploring the entire possible "tree space" is important, because tree space is often complex and numerous islands of equally parsimonious trees may be present. Maddison (1990) defined an island as "a collection of trees, all less than a specified length, each tree connected to every other tree in the island through a series of trees, and each one differing from the next by a single rearrangement of branches." Although trees on the same island differ by only a few branch swaps, those on different islands can be quite different from one another. Results of any study which does not explore tree space thoroughly may be misleading, showing relationships which are fully resolved, but not supported by all the shortest length trees. Running multiple heuristic searches which use a random addition algorithm to generate starting trees is an effective way to map tree space. Output from PAUP* (Swofford, 2000) indicates when multiple islands are or may be present.

Handling a large number of equally parsimonious trees is also problematic. Computer processor speed and memory requirements can become challenged when the number of trees becomes large. Swapping on such a large number of trees can take a prohibitively long time and, depending on which tree you begin with, the next shortest tree may never be reached. One method that has been proposed to circumvent large "plateaus" of trees is channeling (Maddison and Maddison, personal communication). In this technique many searches are performed which only save and swap on a limited number (e.g., 10, 100) of trees longer than a specified length. This makes it possible to rapidly explore more of tree space and possibly discover shorter trees without using extensive (or infinite) computer time. Once a shorter tree is discovered, the algorithm may begin to save a large number of trees again, so performing multiple channel searches with successively shorter tree lengths may be required.

Some authors (Nei and Kumar, 2000) have recently suggested that tree space should *not* be thoroughly explored when using maximum parsimony. They argue that, as the number of taxa increase, parsimony does a poorer and poorer job of finding the "correct" tree, so additional effort to do so is unwarranted (Nei and Kumar, 2000). This notion completely ignores the reason for selecting an optimality criterion in the first place.

Examplars, placeholders and inferred ancestral states

One way to reduce the complexity of a phylogenetic problem is to use representative taxa for large clades, thereby reducing the number of taxa, and the total number of trees, which must be evaluated (Rice et al., 1997). Selecting exemplar taxa to represent larger clades is one method. Suitable strategies to reduce the number of taxa should be carefully evaluated, however, because some are clearly better than others (Yeates, 1995). Rice et al. (1997) conclude that the exemplar method is unsatisfactory because it increases branch lengths on the tree and discards information about inferred ancestral states within a clade. Another method, referred to as inferred ancestral states (Rice et al., 1997), replaces large clades with hypothetical ancestors inferred by optimizing characters to the base of the tree for the clade of interest (see also Donoghue, 1994; Mishler, 1994; Yeates, 1995). This effectively replaces a large clade with a single exemplar while preserving information about polymorphism within that clade. Rice et al. (1997) found that using the inferred ancestral states method greatly reduced the search time required while preserving information about polymorphisms within a clade and not artifictually increasing branch length. They point out, however, that the monophyly of a given clade must be tested independently (Donoghue, 1994) and sensitivity analysis to test the robustness of these conclusions should also be performed (Kellogg and Campbell, 1987).

Successive approximations character weighting

Successive approximations weighting is a method which uses a weight set based on some measure of character fidelity (e.g., consistency index, retention index, etc.) in an attempt to select between a large number of equally parsimonious trees (Farris, 1969; Carpenter, 1988). First, the most parsimonious trees are found. Next, each character is weighted based on how well it performs on the set of most parsimonious trees. Finally, a second parsimony search is performed to find a new set of most parsimonious trees. This technique is repeated until a "reasonable" number of trees is found. A justification for this method, or a point at which one should stop weighting iterations, has not been presented to date. It should be noted that, if successive weighting is to be undertaken, tree space should be explored extensively enough to ensure that all most parsimonious trees in each step are found. If they are not, character weight sets will be biased and a misleading phylogeny could be presented.

Phylogenetic supertrees

The increasing number of phylogenetic studies are expanding our understanding of the tree of life. However, few studies dealing with different taxa ever use the same characters to infer phylogeny. Combining data matrices is often difficult to do in such cases because of the large numbers of missing characters. Cutting and pasting pieces of one tree onto another is not satisfying because there is no global optimality criterion being enforced. Sanderson et al. (1998) reviewed some methods of representing phylogenetic trees as matrices of characters. This method allows one to encode the relationships at each node as a binary character—each taxon is either a member of the clade represented by node x or it is not). As long as there are two taxa shared between studies, the matrices can be combined and maximum parsimony can be used to build the supertree. The percentage of missing characters is thus reduced. An added benefit of this method is that trees constructed with different types of characters, such as allozyme frequency data and nucleotide sequences, can be combined in one analysis, even though the characters themselves cannot.

Conclusions

Pilot studies are essential in the design of any experiment, especially when dealing with groups for which previous systematic hypotheses are lacking. Basing such studies on alpha taxonomy can help identify cases where additional taxa or characters might be included in the analysis. Often the tradeoff between including additional taxa or additional characters is complex. Characters should be added with consideration of the major characteristics of the sequences (e.g., coding *versus* non-coding, single *versus* multiple copy). Many systematists argue that it important to add taxa in such a way as to reduce long branches, either by sampling clades more densely or adding fossil taxa (Hillis 1998; however, see Kim, 1996). Additional taxa may reduce the chance of getting "incorrect phylogenies" (i.e., those with LBA problems) and adding characters may increase support for nodes within a phylogeny.

Analysis of data sets, particularly when they are large and complex, is a critical step of systematics that is often neglected. Search strategies which thoroughly explore maximum parsimony tree space or the maximum likelihood surface are critical when examining large data sets for which heuristic algorithms must be employed.

Acknowledgements
This manuscript was greatly improved through discussions and comments with J. Bonacum and J. Remsen.

Molecular Systematics and Evolution: Theory and Practice
ed. by R. DeSalle, G. Giribet and W. Wheeler
© 2002 Birkhäuser Verlag/Switzerland

Higher-level systematic analysis of birds: current problems and possible solutions

Scott E. Stanley and Joel L. Cracraft

Department of Ornithology, American Museum of Natural History, New York, NY 10024, USA

Summary. Avian systematics has a rich history, as evolutionary biologists have long been interested in this conspicuous and diverse group of vertebrates. Many prominent scientists, and evolutionary biologists in particular, have focused their efforts on birds. Perhaps no other group of vertebrates is so well studied. Yet, despite the attention paid to this group, much about the history of the class Aves remains controversial, both with respect to the origin of birds and the history since that origin. This puts avian systematists in a unique position, with so much information available and so many unanswered questions to pursue.
The fact that avian ordinal relationships are still the center of much controversy speaks to the difficulty of the problem. While many prominent morphologists have worked on avian relationships, relatively few morphological studies have identified characters with informative variation for interordinal relationships. Molecular data offer the hope for phylogenetic information not present (or not discovered) in avian anatomy. Since the first study of avian proteins for the purposes of systematics (Sibley, 1960), several prominent molecular systematists have devoted tremendous time and resources to solving the problems of avian relationships using molecular characters (see Barrowclough, 1992 and Sheldon and Bledsoe, 1993 and references therein). So far, their efforts have not produced adequate resolution, at least not in the minds of most practicing systematists. Here, we first outline what we think we do know about higher order avian systematics and discuss some specific cases of molecular data applied to this question. Next, we consider some of the problems which may be blocking a clearer understanding of avian relationships. We then go on to offer some new directions for systematists working on this difficult group.

The current state of knowledge in avian systematics

After over 200 years of avian systematics we are still very ignorant of avian ordinal relationships. However, there is some consensus regarding a few relationships that we can discuss here. Perhaps the most well supported is the monophyly of the three basal clades of birds: 1) the palaeognaths, or ratites, 2) the Galloanserae and 3) the Neoaves. The relationships between the three groups have been the source of controversy on several occasions. For example, Mindell et al. (1997, 1999), found evidence for a basal split between the Passeriformes and the rest of modern birds, although there was not much support for this result in their data. Sibley and Ahlquist (1988) originally proposed a closer relationship between the palaeognaths and the Galloanserae (a group they named Eoaves). They subsequently rejected this arrangement in favor of the traditional avian rooting with a basal palaeognath *versus* neognath split (Sibley and Ahlquist, 1990), and used the names Eoaves and Neoaves in the place of palaeognath and neognath, respectively. It is now generally agreed that the Galloanserae and the Neoaves are sister taxa to the exclusion of the palaeognath birds (see Cracraft and Clarke, in press, for a review). This arrangement was also supported by an analysis of nuclear DNA sequence data published recently by Groth and Barrowclough (1999). The latter used a crocodilian outgroup and recovered the traditional palaeognath *versus* neognath split, as well as the Galloanserae *versus* Neoaves split within the Neognathae[1]. In a more recent analysis of both mitochondrial and nuclear DNA, van Tuinen et al. (2000) found support for this arrangement, although their data did not provide much information with respect to the relationships within the Neoaves.

Recent work

DNA-DNA hybridization data

DNA-DNA hybridization as applied to avian systematics is nearly synonymous with the name Charles Sibley. While other investigators have applied this technique to avian systematics, Charles Sibley and his colleagues have accumulated the largest amount of DNA-DNA hybridization data. It is beyond the scope of this chapter to discuss in detail this large and influential body of work. Suffice it to say that this research, which culminated in the tome *Phylogeny and Classification of Birds: A Study in Molecular Evolution* (Sibley and Ahlquist, 1990), has engendered much controversy. Yet it is unquestionably the single most important reference for avian systematists to date. The sources of controversy were many, some involving the technique itself (Cracraft, 1987a; Houde, 1987; Templeton, 1985), some involving the methods of analysis (Cracraft, 1987a) and some involving the particular practices of Charles Sibley and Jon Ahlquist (Sheldon and Bledsoe, 1993). In addition, there are several inherent limitations to DNA-DNA hybridization data that make other types of data preferable, including the exclusively phenetic nature of the data, the limited number of taxa possible when examining all pairwise comparisons, and the fact that one cannot add taxa to data sets which have already been collected. Thus, the large tree depicted in Sibley and Ahlquist (1990; Figures 357–368 and 371–385), the so-called tapestry, is composite in nature and does not necessarily represent the shortest network for the taxa included since all possible pairwise comparisons were never investigated. Moreover, the lack of any rigorous framework for assessing the reliability of the results produced by Sibley and Ahlquist (1990) only serves to amplify the ambiguities in the data.

---

[1] They chose the name Plethornithiformes to identify the Neoaves. However, since this clade was first named by Sibley and Ahlquist (1990: Figure 353), we choose to use this name as it has precedence in the literature.

DNA sequence data

With the many problems and inherent limitations of DNA-DNA hybridization data, many workers turned to DNA sequences to address avian interordinal relationships. Most of the work to date has addressed specific problems in avian systematics, thus focusing each taxon sample to include only the groups involved in a particular controversy. A characteristic of most of this work has been the assumption that just a few sequences would be necessary to solve long-standing problems. With respect to higher order avian systematics, a few among these deserve discussion here.

In 1994, three papers appeared which addressed specific questions regarding the relationships between controversial avian groups: 1) New World vultures (Cathartidae) and storks (Avise et al., 1994a), 2) the enigmatic Hoatzin (*Opisthocomus hoatzin*) and its candidate sister taxa (Avise et al., 1994b; but see also Hedges et al., 1995), and 3) the relationships among pelecaniform families (Hedges and Sibley, 1994). All of these relied on relatively small fragments of mitochondrial DNA from only a few taxa. In all cases the results of these analyses conflicted with traditional taxonomy with both strong and weak support. With the first (Avise et al., 1994a), it was subsequently shown that some of the sequences used in the analysis were misidentified (Hackett et al., 1995). With the second and third, the interpretation of the authors tended towards acceptance of the molecular results, in spite of the limited number of both characters and taxa in their analyses, and a rejection of the morphological data as too susceptible to convergence.

More recently, Mindell et al. (1997) approached the problems of avian ordinal relationships using much larger data sets, especially with respect to number of characters. In this case they included complete mitochondrial genomes in an analysis of avian taxa and one crocodilian outgroup. This analysis produced a controversial result, in that the placement of the root node made the Passeriformes basal to the rest of the ingroup taxa, including palaeognaths. However, given the small number of ingroup taxa, virtually nothing was learned about interordinal relationships beyond the placement of the avian root. In the same study, Mindell et al. (1997) also analyzed data sets with more taxa for partial fragments of the mitochondrial 12S rRNA gene. While some of the more closely related groups (e.g., genera within the Galliformes and Anseriformes) were well supported by the data, the resulting trees were generally poorly resolved with respect to interordinal relationships. Subsequently, Mindell et al. (1999) added an additional ratite species and a suboscine songbird to their analysis of complete mitochondrial genomes. The results were similar with respect to the placement of the songbirds at the root. However, in this case they could not reject the traditional arrangement with the basal split occurring between the ratites and the rest of the avian orders.

Problems in avian systematics

Taxon sampling

We interpret these results as evidence that both inadequate character and taxonomic sampling can impede insights into avian interordinal relationships.
Since taxon sampling is known to have a significant effect on the results of
phylogenetic analyses in general (Lecointre et al., 1993; Donoghue, 1994;
Friedlander et al., 1994), it seems premature to conclude anything significant
regarding avian interordinal relationships when studies have very few representatives of the major groups of birds. Our caution is based on the observation that when taxa diverged a relatively long time ago, long branches can confound efforts to recover a phylogenetic signal (Felsenstein, 1978; Olmstead
and Palmer, 1994; Hillis et al., 1996; Siddall, 1998). If the divergence between
avian orders is as old as some suggest (Cracraft, 1986; Chiappe, 1995; Hedges
et al., 1996; Cooper and Penny, 1997; Mindell et al., 1999), interordinal divergences are likely to be deep, and taxon sampling will be extremely important.
Our preliminary results in a large-scale effort, using both mitochondrial and
nuclear sequences to resolve some of the avian ordinal relationships, suggest
that many taxa from within each order are required even to recover monophyly of orders that are also well supported by morphology. Recent work by van
Tuinen et al. (2000) supports this position. Beyond recovering the
Palaeognath/ Neognath split and the Galloanserae/Neoaves split, their analysis
of over 4000 bps of nuclear and mitochondrial data did not identify any well-
supported clades within those groups. As a result of these and other studies, we
suggest that dense taxon sampling from within avian orders will be critical
before interordinal relationships can be resolved.

Short internode: real or imagined?

Perhaps the most striking feature of the family-level component of the DNA
hybridization "tapestry" (Figs 354–356 in Sibley and Ahlquist, 1990) is the
short internodes between most of the avian families and orders. This result and
others like it (Avise et al., 1994a,b; Hedges and Sibley, 1994; Mindell et al.,
1997), have convinced many ornithologists that the major avian lineages radiated rapidly. This scenario is also supported by the lack of morphological characters discovered to date to define interordinal groupings, and the sudden
appearance of many avian groups in the fossil record (see Benton, 1999 and
references therein). Moreover, it is tempting to use the reality of short internodes between avian orders as an explanation for the difficulty experienced by
those who have tried to resolve avian interordinal relationships.
    At this point, we are reluctant to accept the fact that the short internodes
recovered in molecular analyses are a definitive indication of a rapid avian
radiation. We attribute our reluctance to the possibility that the lack of resolu-

tion and short internodes in avian phylogenies is in fact an artifact associated with DNA data and very old divergences. There are two different patterns for the origin of avian orders which are currently debated in the literature. In one (Fig. 1A), a rapid radiation of neornithine orders just after the Cretaceous/ Tertiary (K/T) boundary is proposed (Feduccia, 1995, 1996; Benton, 1999), as indicated by the sudden appearance of fossil taxa that can be assigned to modern orders. In the second scenario (Fig. 1B), based on molecular clock calibrations, it is assumed that extant avian orders diverged from one another well before the K/T boundary (Hedges et al., 1996; Cooper and Penny, 1997; Easteal, 1999) and the limitations of the fossil record do not allow us to document their presence. There are two additional scenarios worthy of mention here. In one, modern avian orders radiate at or shortly after the K/T boundary, but do so less rapidly than a strict interpretation of the molecular trees would suggest (Fig. 1C). In the other (Fig. 1D), avian orders radiated rapidly, as in Figure 1a, but much earlier in the Cretaceous.

With respect to these latter two scenarios, we can comfortably reject the first one (Fig. 1C), in which avian ordinal divergence events were well spaced

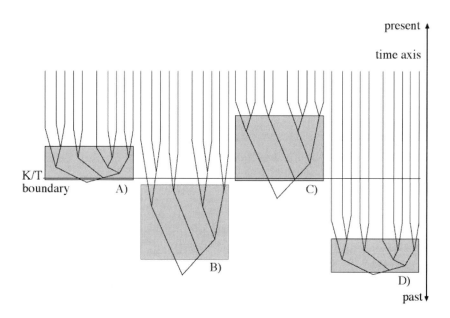

Figure 1. Four hypothetical alternatives for the pattern and timing of avian ordinal divergences (see also Easteal, 1999; Figure 1). The solid horizontal line represents the boundary between the Cretaceous and Tertiary periods (the so-called K/T boundary), approximately 65 million years ago. The shaded boxes represent the time period in which the avian orders diverged in the four examples. (A) Avian orders diverged rapidly just after the K/T boundary. (B) Avian orders diverged in the Cretaceous, well before the K/T boundary, and those divergences were spaced out in time. (C) Avian orders diverged just after the K/T boundary, but those divergences were also spaced out in time. (D) Avian orders radiated rapidly as in A), but did so in the Cretaceous.

throughout the Tertiary, since fossil representatives of many avian orders were present soon after the K/T boundary in the Paleogene. It is the last scenario (Fig. 1D) that we find intriguing, particularly in contrast to Figure 1B. If the molecular clock calibrations are correct, avian interordinal divergences are older than traditionally believed. If the interpretation of short internodes on molecular phylogenies is also correct, as depicted in Figure 1D, then recovering the structure in this portion of the avian tree will indeed be a difficult task. If, however, avian orders are simply old, but did not radiate rapidly (Fig. 1B), then the quest for additional taxa and characters should help resolve many issues in avian systematics.

We make this distinction because the interpretation of molecular phylogenies with short internodes as evidence of a rapid radiation assumes that estimates for branch lengths among internal nodes are accurate. More to the point, it assumes that estimates of internal branch length are not biased in favor of short branches, as compared to estimates for terminal branches. It is possible that when branch lengths are estimated for trees with only a few distantly related taxa, mutations will have erased changes that occurred along deeper internal branches (Fig. 2). This might lead to a compression effect in which the branch lengths for deeper internal branches are underestimated more than branch lengths for the recent terminal branches (Fig. 2B), giving the impression of a rapid radiation of the basal lineages. Depending on how severe this

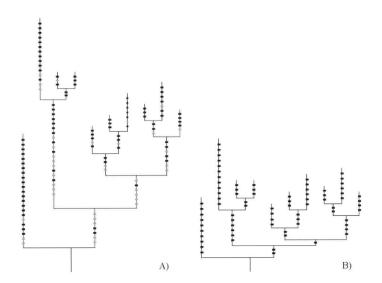

Figure 2. The phylogram in (A) depicts the actual history of ten taxa in this hypothetical example. The branch lengths are proportional to the number of mutations, designated as both black and gray hatch marks, that have occurred along each. The gray hatch marks represent mutations that were overwritten by subsequent mutations. The phylogram in (B) is the one which would be estimated by the resulting sequence data. Note that branch lengths in (B) are underestimated due to multiple mutations, and that this effect is greater for internal branches, leading to a compression effect deeper in the tree.

compression effect is, the scenario depicted in Figure 1B might produce phylograms which resemble the one in Figure 1A.

As a first step in determining the extent to which short internodes might be an artifact rather than an indication of rapid radiations, a computer program written by Mark Siddall was used to generate sequences with an evolutionary history lacking any such rapid radiations. The model of sequence evolution used was a Jukes-Cantor model, and the starting sequence of 1000 bp was generated using equal frequencies of all four bases. The simulation had two main features. First, mutations occurred via mutational bouts where a certain number of bases were chosen at random and mutated to a different base. The resulting base had an equal probability of being each of the three alternative bases. The number of bases mutated during any mutational bout was determined by the mutation rate, which was set so that a certain percentage (from 1 to 75 percent) of the bases in a sequence experienced mutations, leading to the same amount of percent sequence divergence along all lineages between bouts. Second, after each mutational episode, one of the sequences was chosen at random to "speciate", producing two daughter sequences. This process was repeated until there were 100 sequences and the true tree for these sequences was retained. Using that tree, we then estimated the branch lengths with both parsimony and neighbor-joining based on a Jukes-Cantor correction. We did this for 10 different mutation rates and calculated the ratio of internal to terminal branch lengths for each simulation to see if this ratio would change with mutation rates. Since the relative branch lengths were kept constant between simulations (i.e., there was no bias in branch lengths for particular areas of the tree in any of the simulations), we could attribute any differences in this ratio to differences in mutation rates. Specifically, we were interested in determining whether or not this ratio would decrease as mutation rates increased, which would be consistent with the compression phenomenon described above. This is precisely the result we observed, and the effect was worse for neighbor-joining than for parsimony (Fig. 3). Thus, short internodes in phylogenies may be a consequence of biases in the methods we use to estimate branch lengths and might not necessarily reflect rapid radiations of the taxa involved.

Outgroups in avian systematics

Outgroup selection is another important issue in avian systematic studies for two disparate reasons. First, an inappropriate choice of outgroup has obvious consequences for delimiting monophyletic groups. In their attempts to determine whether or not the problematic Hoatzin (*Opisthocomus hoatzin*) was more closely related to Galliformes or to cuckoos, Avise et al. (1994b) used a passeriform species as an outgroup. Yet, the conclusions were compromised because of their choice of outgroup (Hackett et al., 1995): Galliformes, one of the ingroups, are more basal within neognaths than are Passeriformes. The choice of an inappropriate outgroup, as is probably the case in this example,

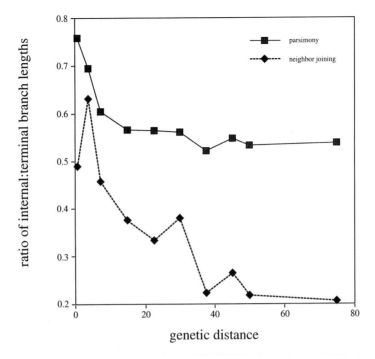

Figure 3. The graph depicts the ratio of internal to terminal branch lengths estimated for 10 simulat-ed data sets with different mutation rates, described here as the genetic distance generated in each sim-ulation (see text). The squares represent reconstructed branch lengths using parsimony and the dia-monds represent reconstructed branch lengths using neighbor-joining with a Jukes-Cantor correction. Note that in each case, as genetic distance increases, the ratio decreases, showing that the recon-structed branch lengths are biased toward shorter branch lengths for internal branches relative to ter-minal branches. This leads to a compression effect for internal branches as mutation rates increase.

can sometimes arise because of uncertainty in our knowledge of relationships. The problem can also arise through ignoring outgroups altogether.

For many years the choice of the root for the Neornithes has been a vexing problem. Sibley and Ahlquist (1990) presented rooted trees, but did not include any outgroups and simply relied on midpoint rooting. This created other prob-lems because it rendered the Neognathae non-monophyletic. For morphologi-cal studies the fossil taxa *Archaeopteryx*, *Hesperornis* and *Ichythyornis* were generally used as outgroups. Now, a host of new fossil material for taxa basal to neornithines exist and can be used to address the problem of rooting the avian tree using morphological data. For studies using molecular data, on the other hand, the closest living outgroup to modern birds are the crocodilians, which diverged some 230–270 million years ago (Benton, 1993).

Using such distant outgroups has its disadvantages. The controversial results of Mindell et al. (1997, 1999), placing passeriform at the neornithine root may be indicative of the problems associated with using overly distant outgroups, or of inadequate taxon sampling (see above), but probably of both.

The current consensus, based on a variety of morphological and molecular studies, supports the position of a neornithine root between the Palaeognathae and Neognathae (see review in Cracraft and Clarke, in press), yet additional data are always welcome. In the meantime, palaeognaths provide an obvious choice for rooting neognaths. Within the latter, most of the evidence points to the Galloanserae as being the sister-group of the Neoaves and therefore being an appropriate outgroup, along with palaeognaths. Where the root of the Neoaves goes, however, is highly uncertain and seems likely to remain a very difficult problem.

New methods of rooting molecular phylogenies may be required. One such source may come in the form of duplicate, or paralogous, genes that diverged prior to the origin of the Neoaves (Fig. 4). The use of paralogous genes as outgroups has proven useful in other analyses where ancient groups were involved (Iwabe et al., 1989; Donoghue and Mathews, 1998). Assuming that sufficient time has passed between the crocodilian-neornithine split and the divergence of the Neoaves, gene duplication events may have occurred and, when discovered, may prove useful to avian systematists through the discovery of clade-specific gene duplication events within the Neornithes. If, for example, a gene

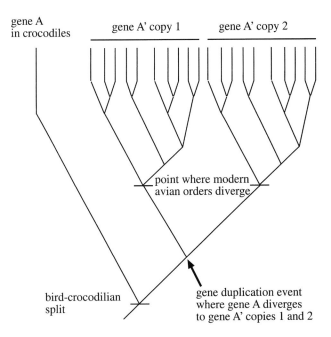

Figure 4. This tree represents a hypothetical example of a gene duplication event that would provide a means to root the avian ordinal tree with an outgroup that is more closely related to a gene in birds than is the orthologous copy in crocodilians. In this example, a gene duplication takes place for gene A at some point after the bird-crocodile split, but before the point where modern avian orders diverged, producing copies 1 and 2 of gene A'. Here, either copy of gene A' can be used to root a tree made with the other. This is preferable to using gene A in crocodilians to root either copy of gene A' birds, since it will represent a more closely related outgroup.

duplication event occurred within neognaths, not only can the paralogous copies which result be used as outgroups for each of the corresponding genes trees, the duplication event itself may be a useful synapomorphy for those taxa within the Neognathae that contain both copies.

Molecules versus *morphology*

The belief that there is a conflict between molecular and morphological data arose as soon as molecular data became widely available. An expectation of conflict, as seems to underlie some investigators' approach to this issue, is perplexing given that there has been only a single history which is being reconstructed with both types of data. There have been different responses to this apparent conflict. Many molecular systematists have adopted the opinion that conflict demonstrates the inferiority of morphological data (Sibley and Ahlquist, 1990; Hedges and Sibley, 1994; Siegel-Causey, 1997). Thus, Hedges and Sibley (1994) recommended that in the face of a "robust phylogeny" based on molecular data, the morphological data should be reevaluated, since morphological characters are more likely to be shared-primitive or convergent (i.e., homoplasious).

But there are a number of analytical issues to be considered with such claims. First, little, if any, direct evidence has been produced to support claims of superiority of molecular evidence. Indeed, most claims are little more than special pleading rather than an objective, empirical investigation of whether certain kinds of data are inherently less informative. Opinions are often biased by false assumptions and comparisons. For example, differences between topologies generated using different analytical methods and both molecular and morphological data are not evidence for the superiority of molecular data over morphological data, or *vice versa*. Before conflict between morphology and molecules can be addressed directly, morphological and molecular data for the same taxa need to be included in a single phylogenetic analysis. This follows because homoplasy cannot be identified except in the context of overall character congruence (Farris, 1983).

Another argument underlying the conclusion that molecular trees are more "accurate" than those based on morphology is the notion that morphological data are "known" to be very homoplastic due to functional convergence. Such an argument, however, ignores the intrinsic limitations of molecular data, namely, that DNA sequences are limited in the possible character states they can assume (Mishler et al., 1988), thus producing "cyclic homoplasy," particularly in rapidly evolving sequences like mitochondrial genes. In fact, in a comparison of morphological and molecular data, Sanderson and Donoghue (1989) found no evidence that morphological characters tended to be more homoplasious than molecular characters.

A second issue in need of more discussion is whether apparent conflict is more perceived than real. Conflicting molecular and morphological trees both frequently have very poorly supported nodes. In many instances poorly supported short internodes allow for the possibility of numerous extremely different topologies. Virtually all molecular studies of birds using moderate to large taxon sampling are characterized by short internodes having poor support. This includes topologies derived from DNA hybridization, immunological distances and DNA sequence data.

Ambiguities in rooting of molecular trees also lead to apparent conflict. For example, some conflicting molecular and morphological results for ratites are most easily explained by different placements of roots (Lee et al., 1997). Other examples have already been mentioned, including the three different locations for the root of the neornithine tree (1) between Palaeognathae and Neognathae, (2) between Palaeognathae and Galloanserae on the one hand and Neoaves on the other, or (3) on the passeriform branch itself. All of these are largely conflicts, and artifacts, of rooting.

Future directions

In 1992 George Barrowclough suggested we would have all of the problems of avian systematics solved in 5 years (Barrowclough, 1992). Clearly, we have fallen short of that goal. The relationships between avian orders seem to present some of the most difficult problems for systematists. Here we have stressed the importance of taxonomic sampling in systematics in general, and avian systematics in particular, and we suggest that dense taxon sampling will be one of the keys to improvements in our understanding of avian ordinal relationships. There are several good examples of dense taxon sampling applied at the intraordinal and intrafamilial levels (e.g., Burns, 1998; Nunn and Stanley, 1998; Groth, 1998; Lanyon and Omland, 1999), and we see these as steps in the right direction.

Avian systematists also need to look for characters that are evolving at slow rates. If internodes between avian orders are extremely short, or even if those divergences are simply very ancient, then many slowly evolving characters will be required if we are to identify these older avian clades. Slowly evolving characters will also be required when taxon sampling is limited by the fact that there are simply few extant taxa. If many avian orders are composed of closely related species that together are distantly related to their sister taxa, then only informed character sampling is likely to help resolve these relationships.

It is also important to keep in mind the difference between a character that evolves slowly because of a slow mutation rate and one that evolves slowly because of extreme selective constraints. Ideally we would target characters that have slow mutation rates and low levels of selective constraint over ones that are simply highly conserved. This might appear to contradict the notion that conserved characters are best for tracking older divergences. However, "conservative" is not really the property one seeks when addressing questions in higher-level systematics. A useful hypothetical example of why a conservative gene, *per se*, is not desirable for tracking older divergences involves a protein-coding gene which is highly conserved but has a high mutation rate. In this case, synonymous sites would be evolving at a high rate, but replacement sites will be invariant, or nearly so. If this hypothetical gene were used to reconstruct the phylogeny for a group of distantly related taxa, it would provide little or no phylogenetically informative variation for distantly related groups. Synonymous sites would be evolving so rapidly that multiple mutations at a given site would obscure phylogenetic relationships while the replacement sites would be invariant, thus leaving no phylogenetic information. Rather than targeting highly conserved genes for addressing phylogenetic questions when distantly related taxa are involved, systematists would be wise to search for genes with low mutation rates and a high proportion of sites that are free to vary.

In vertebrates, the closest approximation to this ideal marker is likely to be nuclear genes. In a recent study using DNA sequence data from the nuclear gene RAG-1, Groth and Barrowclough (1999) demonstrated the utility of nuclear sequences for higher-order avian systematics, and several labs are currently investigating this approach with both more taxa and additional nuclear genes. The search for additional genes is particularly important, especially with respect to reduced mutation rates. In birds, genes located on the W chromosome may prove useful, as this chromosome appears to evolve more slowly than autosomal genes or genes on the Z chromosome (Ellegren and Fridolfsson, 1997; Kahn and Quinn, 1999). Additionally, molecular markers such as large insertions or deletions in genes (e.g., Groth and Barrowclough, 1999), or insertion locations of mobile elements, may prove to be extremely useful in so far as these markers might represent unique events in avian history. If this is the case, then these types of characters may help define some clades that would otherwise be difficult to identify. The problem is that such markers may be so rare that few will be available to the avian systematist and they will likely only be discovered very gradually as more is learned about avian genome structure. However, even a few such characters may help provide some stable structure to avian ordinal trees.

Conclusions

Studies of higher avian systematics conducted to date have suffered from both limited character and taxon sampling. We suggest here that improvements in both are required before significant advances cane be made in our understanding of higher-order avian systematics. In addition, new sources of molecular characters need to be explored to discover more slowly evolving markers. Given recent advances in sequencing technology, it is now possible to collect sequence data at much higher rates. As a result, larger data sets with many more taxa are now a reasonable goal for avian systematists. We also call upon avian systematists working on morphological characters to contribute data that will be informative for ordinal-level analyses and we advocate the use of these data in combined approaches with molecular data. Finally, we recommend that avian systematists pool their resources, efforts and data in collaborative studies, in order to generate data sets of the size that may be required to solve the problems they currently face.

Acknowledgements
We thank Mark Siddall for writing the software used to do the simulations on the biases associated with branch length reconstructions. Julie Feinstein, Kay Earls and Anwar Janoo provided helpful comments on the manuscript. This paper is a contribution from the Monell Molecular Laboratory and the Cullman Research Facility in the Department of Ornithology, AMNH, and has received generous support from the Lewis B. and Dorothy Cullman Program for Molecular Systematics Studies, a joint initiative of The New York Botanical Garden and the AMNH.

Relative quality of different systematic datasets for cetartiodactyl mammals: assessments within a combined analysis framework

John Gatesy

Department of Molecular Biology, University of Wyoming, Laramie, WY 82071, USA

Summary. High congruence, support, stability, resolution and decisiveness are seen as positive attributes by many systematists. Within a cladistic context, the consistency index, the retention index, branch support, data decisiveness, the number of nodes resolved in a strict consensus tree and the incongruence length difference are direct measures of these qualities. Phylogenetic analyses of 29 datasets for cetartiodactyl mammals show that for a particular character partition, these indices can vary radically in separate *versus* combined analysis of datasets. The quality of any single dataset is of little importance in comparison to a thorough sampling of the available character space.

Introduction

Humans like to rank things. Not surprisingly, phylogeneticists commonly spar over the relative utility of different systematic datasets. These discussions are sometimes contentious and oftentimes contradictory. For example, certain molecular systematists would prefer to limit the use of gross anatomical evidence in favor of DNA sequence data. They consider morphological characters to be more prone to selective convergence and homoplasy generally (e.g., Hedges and Maxson, 1996; Givnish and Sytsma, 1997). At the other extreme, some morphologists indiscriminately dismiss simple nucleotide characters whose functional properties may be poorly understood in favor of "a few well-analyzed morphological characters or character complexes, whose history can be followed by paleontological and/or ontogenetic methods and whose adaptive and functional significance can be reasonably inferred" (Luckett and Hong, 1998). Alternatively, Shimamura et al. (1997) and Nikaido et al. (1999) have countered that SINE retrotransposons are superior to both DNA sequence data and gross anatomical characters because insertions of these transposons are apparently irreversible and "the probability that homoplasy will obscure phylogenetic relationships is, for all practical purposes, zero." (Nikaido et al., 1999). Claims and counterclaims on the relative merits of different character sets are rampant.

Preference for one dataset over another has been based on a variety of criteria. The factors listed below have all been implemented in arguments for or against the utility of a particular character partition: overall rate of evolution/variability, mode of inheritance (i.e., biparental *versus* maternal *versus* paternal), complexity of characters, amount of homoplasy, support for well-substantiated "known" phylogenies, probability of confusion among paralogous gene copies, resolving power, apparent irreversibility of character transformations, stability, the possibility of confounding gene conversion events, rates of insertion and deletion of nucleotides, ambiguity in sequence alignment, saturation of substitutions at third codon positions, independence of characters, size of a dataset, ease and cost efficiency of data collection, rate heterogeneities among different lineages, the likelihood of fixation for ancestral polymorphisms, chance of horizontal gene transfer, stationarity of nucleotide base composition, transition/transversion bias, differential functional constraints and susceptibility to convergent selection pressures. This list is certainly not comprehensive.

Many of the factors listed above, such as uneven rates of nucleotide substitution among lineages, rapid evolution at third codon positions, differential functional constraints and transition/transversion bias, are biological phenomena that may encourage homoplastic change. In certain cases, these factors *could* have a negative impact on phylogenetic support, stability, congruence, decisiveness and resolution. In contrast, some measures of dataset quality *are* indices of support, stability, congruence, decisiveness and resolution. The ensemble consistency index (CI; Kluge and Farris, 1969), the ensemble retention index (RI; Farris, 1989), data decisiveness (DD; Goloboff, 1991), branch support (BS; Bremer, 1994), the incongruence length difference (ILD; Farris et al., 1994), and the number of nodes resolved in a strict consensus tree (Schuh and Polhemus, 1980) fit into this class of indices. These and similar measures of dataset quality will be considered further in this chapter.

I first use separate analyses of 29 datasets for Cetartiodactyla (whales and even-toed ungulates) to assess homoplasy, congruence, stability, support, decisiveness and resolution in these character partitions. I then measure these same qualities within a combined analysis framework and compare dataset quality in separate *versus* combined analysis. I suggest that, within a combined cladistic analysis framework, extensive sampling of the available character space is much more important than a determination of the relative quality of different datasets. But, the differential utility of various datasets in combined analysis can be used to make educated guesses as to which character partitions may be useful in future phylogenetic analyses.

Methods

Many different datasets have been applied to higher-level problems in mammalian systematics. One apparent consensus from these studies is the grouping of the eutherian orders, Artiodactyla (even-toed ungulates) and Cetacea (whales), to the exclusion of other extant mammals. Numerous data support Cetartiodactyla (reviewed in Gatesy, 1998; O'Leary and Geisler, 1999).

Relationships within Cetartiodactyla are more contentious. Molecular data generally favor the placement of Cetacea within a paraphyletic Artiodactyla (e.g., Milinkovitch et al., 1998; Ursing and Arnason, 1998a; Nikaido et al., 1999), while morphological characters support a sister group relationship between Cetacea and a monophyletic Artiodactyla (e.g., Geisler and Luo, 1998; O'Leary and Geisler, 1999). The diversity of datasets for Cetartiodactyla is impressive, and includes gross anatomical characters, evidence from the fossil record, numerous nuclear (nu) gene sequences, complete mitochondrial (mt) genomes, amino acid sequences from several nu proteins, and insertions of SINE retrotransposons (Fig. 1).

Twenty-nine cetartiodactyl character sets from the literature were concatenated. The five taxa in this large combined matrix of over 18,000 characters were: Bovidae (Ruminantia, Artiodactyla), Hippopotamidae (Ancodonta, Artiodactyla), Suidae (Suina, Artiodactyla), Mysticeti (Cetacea) and an outgroup. The outgroup taxon in 28 of the 29 datasets was a member of Perissodactyla (odd-toed ungulates). For α-lactalbumin, the outgroup was a member of Rodentia (rodents). Most taxa in the combined matrix were composites of different species. Taxonomic exemplars for all mt protein coding genes but cytochrome *b* were: *Bos taurus* (Bovidae; Anderson et al., 1982),

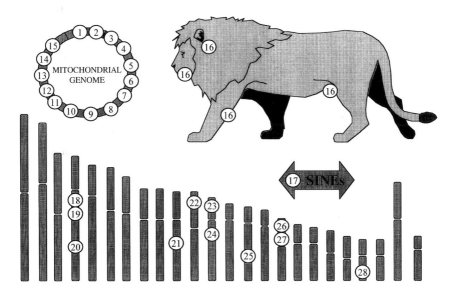

Figure 1. The distribution of datasets in character-space. Nu DNA and amino acid sequences sampled in this study are mapped onto the chromosomes of *Homo sapiens*. 1 = 12S rDNA, 2 = 16S rDNA, 3 = NADH 1, 4 = NADH 2, 5 = CO1, 6 = CO2, 7 = 5'ATPase 8, 8 = 3' ATPase 8 + ATPase 6, 9 = CO3, 10 = NADH 3, 11 = NADH 4L, 12 = NADH 4, 13 = NADH 5 + 5' NADH 6, 14 = 3' NADH 6, 15 = cytochrome *b*, 16 = morphology (dental, basicranial, postcranial, and soft anatomical characters), 17 = SINE retrotransposons, 18 = β-casein, 19 = κ-casein, 20 = γ-fibrinogen, 21 = IRBP, 22 = β-hemoglobin, 23 = vWF, 24 = α-lactalbumin, 25 = pancreatic ribonuclease, 26 = protamine P1, 27 = α-hemoglobin, 28 = α-crystallin.

Hippopotamus amphibius (Hippopotamidae; Ursing and Arnason, 1998a), *Sus scrofa* (Suidae; Ursing and Arnason, 1998b), *Balaenoptera physalus* (Mysticeti; Arnason et al., 1991) and *Equus caballus* (outgroup; Xu and Arnason, 1994). These 12 mt protein coding genes were aligned manually using SeqApp (Gilbert, 1992); the alignments required the introduction of only a few internal gaps. Taxonomic exemplars for mt 12S ribosomal (r) DNA, mt 16S rDNA, mt cytochrome *b*, all nu loci and SINEs were as in the WHIPPO-1 matrix of Gatesy et al. (1999a). Sequence alignments also were as in Gatesy et al. (1999a) for these datasets. Insertions of SINE retrotransposons were coded as in Nikaido et al. (1999).

Morphological data were primarily from O'Leary and Geisler (1999). Taxonomic exemplars were: *Ovis* (Bovidae), *Hippopotamus amphibius* (Hippopotamidae), *Sus* (Suidae), *Balaenoptera* (Mysticeti) and *Equus* (outgroup). The 123 characters of O'Leary and Geisler (1999) were augmented by two traits taken from the literature by Gatesy (1997), position of the testes (scrotal-0, inguinal-1, intra-abdominal-2) and ability of offspring to suckle underwater (suckle exclusively on land-0, both on land and underwater-1, exclusively underwater-2). These two characters were treated as ordered, and selected morphological characters from O'Leary and Geisler (1999) also were ordered as recommended by these authors.

Individual genes or gene products were considered different datasets, as were the morphological characters and the SINE retrotransposons. Individual and combined datasets were analyzed cladistically with all character transformations equally weighted (Kluge, 1997) and with nucleotide and amino acid ambiguities coded as in Gatesy et al. (1999a and b). In most cases, gaps were treated as missing data. Analyses were done with PAUP or PAUP* (Swofford, 1993, 2000), and searches were either branch and bound or exhaustive. All indices of dataset quality were also determined using PAUP or PAUP*.

A combined matrix of 79 mammalian taxa, including 52 cetartiodactyls, was also compiled. This dataset of over 8,000 characters is simply a combination of the WHIPPO-2 dataset of ten genes from Gatesy et al. (1999a) and the morphological dataset of O'Leary and Geisler (1999). The nine extant eutherian taxa in the matrix of O'Leary and Geisler (1999) were merged with the molecular information from three mt and seven nu loci in the WHIPPO-2 dataset. In parsimony analysis of this dataset, gaps were treated as missing data, selected morphological characters were ordered as in O'Leary and Geisler (1999), all character transformations were equally weighted, and nucleotide ambiguities were coded according to IUPAC conventions. The PAUP* search was heuristic with 1000 replications of random taxon addition. The "amb-" option was in effect, and branch swapping was by tree bisection reconnection (Swofford, 2000). Optimal trees were rooted with Xenarthra. The two combined datasets, for five taxa and for 79 taxa, are available from the author. Below, indices and statistics determined for the matrix of five taxa are summarized.

Separate analyses

Each of the 29 individual datasets in the five-taxon matrix was analyzed separately, and the following information was recorded:

1. *Overall number of characters.*
2. *Number of informative characters for the five taxa analyzed here.*
3. *Variability, percentage of characters that are phylogenetically informative.*
4. *Minimum tree length.*
5. *Number of minimum-length topologies.*
6. *The strict consensus of all minimum-length topologies.*
7. *CI (Kluge and Farris, 1969) and CI disregarding uninformative characters (CIdu) for minimum-length trees.*

   The CI has been advocated as a measure of homoplasy, that is, a measure of "deviation from hierarchy" (Goloboff, 1991). The CI and the RI (see below) also can be interpreted as measures of character congruence. If characters are perfectly congruent, the CI and the RI are 1.0, the maximum possible. If characters are highly incongruent, the RI and the CI will be low.

8. *RI (Farris, 1989) for minimum-length trees.*

   The RI is a measure of the amount of homoplasy "expressed as a fraction of possible homoplasy" (Farris, 1989).

9. *A modification of DD (Goloboff, 1991), unscaled DD (uDD).*

   The length of shortest trees was subtracted from the average length of all possible trees as in the calculation of data decisiveness, but this difference was not scaled to the sum of minimum lengths for all characters as in Goloboff (1991).

   uDD = average length of all possible trees – length of shortest tree(s)

   Goloboff (1991) defined decisiveness as, "…information allowing a choice or a decision between different classifications…A dataset with high decisiveness offers good grounds to prefer some trees over others." In addition to measuring the difference in length between average tree length and minimum tree length, DD takes into account the amount of homoplasy in a dataset (De Laet and Smets, 1999). This is because DD, like CI, is scaled to the amount of variation in the character set. For uDD, this scaling factor is eliminated so that redundancy with the CI is reduced. uDD is zero when all possible trees are of equal length, and is high when the difference in character steps between minimum-length trees and all other trees is great, and when there are few minimum-length trees.

10. *Total support (TS; Källersjö et al., 1992).*

    TS equals the sum of BS scores (Bremer, 1994) for all nodes supported by the dataset.

The 29 datasets were ranked for ten criteria: number of characters, number of informative characters, variability, number of minimum-length trees, nodes resolved in the strict consensus of minimum-length trees, CI, CIdu, RI, uDD and TS. These measures assess dataset size, resolution, congruence, homoplasy, decisiveness, support and stability.

Simultaneous analysis

Next, all 29 datasets in the matrix of five taxa (18,623 characters) were combined in simultaneous analysis. The following information was recorded:
1. *Minimum tree length.*
2. *Number of minimum-length topologies and the strict consensus of those topologies.*
3. *CI and CIdu for minimum-length trees.*
4. *RI for minimum-length trees.*
5. *uDD.*
6. *BS (Bremer, 1994) for each node supported by the combined analysis.*
   In this paper, BS is defined as the difference in length between the shortest topology that lacks the node of interest and the shortest topology that contains the node of interest (see Gatesy et al., 1999b). With this definition, BS scores may be positive, zero, or negative.
7. *Hidden branch support (HBS; Gatesy et al., 1999b) for each node supported by the combined analysis of all 29 datasets.*
   For a particular node supported by a combined matrix of several datasets, HBS equals the BS score at that node for the total combined analysis, minus the sum of BS scores for that node in separate analyses of the individual datasets that compose the combined matrix.

Quality of character partitions within a simultaneous analysis framework

For each of the 29 datasets in the five-taxon matrix, the following indices were measured within the context of the total character set.
1. *For each individual dataset, the number of extra steps required to fit that dataset onto the topology supported by the total combined dataset.*
   The sum of these extra steps for all 29 datasets is the ILD (Farris et al., 1994).
2. *CI and CIdu of each dataset for the minimum-length tree(s) supported by the combined dataset.*
3. *RI of each dataset for the minimum-length tree(s) supported by the combined dataset.*
4. *uDD for each dataset in the combined analysis.*
   For each dataset, the length of that dataset on the shortest tree(s) supported by the combined dataset was subtracted from the average length of trees for the individual dataset.
5. *Partitioned Bremer support (PBS; Baker and DeSalle, 1997) for each individual dataset for each node supported by the combined analysis of all 29 datasets.*
   PBS describes the contribution of each dataset to a BS score for the total combined dataset.

6. *The sum of PBS scores (ΣPBS) for each of the 29 datasets in the combined matrix.*

   ΣPBS for a particular dataset for all nodes supported by the combined dataset has been suggested as a measure of dataset influence in combined analysis (Baker and DeSalle, 1997).

7. *Nodal dataset influence (NDI; Gatesy et al., 1999b) for each individual dataset for each node supported by the combined analysis of all 29 datasets.*

   For a particular node and a particular data partition within a combined dataset, NDI is the BS score for that node in the combined analysis of all datasets, minus the BS score at that node for the combined dataset without the partition of interest.

8. *Dataset influence (DI; Gatesy et al., 1999b) for each of the 29 datasets in the combined analysis.*

   DI is the sum of all NDI scores for a particular data partition for all nodes supported by the combined dataset. DI and ΣPBS in combined analysis are analogous to TS in separate analysis, but include estimates of hidden stability and instability that may emerge in combined analyses (Baker and DeSalle, 1997; Gatesy et al., 1999b).

9. *Partitioned hidden branch support (PHBS; Gatesy et al., 1999b) for each dataset at each node supported by the combined matrix.*

   PHBS summarizes the contribution of a data partition to HBS that emerges in combined analysis at a particular node.

10. *Differences in the number of minimum-length topologies that came with the removal of each individual dataset from the combined matrix of 29 datasets.*

11. *Differences in the resolution of the strict consensus of minimum-length topologies that came with the removal of each individual dataset from the combined matrix of 29 datasets.*

Each of the 29 datasets was ranked for nine criteria within the simultaneous analysis framework: CI, CIdu, RI, extra steps in combined analysis, uDD, ΣPBS, and effect of removal of the dataset from the combined analysis on the number of minimum-length trees, number of nodes resolved, and stability to relaxation of the parsimony criterion (DI). These measures assess homoplasy, congruence, decisiveness, support, stability, and resolution.

A comparison was then made between indices of dataset quality in separate analysis and analogous indices in the context of all 29 datasets. Comparisons between ranks for separate *versus* simultaneous analysis were made for CI, CIdu, RI, uDD, TS *versus* ΣPBS or DI, number of equally short trees in separate analysis *versus* change in number of minimum-length trees with the deletion of a particular dataset from the combined matrix, number of nodes resolved in separate analysis *versus* change in number of nodes resolved with the deletion of a particular dataset from the combined matrix. Correlations between ranks in separate *versus* combined analysis were measured using the

Spearman rank correlation coefficient. Significance of correlation coefficients ($p < 0.05$) was estimated with standard statistical tables (Zar, 1974).

There has been some concern in the literature about the "swamping" of characters in small datasets by abundant characters in larger datasets (e.g., Luckett and Hong, 1998). That is, the signal in a small, internally consistent dataset could in some cases be overridden by a huge, homoplasious dataset. Therefore, a comparison of the number of informative characters in a dataset to the amount of stability/support provided by that dataset in combined analysis (ΣPBS) was made (Baker et al., 1998). Significance of the correlation coefficient between relative ranks of the 29 datasets was determined as above.

A similar comparison was made between the consistency of characters within each dataset *versus* character incongruence with other datasets in combined analysis. If consistency of a dataset in separate analysis (internal consistency) is indicative of consistency of that dataset in combined analysis (global consistency), measures of homoplasy in separate analysis should be correlated with measures of *additional* homoplasy required in combined analysis. So, the CI for a particular dataset, a measure of homoplasy in separate analysis, was compared to the number of extra steps required to fit a dataset to the combined data tree. The CI in separate analysis also was compared to the number of extra steps for a dataset in combined analysis divided by the total number of steps for that dataset in combined analysis. Significance of correlation coefficients was determined as above.

Results

The shortest topology for the total combined character set of five taxa is shown in Figure 2. As in most recent analyses of molecular data (e.g., Gatesy, 1998; Milinkovitch et al., 1998; Ursing and Arnason, 1998a; Nikaido et al., 1999), a sister group relationship between Hippopotamidae and Cetacea (Mysticeti) is supported (BS = 41, BS = 38 if gaps are coded as a fifth character state) along with a grouping of Hippopotamidae, Mysticeti and Bovidae to the exclusion of Suidae (BS = 72, BS = 104 with gaps).

There is abundant hidden support for these relationships. Over 92% of the BS in simultaneous analysis emerges through combination of the 29 individual datasets (total HBS = +104; Fig. 2). Within this combined analysis framework, conflict at the two supported nodes is due solely to mt DNA datasets (cytochrome *b*, ATP6, ATP8, COII, NAD2, NAD3, NAD4, NAD6) and nu amino acid datasets (β-hemoglobin, pancreatic ribonuclease) according to PBS and NDI (Fig. 2). For each data partition, NDI scores and PBS scores are identical for each node supported by the combined dataset. This result obtained because the minimum-length topology and the shortest topologies without a given supported node did not change with the removal of any single character partition from the combined matrix. Although several mt DNA datasets conflict with the two nodes supported by the combined matrix

(Fig. 2), an analysis of all 15 mt DNA datasets combined strongly supports the total data tree as in recent analyses of the mt genome (Ursing and Arnason, 1998a).

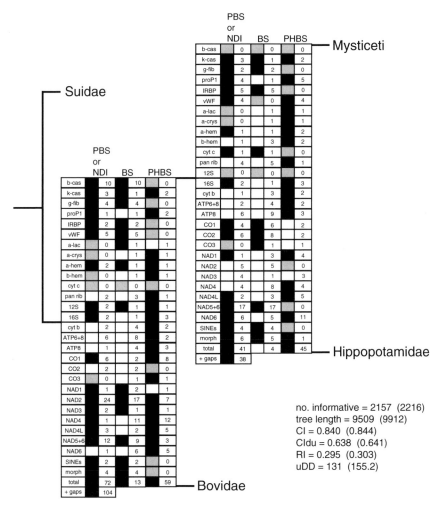

no. informative = 2157 (2216)
tree length = 9509 (9912)
CI = 0.840 (0.844)
CIdu = 0.638 (0.641)
RI = 0.295 (0.303)
uDD = 131 (155.2)

Figure 2. Topology supported by combined analyses of the 29 datasets. Number of informative characters (no. informative), tree length, CI, CIdu, RI, and uDD are shown for all of the data excluding gap characters and including gap characters (in parentheses). PBS, NDI, BS and PHBS are shown for all 29 datasets at nodes (gap characters excluded). Positive scores are marked by black boxes to the left, negative scores are marked by white boxes, and scores of zero are marked by gray boxes. Datasets are b-cas = β-casein, k-cas = κ-casein, g-fib = γ-fibrinogen, proP1 = protamine P1, IRBP, vWF, a-lac = α-lactalbumin, a-crys = α-crystallin, a-hem = α-hemoglobin, b-hem = β-hemoglobin, cyt c = cytochrome c, pan rib = pancreatic ribonuclease, 12S = 12S rDNA, 16S = 16S rDNA, cyt b = cytochrome b, ATP6 + 8 = 3' ATPase 8 + ATPase 6, ATP8 = 5'ATPase 8, CO1, CO2, CO3, NAD1 = NADH 1, NAD2 = NADH 2, NAD3 = NADH 3, NAD4 = NADH 4, NAD4L = NADH 4L, NAD5 + 6 = NADH 5 + 5' NADH 6, NAD6 = 3' NADH 6, SINEs = SINE retrotransposons, and morph = morphology.

Table 1. Datasets and tree statistics for these datasets in separate analysis

Dataset	#chars.	#inf.	inf./chars.	#trees	#nodes
nuclear DNA					
β-casein	1100 (7)	52 (13)	0.047 (26.5)	2 (26)	1 (25.5)
κ-casein	489 (16)	23 (23)	0.047 (26.5)	1 (12)	2 (11.5)
γ-fibrinogen	590 (13)	37 (17)	0.063 (23)	1 (12)	2 (11.5)
protamine P1	452 (17)	40 (16)	0.088 (18)	3 (29)	0 (29)
IRBP	1247 (4)	51 (14)	0.041 (28)	1 (12)	2 (11.5)
vWF	1242 (5)	81 (10)	0.065 (22)	2 (26)	1 (25.5)
α-lactalbumin	279 (21)	28 (20.5)	0.100 (16)	1 (12)	2 (11.5)
nuclear amino acids					
α-crystallin	173 (22)	4 (29)	0.023 (29)	1 (12)	2 (11.5)
α-hemoglobin	141 (25)	10 (25)	0.071 (21)	1 (12)	2 (11.5)
β-hemoglobin	146 (24)	8 (26)	0.055 (25)	1 (12)	2 (11.5)
cytochrome c	104 (28)	6 (27.5)	0.058 (24)	1 (12)	1 (25.5)
pancr. ribonuclease	124 (27)	11 (24)	0.089 (17)	1 (12)	2 (11.5)
mitochondrial DNA					
12S rDNA	406 (18)	34 (18.5)	0.084 (19)	2 (26)	1 (25.5)
16S rDNA	564 (14)	45 (15)	0.080 (20)	1 (12)	2 (11.5)
cytochrome b	1140 (6)	158 (6)	0.139 (11)	1 (12)	2 (11.5)
ATPase 6 + 3'ATPase 8	681 (12)	177 (5)	0.260 (2)	1 (12)	2 (11.5)
5'ATPase 8	161 (23)	28 (20.5)	0.174 (4)	1 (12)	2 (11.5)
CO1	1551 (2)	231 (2)	0.149 (9)	1 (12)	2 (11.5)
CO2	696 (11)	98 (9)	0.141 (10)	1 (12)	2 (11.5)
CO3	804 (10)	110 (8)	0.137 (13)	1 (12)	2 (11.5)
NADH 1	957 (9)	113 (7)	0.118 (14)	1 (12)	2 (11.5)
NADH 2	1044 (8)	181 (4)	0.173 (5)	2 (26)	1 (25.5)
NADH 3	357 (19)	55 (12)	0.154 (8)	1 (12)	2 (11.5)
NADH 4	1383 (3)	218 (3)	0.158 (6)	1 (12)	2 (11.5)
NADH 4L	297 (20)	34 (18.5)	0.114 (15)	2 (26)	1 (25.5)
NADH 5 + 5'NADH6	1822 (1)	285 (1)	0.156 (7)	1 (12)	2 (11.5)
3'NADH 6	528 (15)	73 (11)	0.138 (12)	1 (12)	2 (11.5)
Other data					
SINEs	20 (29)	6 (27.5)	0.300 (1)	1 (12)	2 (11.5)
Morphology	125 (26)	27 (22)	0.216 (3)	1 (12)	2 (11.5)

(continued on next page)

Tree statistics/indices for separate analyses of individual datasets are in Table 1. Tree statistics/indices for individual datasets in the context of all 29 datasets are in Table 2. Ranks in separate *versus* combined analysis for CI, RI, uDD and stability (TS *versus* ΣPBS or DI) are shown in graphical format in Figure 3. For each data partition, ranks for CI in separate analysis and number of extra steps in combined analysis are portrayed in Figure 4, as are ranks for number of informative characters and contribution to stability in combined analysis (ΣPBS).

Table 1. (continued)

Dataset	TL	CI	CIdu	RI	uDD	TS
nuclear DNA						
β-casein	456	0.941 (4)	0.700 (7)	0.481 (8)	12.7 (5)	10 (4.5)
κ-casein	194	0.933 (5)	0.683 (9)	0.435 (11)	4.7 (20.5)	2 (22)
γ-fibrinogen	225	0.911 (9)	0.677 (10)	0.459 (10)	8.5 (9.5)	6 (13)
protamine P1	191	0.859 (15)	0.663 (14)	0.325 (21)	3.1 (25)	0 (29)
IRBP	405	0.926 (7)	0.659 (16)	0.412 (12)	9.5 (7)	7 (10)
vWF	466	0.888 (10.5)	0.636 (25)	0.358 (15)	10.9 (6)	5 (15)
α-lactalbumin	121	0.860 (14)	0.638 (24)	0.393 (13)	5.0 (19)	2 (22)
nuclear amino acids						
α-crystallin	9	0.778 (29)	0.667 (13)	0.500 (5.5)	1.2 (28)	2 (22)
α-hemoglobin	61	0.918 (8)	0.750 (4)	0.500 (5.5)	2.3 (26.5)	2 (22)
β-hemoglobin	59	0.932 (6)	0.692 (8)	0.500 (5.5)	2.3 (26.5)	2 (22)
cytochrome c	7	1.000 (1.5)	1.000 (1.5)	1.000 (1.5)	0.8 (29)	1 (27.5)
pancr. ribonuclease	72	0.958 (3)	0.833 (3)	0.727 (3)	5.3 (17)	7 (10)
mitochondrial DNA						
12S rDNA	149	0.846 (16)	0.635 (26.5)	0.324 (22)	3.3 (24)	1 (27.5)
16S rDNA	186	0.839 (18)	0.670 (12)	0.333 (17)	3.9 (23)	2 (22)
cytochrome b	615	0.813 (24)	0.623 (29)	0.272 (29)	7.1 (12)	4 (16.5)
ATPase 6 + 3'ATPase 8	437	0.826 (22)	0.643 (21.5)	0.309 (24)	8.4 (11)	8 (7.5)
5'ATPase 8	134	0.888 (10.5)	0.717 (6)	0.464 (9)	6.1 (15)	6 (13)
CO1	823	0.802 (28)	0.643 (21.5)	0.294 (26)	13.5 (4)	8 (7.5)
CO2	365	0.819 (23)	0.647 (19)	0.327 (19.5)	9.3 (8)	10 (4.5)
CO3	423	0.811 (26)	0.633 (28)	0.273 (28)	4.3 (22)	2 (22)
NADH 1	505	0.842 (17)	0.635 (26.5)	0.292 (27)	6.9 (13.5)	4 (16.5)
NADH 2	706	0.834 (21)	0.672 (11)	0.354 (16)	19.9 (2)	17 (2.5)
NADH 3	192	0.807 (27)	0.648 (18)	0.327 (19.5)	5.3 (17)	2 (22)
NADH 4	815	0.812 (25)	0.639 (23)	0.298 (25)	14.3 (3)	17 (2.5)
NADH 4L	173	0.879 (12)	0.661 (15)	0.382 (14)	5.3 (17)	2 (22)
NADH 5 + 5'NADH6	1191	0.835 (20)	0.645 (20)	0.312 (23)	21.7 (1)	26 (1)
3'NADH 6	302	0.838 (19)	0.652 (17)	0.329 (18)	6.9 (13.5)	7 (10)
Other data						
SINEs	20	1.000 (1.5)	1.000 (1.5)	1.000 (1.5)	4.7 (20.5)	6 (13)
Morphology	123	0.870 (13)	0.719 (5)	0.500 (5.5)	8.5 (9.5)	9 (6)

The relative rank of each statistic is given for each of the 29 datasets in parentheses (#chars. = total number of characters, #inf. = number of informative characters, inf./chars. = variability, #trees = number of minimum-length trees, #nodes = number of resolved nodes, TL = tree length). When there were ties, ranks were averaged.

Rank correlations of dataset quality in separate *versus* combined analysis were positive and significant at p < 0.05 for CI, CIdu, RI, uDD, CI in separate analysis *versus* extra steps in combined analysis, and CI in separate analysis *versus* extra steps in combined analysis scaled to total number of steps in combined analysis. Rank correlations of dataset quality in separate *versus* combined analysis were positive and not significant at p < 0.05 for measures of tree stability (TS *versus* ΣPBS or DI). Because the removal of any individual data partition from the combined dataset had no effect on resolution or the number

Table 2. Datasets and tree statistics for these datasets in combined analysis

Dataset	Extra	CI	CIdu	RI	uDD
nuclear DNA					
β-casein	0 (5.5)	0.941 (3)	0.700 (5)	0.481 (4)	12.7 (3)
κ-casein	0 (5.5)	0.933 (4)	0.683 (6)	0.435 (6)	4.7 (10.5)
γ-fibrinogen	0 (5.5)	0.911 (6)	0.677 (7)	0.459 (5)	8.5 (7.5)
protamine P1	1 (13.5)	0.854 (13)	0.654 (11)	0.300 (16)	2.1 (17.5)
IRBP	0 (5.5)	0.926 (5)	0.659 (10)	0.412 (7)	9.5 (6)
vWF	0 (5.5)	0.888 (8)	0.636 (15)	0.358 (9)	10.9 (5)
α-lactalbumin	1 (13.5)	0.852 (14)	0.625 (20)	0.357 (10)	4.0 (12)
nuclear amino acids					
α-crystallin	1 (13.5)	0.700 (29)	0.571 (28)	0.250 (21)	0.2 (24)
α-hemoglobin	1 (13.5)	0.903 (7)	0.714 (4)	0.400 (8)	1.3 (19.5)
β-hemoglobin	3 (19.5)	0.887 (9)	0.562 (29)	0.125 (27)	−0.7 (25.5)
cytochrome c	0 (5.5)	1.000 (1.5)	1.000 (1.5)	1.000 (1.5)	0.8 (23)
pancr. ribonuclease	7 (26)	0.873 (10)	0.600 (27)	0.091 (29)	−1.7 (28)
mitochondrial DNA					
12S rDNA	0 (5.5)	0.846 (15)	0.635 (16)	0.324 (12)	3.3 (13.5)
16S rDNA	1 (13.5)	0.834 (17)	0.663 (8.5)	0.311 (14)	2.9 (15)
cytochrome b	5 (22.5)	0.806 (25)	0.613 (25)	0.241 (23)	2.1 (17.5)
ATPase 6 + 3'ATPase 8	10 (27.5)	0.808 (24)	0.614 (24)	0.218 (26)	−1.6 (27)
5'ATPase 8	10 (27.5)	0.826 (20)	0.603 (26)	0.107 (28)	−3.9 (29)
CO1	2 (17.5)	0.800 (26)	0.640 (13.5)	0.286 (18)	11.5 (4)
CO2	2 (17.5)	0.815 (22)	0.640 (13.5)	0.306 (15)	7.3 (9)
CO3	1 (13.5)	0.809 (23)	0.630 (18)	0.264 (19)	3.3 (13.5)
NADH 1	6 (24.5)	0.832 (18)	0.618 (22)	0.239 (24)	0.9 (21.5)
NADH 2	5 (22.5)	0.828 (19)	0.663 (8.5)	0.326 (11)	14.9 (2)
NADH 3	4 (21)	0.791 (28)	0.624 (21)	0.255 (20)	1.3 (19.5)
NADH 4	15 (29)	0.798 (27)	0.617 (23)	0.229 (25)	−0.7 (25.5)
NADH 4L	3 (19.5)	0.864 (12)	0.631 (17)	0.294 (17)	2.3 (16)
NADH 5 + 5'NADH6	0 (5.5)	0.835 (16)	0.645 (12)	0.312 (13)	21.7 (1)
3'NADH 6	6 (24.5)	0.821 (21)	0.626 (19)	0.247 (22)	0.9 (21.5)
Other data					
SINEs	0 (5.5)	1.000 (1.5)	1.000 (1.5)	1.000 (1.5)	4.7 (10.5)
Morphology	0 (5.5)	0.870 (11)	0.719 (3)	0.500 (3)	8.5 (7.5)

(continued on next page)

of minimum-length trees, Spearman rank correlations were undefined for number of resolved nodes and number of minimum-length trees in separate *versus* combined analysis.

CI, CIdu, RI, uDD, stability and CI *versus* extra steps in combined analysis were positively correlated in separate *versus* combined analysis. Within this general pattern there were some further regularities. For CI, RI, uDD and stability, *all* of the nu DNA datasets ranked higher in combined analysis than in separate analysis. An example is the protamine P1 dataset (Queralt et al., 1995; Gatesy, 1998). Protamine P1 ranked rather poorly in separate analysis, but the

Table 2. (continued)

Dataset	ΣPBS or DI	#trees or nodes
nuclear DNA		
β-casein	10 (4)	0 (15)
κ-casein	6 (9)	0 (15)
γ-fibrinogen	6 (9)	0 (15)
protamine P1	5 (12)	0 (15)
IRBP	7 (7)	0 (15)
vWF	9 (6)	0 (15)
α-lactalbumin	0 (21)	0 (15)
nuclear amino acids		
α-crystallin	0 (21)	0 (15)
α-hemoglobin	3 (16)	0 (15)
β-hemoglobin	−1 (23)	0 (15)
cytochrome *c*	1 (19)	0 (15)
pancr. ribonuclease	−6 (27)	0 (15)
mitochondrial DNA		
12S rDNA	2 (17.5)	0 (15)
16S rDNA	4 (14.5)	0 (15)
cytochrome *b*	−3 (25.5)	0 (15)
ATPase 6 + 3'ATPase 8	−8 (29)	0 (15)
5'ATPase 8	−7 (28)	0 (15)
CO1	10 (4)	0 (15)
CO2	4 (14.5)	0 (15)
CO3	0 (21)	0 (15)
NADH 1	2 (17.5)	0 (15)
NADH 2	19 (2)	0 (15)
NADH 3	−2 (24)	0 (15)
NADH 4	−3 (25.5)	0 (15)
NADH 4L	5 (12)	0 (15)
NADH 5 + 5'NADH6	29 (1)	0 (15)
3'NADH 6	5 (12)	0 (15)
Other data		
SINEs	6 (9)	0 (15)
Morphology	10 (4)	0 (15)

The relative rank of each statistic is given for each of the 29 datasets in parentheses (Extra = extra steps required in combined analysis, #trees or nodes = change in number of minimum-length trees or resolved nodes with the removal of a particular dataset from the combined dataset). When there were ties, ranks were averaged.

relative ranking of this dataset improved in combined analysis for CI (15 to 13), CIdu (14 to 11), RI (21 to 16), uDD (25 to 17.5) and stability (29 to 12) (Tabs 1 and 2; Fig. 3). This pattern held for the SINE retrotransposons and the morphological character set as well; ranking in combined analysis was higher than or, at worst, equal to ranking in separate analysis (Fig. 3). In contrast, many of the nu amino acid datasets and mt DNA datasets are characterized by

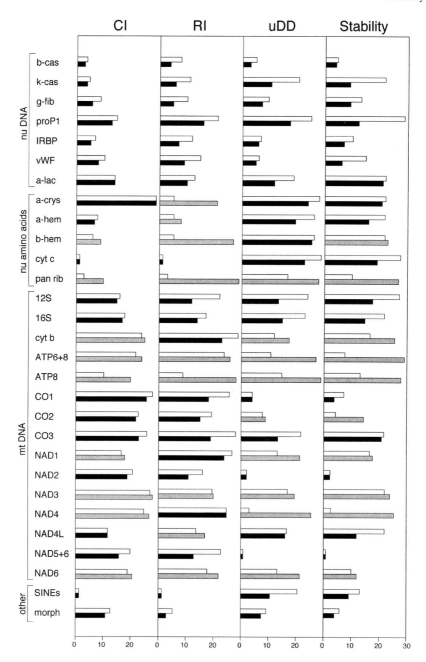

Figure 3. A comparison of relative ranks in separate analysis (white bars) *versus* relative ranks in combined analysis (black bars and gray bars) for CI, RI, uDD and stability (TS *versus* ΣPBS) for each of the 29 datasets. Datasets are abbreviated as in Figure 2. Relative ranks that are higher or equal in combined analysis relative to separate analysis are represented by black bars. Relative ranks that are lower in combined analysis relative to separate analysis are represented by gray bars. There is a significant ($p < 0.05$) positive correlation in separate *versus* combined analysis for CI, RI and uDD.

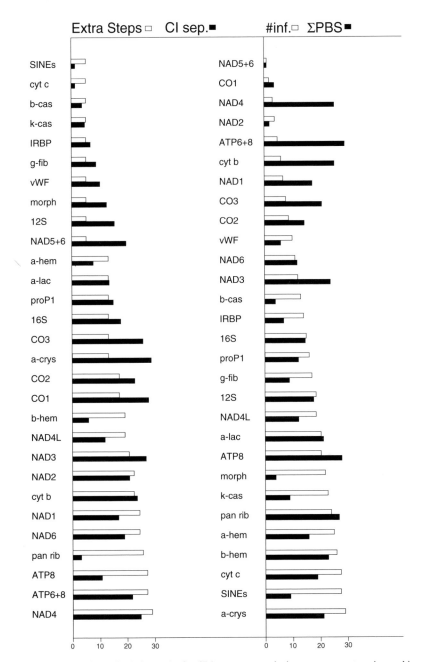

Figure 4. A comparison of relative ranks for CI in separate analysis *versus* extra steps in combined analysis and for number of informative characters *versus* contribution to stability in combined analysis (ΣPBS) for all 29 datasets. There is a significant ($p < 0.05$) positive correlation between CI in separate analysis *versus* extra steps in combined analysis.

lower ranks in combined relative to separate analysis (Tabs 1 and 2; Fig. 3). Pancreatic ribonuclease and mt ATPase 8 are extremes in this regard. RI for nu pancreatic ribonuclease is 0.727 in separate analysis but only 0.091 in combined analysis, and TS is +7 in separate analysis but ΣPBS is only -6 in the context of the other data. Similarly, RI for mt ATPase 8 is 0.464 in separate analysis but only 0.107 in combined analysis, and uDD for this dataset plummets from +6.1 in separate analysis to −3.9 in combined analysis (Tabs 1 and 2).

In terms of homoplasy, congruence with other datasets, stability and decisiveness, β-casein, κ-casein, γ-fibrinogen, IRBP, vWF, SINE retrotransposons and morphology consistently ranked highly among the 29 datasets in the combined matrix (Figs 3 and 4). None of these partitions are nu amino acid sequences or mt genes. The nu amino acid datasets are generally too small, from 4 to 11 informative characters, to have much influence in the combined analysis in terms of stability and decisiveness (but note the highly influential yet tiny SINE retrotransposon dataset; Figs 3 and 4; Nikaido et al., 1999). Some of the mt DNA datasets, such as CO1, NADH 2, and NADH 6, do provide extensive stability and decisiveness in the combined analysis (Tab. 2; Fig. 3).

There was a positive correlation between number of informative characters in a dataset and contribution to stability in combined analysis (ΣPBS), but the correlation was not significant at $p < 0.05$ (Fig. 4). Eleven of the 15 mt DNA datasets have a lower rank for contribution to stability in combined analysis than for number of informative characters. The morphological partition, the SINE retrotransposons, six of the seven nu DNA datasets, and four of the five nu amino acid datasets show the opposite pattern (Fig. 4). The SINE retrotransposon dataset is very small, six characters, but strongly contributes to the combined data result (ΣPBS = +6), and the morphological dataset, only 27 informative characters, also offers much stability in combined analysis (ΣPBS = +10). In contrast, some of the large mt DNA datasets such as NADH 4, 218 characters, detract from the stability of the total data tree (ΣPBS = −3). Per informative character, the mt DNA partitions are not very influential, and the overall quality of many mt DNA datasets decays in combined analysis.

Figure 5 shows the results for cladistic analysis of the combined 79-taxon matrix. Six minimum-length trees were discovered that had a length of 17,868 steps. The strict consensus of these topologies is consistent with relationships supported by analysis of the five-taxon matrix (Fig. 2). Cetacea and Hippopotamidae cluster to the exclusion of all other extant eutherian mammals, and Ruminantia is more closely related to Cetacea and Hippopotamidae than is Suidae (Fig. 5). More complete taxonomic sampling did not alter relationships favored by the matrix of five taxa.

Discussion

Phylogenetic accuracy cannot be measured for most systematic problems (Hillis et al., 1994). Exceptions are problems that systematists make for them-

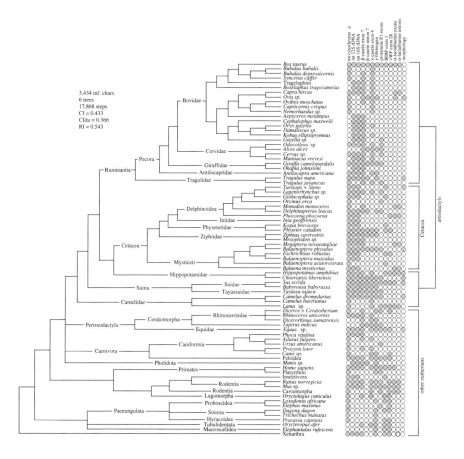

Figure 5. Strict consensus of minimum-length trees for the combined matrix of 79 taxa. Tree statistics (number of steps, number of minimum-length trees, CI, CIdu, RI, inf. chars. = number of informative characters) and delimitations of higher-level taxa are shown. Datasets sampled for each taxon are marked by shaded circles. Some taxa (e.g., *Equus* sp.) are hybrid terminals that are made up of character partitions from different closely related species. These hybrid terminals are as in Gatesy et al. (1999a).

selves, such as experimental phylogenies of viruses (Hillis et al., 1992) and phylogenies that are simulated on computers (e.g., Huelsenbeck and Hillis, 1993). Therefore, assessments of homoplasy, congruence among datasets, stability, support, decisiveness, ambiguity and resolution have been emphasized in empirical phylogenetic studies.

Unlike accuracy, these qualities are readily quantified in *real* datasets and are attributes that are of obvious importance to many systematists. For example, if there is much homoplasy in a character set, the data may not be especially hierarchical, and there is an abundance of conflicting evidence. All else being equal, perfectly hierarchical, highly decisive data are superior to homoplastic, incongruent data. Furthermore, if there are millions of equally parsimonious trees for a dataset, and no clades are resolved in the strict consensus of those topologies, a systematist has not appreciably narrowed the spectrum of viable topologies for the group of interest, and phylogenetic conclusions are vague. Lack of resolution is the result of conflicting character support (see above) and/or absence of character support. All else being equal, high resolution is preferable to a general lack of resolution; if there is no resolution, no groups are supported by a dataset. Similarly, if the strict consensus of shortest trees for a particular dataset is not stable to only a slight relaxation of the parsimony criterion (i.e., BS is low for all supported nodes), systematic results are not robust. Stability to the addition of only a few novel characters is not guaranteed, and the information for choosing one tree topology over another, decisiveness (Goloboff, 1991), may be poor. For a particular dataset, if BS is zero or negative for all possible clades, resolution is nonexistent (see above), character conflict is high (see above), and/or character support is lacking (see above). Again, all else being equal, stability is better than instability, and stability is at some level intertwined with homoplasy, congruence, ambiguity, resolution, support and decisiveness.

Nixon and Carpenter (1996) correctly noted that, "…simultaneous analysis of combined data better maximizes cladistic parsimony than separate analyses, hence is to be preferred. The position that datasets should be analyzed separately is clearly based on a rejection of the principle of parsimony in cladistics" (also see Miyamoto, 1985; Kluge, 1989). The quality of a particular dataset is often assessed in isolation from other datasets, but the positive attributes of any individual dataset may erode or be enhanced in a combined analysis of many datasets. For example, a single dataset that is perfectly internally consistent, is decisive, produces a single well-supported topology, but is sharply contradicted by 28 independent mutually congruent datasets, might not be considered especially useful. An undetected gene duplication event, interspecific hybridization, gene conversion, extreme directional selection, or other biological processes could produce such *internally* consistent but *globally* inconsistent datasets. Therefore, the utility of any particular dataset should be determined within the context of all relevant character data (Kluge, 1989; Eernisse and Kluge, 1993; Farris et al., 1994; Baker and DeSalle, 1997; Baker et al., 1998).

Phylogenetic analyses of cetartiodactyl mammals show that for the 29 character partitions analyzed here, there is a general positive correlation between dataset quality in separate *versus* combined analysis (Figs 3 and 4). Nevertheless, the relative quality of some datasets differs radically in separate *versus* combined analysis (Tabs 1 and 2; Figs 3 and 4). These results are not surprising, but should give pause to those who overstate the phylogenetic implications of any one character partition in isolation from the remaining evidence.

For example, Luckett and Hong (1998) presented a detailed analysis of two morphological characters in cetartiodactyls: presence/absence of elongate trilobed lower fourth deciduous premolars and presence/absence of a double-trochleated astragalus. Both characters are present in all extant and extinct artiodactyls that were examined (Luckett and Hong, 1998) and apparently are absent from all cetaceans. This led Luckett and Hong (1998) to suggest that Artiodactyla, excluding Cetacea, is monophyletic. They pointed out that, "This corroboration of 'traditional' hypotheses presents few surprises for evolutionary morphologists, who have investigated in detail the comparative aspects of artiodactyl and cetacean systematics for more than 170 years" and concluded that, "Common sense indicates that the two rare or unique synapomorphic character complexes of the tarsus and deciduous dentition, shared by all known fossil and extant hippopotamids and artiodactyls, provide robust corroboration for artiodactyl monophyly…"

This sort of argument may have been par for the course, say 170 years ago, but does not hold much weight in the 21st century. A dataset composed of two phylogenetic characters cannot "provide robust corroboration" for anything. Luckett and Hong (1998) were insightful in noting that, "…the primary source of error in both morphological and molecular analyses of evolution is *homoplasy* [their italics]." Within the context of 18,621 characters that apparently lack "common sense," the two "uniquely derived morphological traits" shared by all artiodactyls were apparently gained in the common ancestor of Cetartiodactyla and then lost on the lineage that leads to Cetacea. That is, they are *homoplastic* and are not uniquely derived.

In another study, Nikaido et al. (1999) argued that insertions of SINE retrotransposons provide the definitive evidence for cetartiodactyl phylogeny. However, the relationships supported by the 20 SINE characters (Nikaido et al., 1999) were overwhelmingly and robustly supported in several earlier studies (Gatesy, 1998; Ursing and Arnason, 1998a; Gatesy et al., 1999a), missing data in the transposon matrix weaken the support provided by this dataset, and claims that insertions of transposons are homoplasy-free have been shown to be false (Pecon Slattery et al., 2000). This is not to say that insertions of transposons are poor systematic characters. In the present analysis, the SINE transposons ranked highly for measures of congruence and consistency, and ranked above average for stability and decisiveness in the combined analysis framework (Tab. 2, Figs 3 and 4). Indeed, the strength of the SINE transposons is not their presumed infallibility. The strength of the SINEs is high congruence with independent character information; the transposon dataset is perfectly consistent with the total data tree that is based on over 18,000 characters (Fig. 2).

On a much grander scale, certain systematists have argued that complete mt genomes offer powerful evidence for phylogenetic relationships because of the large number of characters that they provide (e.g., Ursing and Arnason, 1998a). Other researchers have suggested that complete mt genome datasets can be highly misleading (Naylor and Brown, 1998). To my knowledge, complete mt genome data have *never* been merged with other datasets in combined phylogenetic analyses. Hidden suppport or conflicts that emerge with the combination of datasets in simultaneous analysis cannot be assessed unless data partitions are analyzed together (Barrett et al., 1991). Therefore, previous interpretations of the relative utility of mt genomes, either positive or negative, may be misleading (e.g., Sullivan and Swofford, 1997; Naylor and Brown, 1998; Ursing and Arnason, 1998a).

The simultaneous analysis here suggests that the mt genome, a combination of all 15 mt DNA datasets (12,391 sites), is among the weakest character partitions in terms of homoplasy (CI = 0.818, CIdu = 0.633, RI = 0.272). Among the nu DNA datasets, morphology, and the SINEs, only α-lactalbumin is more internally inconsistent than the mt genome according to CIdu. In combined analysis, all of these datasets have higher RI and CI than the mt genome (Tabs 1 and 2), and a concatenated nu DNA dataset (caseins + γ-fibrinogen + protamine P1 + α-lactalbumin + vWF + IRBP = 5,399 aligned sites) is characterized by less homoplasy than the mt DNA according to CI (0.909), CIdu (0.660) and RI (0.397). However, the mt data do not conflict with the total data tree (zero extra steps in combined analysis) and provide more stability in combined analysis (ΣPBS = +57) than in separate analysis (TS = +50).

For the four cetartiodactyl taxa sampled, mt genomes are fairly effective phylogenetic markers. In the context of 14 other datasets, the mt DNA data are highly congruent and provide abundant stability and decisiveness (uDD = +66.2) despite a high level of homoplasy. However, in terms of value per basepair sequenced or value per phylogenetically informative character, the huge mt genome dataset does not compare favorably to the less homoplasious nu DNA, SINE, and morphological datasets (Tabs 1 and 2; Figs 3 and 4). These partitions provide proportionately more stability and decisiveness in combined analysis. Compared to the mt genome, the combined nu DNA dataset has less than one-fifth the number of informative characters and less than one-half the number of nucleotide sites, but provides almost as much stability (ΣPBS = +43) and decisiveness (uDD = +52.5) as the entire mt genome.

A diversity of phylogenetic characters can be collected for any systematic problem. Two morphological traits, 20 transposition events, or even a complete mt genome of over 12,000 nucleotides is a highly skewed sampling of the available character-space (Fig. 1). Presumably stable morphological traits, such as a double-trochleated astragalus, can be lost in the course of evolution. Insertions of SINEs can be convergent or retained as inconsistently sorted polymorphisms (Pecon Slattery et al., 2000). Similarly, even a giant molecular dataset of over 12,000 nucleotides may be highly homoplastic, and the possibility of incomplete lineage sorting, gene duplication, or introgression makes phylogenetic interpretations of a single non-recombining locus problematic. One approach to counteracting such dataset-specific biases is to broadly sample the available character-space and merge different datasets in simultaneous analysis (Barrett et al., 1991; Wheeler et al., 1993; Nixon and Carpenter, 1996; Baker et al., 1998).

Systematists have identified a variety of factors that may negatively affect congruence, support and resolution. For mt DNA in mammals, these biases include an overall rapid rate of nucleotide substitution, a high transition/transversion ratio, saturation of substitutions at third codon positions, nonstationarity of base composition, lack of independence among substutions in stem regions of mt rDNAs, and structural constraints of mt proteins (Kraus et al., 1992; Collins et al., 1994; Honeycutt et al., 1995; Naylor and Brown, 1998). An empirical approach to overcoming such biases is to combine the mtDNA sequences with datasets that do not have these same biases. For example, nu DNA sequences in mammals generally do not evolve as rapidly as mt DNA sequences, third codon positions of nu genes are not necessarily saturated even among distantly related species, nu introns cannot have the same structural constraints as mt DNA sequences that encode functional proteins, and transition/transversion ratios for nu genes are oftentimes not as high as in mt DNA. Alternatively, morphological characters or SINE retrotransposons cannot have any of the aforementioned biases of nucleotides in mt DNA, simply because gross anatomical characters and SINEs are not individual nucleotides. If datasets without the apparent biases of mt DNA support the same controversial relationships as mt DNA, perhaps such controversial relationships are not controversial. Alternatively, if other datasets sharply contradict the mt DNA sequences in combined analysis, then the controversial relationships supported by mt DNA alone may not be as well supported as was previously thought. Either way, the potential biases of mt DNA data are tested empirically by independent characters that do not have all of the same prejudices as mt DNA. Assuming that mt genomes are misleading before merging these data with apparently contradictory evidence in combined analysis could be considered an appeal to authority (e.g., Sullivan and Swofford, 1997; Naylor and Brown, 1998). The interactions among different datasets in combined analysis are not always predictable based on separate analyses of individual datasets (note the hidden support in Fig. 2).

The utility or quality of any particular dataset should not be the primary concern of a systematist. Instead, sampling the available character-space broadly and including a wide-ranging representation of relevant taxa is more important. However, a large combined matrix is by definition composed of many smaller datasets. In order to best utilize their time and money, systematists need to choose component datasets in a combined matrix with care (Nixon and Carpenter, 1996). There are several concerns. First, the cost required to collect a dataset is relevant. In many cases, the collection of morphological evidence is much less expensive than the collection of molecular evidence, and per informative character, DNA sequence data are much cheaper than SINE retrotransposon data. Second, dense taxonomic sampling is critical (e.g., Gauthier et al., 1988). Most species are extinct; O'Leary and Geisler (1999) estimated that over 85% of cetartiodactyl genera are only known from fossil material. It may be impossible to score molecular characters from the majority of these extinct taxa. Therefore, morphological data again have an advantage over molecular information; at least some gross anatomical characters can be recorded from fossils. Third, a broad spectrum of datasets should be combined to offset any biases inherent in any single dataset. So, given the available character-space (Fig. 1), it would be silly to sample 15 mt genes.

Instead, a combined dataset that includes morphological characters, a few mt genes, a few nu genes, transposons, behavior and chromosomal rearrangements might be easier to defend as a representative sample of the available character-space. Fourth, the component datasets of a combined matrix should be informative for the systematic problem of interest.

This last concern is by far the most problematic. There is no guarantee that a given character partition is going to be effective for resolving a particular phylogenetic tree. Molecular systematists commonly survey published datasets to identify genes that have evolved at an "appropriate" rate for the systematic problem at hand. For example, mt D-Loop sequences have been shown to evolve at an extremely rapid rate in many mammalian taxa. Because of alignment ambiguity and a saturation of nucleotide substitutions, this region of DNA has been ignored in most analyses of higher-level mammalian relationships (e.g., Ursing and Arnason, 1998a). However, D-Loop sequences have been utilized successfully in many phylogenetic studies of closely related species and subspecies of mammals (e.g., Arctander et al., 1999). Conversely, slowly evolving molecules that may be quite effective for resolving ancient branching events within Mammalia (e.g., IRBP; Stanhope et al., 1996) may be relatively useless in an analysis of very closely related taxa because of a lack of informative variation. At least some sequence variation is necessary, but too much variation may imply conflicting character evidence.   There is no guarantee that, for a particular gene, the rate of evolution in one lineage is the same as the rate in another lineage. For example, the simultaneous analyses here suggest that κ-casein is a good locus for examining cetartiodactyl relationships. In combined analyses with 28 other datasets, κ-casein is congruent and decisive (Figs 3 and 4). Furthermore, κ-casein is effective for resolving relationships among pecoran artiodactyls, species that have diversified within the last 20 million years (Cronin et al., 1996; Gatesy, 1998). Nevertheless, this locus is not especially useful for discerning relationships among cetaceans. Within Cetacea, κ-casein has evolved at a much slower rate than within Pecora, especially at synonymous sites; only a few nucleotide positions in this molecule are phylogenetically informative among cetacean genera (Gatesy, 1998). The radiation of Cetacea is more ancient than the diversification of Pecora (Carroll, 1988), but κ-casein provides only trivial character support in analyses of higher-level cetacean phylogeny.

The support, stability, congruence, decisiveness and resolving power of a particular gene may be high in some taxa, but not in others. For the four cetartiodactyl taxa analyzed here, mt 12S rDNA is one of the more congruent partitions. Zero extra steps are required to fit this character set to the total data tree, and in the combined analysis, CI (0.846) and RI (0.324) are high relative to the other mt DNA datasets (Tab. 2, Figs 3 and 4). Nevertheless, a combined analysis of 12 cetartiodactyl species, including the four taxa analyzed here (Gatesy et al., 1999b), showed that mt 12S rDNA was by far the most incongruent of the 17 datasets in the combined matrix according to the ILD test (Farris et al., 1994), and the ΣPBS from mt 12S rDNA was only –0.75.

Choosing a gene for a particular systematic problem on the basis of previous studies is simply an educated guess.

Within a cladistic framework, the quality of a particular character set is best judged within a combined analysis framework, because a dataset that is internally consistent, decisive, and supports a stable, well-resolved result in separate analysis may be ineffectual or contrary within the context of independent data. Unfortunately, variability, congruence, support, stability, resolving power and decisiveness of a dataset are not always predictable among different taxa. So assessments of dataset quality or utility can only be used to make educated guesses about which partitions should be sampled in a future systematic study. Instead of worrying about the performance of any single dataset in separate *or* combined analysis, more effort should be invested in sampling a diversity of character sets and taxa. In this way, the final analysis will not be dependent on any single educated guess or on any single dataset that may have a variety of unique and inconsistent biases.

Acknowledgements
C. Hayashi, S. Jansa and A. de Queiroz commented on earlier drafts of the manuscript. This work was funded by an NSF systematics grant (DEB 998 5847).

Phylogeny of the holometabolous insect orders based on 18S ribosomal DNA: when bad things happen to good data

Michael F. Whiting

Department of Zoology, Brigham Young University, Provo, UT 84602, USA

Summary. The purpose of this chapter is two-fold. First, all available 18S rDNA sequences for the Holometabola to reappraise their phylogenetic relationships will be compiled. Second, these data and analyses will be used to highlight general problems in using molecular data to infer higher-level phylogeny.

Introduction

When one speaks of biodiversity, one is really talking about the insects with complete metamorphosis, the Holometabola. The Holometabola account for more than 80% of insect species and more than 50% of all animal species (Kristensen, 1995; Wilson, 1988). The monophyly of the Holometabola, and of each of the 11 orders included, is relatively well supported by morphological and molecular evidence (Kristensen, 1995; Whiting et al., 1997). The only exception may be the possible paraphyly of the order Mecoptera (scorpionflies), since molecular (Whiting et al., 1997) and morphological (Bilinski et al., 1998; Schlein, 1980) evidence suggests that the order Siphonaptera (fleas) may be nested within Mecoptera, most likely as sister group to Boreidae. With the exception of Lepidoptera + Trichoptera (Amphiesmenoptera), ordinal sister group relationships are not well supported and some have been highly controversial in the recent literature (e.g., Diptera + Strepsiptera).In the past few years, some effort has been placed on using DNA sequence data to decipher phylogenetic relationships among these insects (Carmean et al., 1992; Pashley et al., 1993; Whiting et al., 1997), though many relationships are still ambiguous.

What is most striking about holometabolan phylogeny is how little we actually know about it. For instance, neither molecular nor morphological evidence robustly supports placement for Hymenoptera. While a sister group relationship between Coleoptera and Neuroptera appears widely accepted, it is based on a single character of the female ovipositor (Achtelig, 1975; Mickoleit, 1973) and this hypothesis receives very limited support from molecular data (Whiting et al., 1997). The monophyly of Mecopterida (Amphiesmenoptera +

Diptera + Strepsiptera + Mecoptera + Siphonaptera) is questionable, and relationships among the neuropterid orders (Neuroptera, Megaloptera and
Raphidioptera) are arguable (Kristensen, 1995), though it seems likely that
Neuropterida is a monophyletic group. While current molecular data have
helped resolve some questions in insect phylogeny, it has mostly highlighted
the sparseness of evidence supporting phylogenetic relationships.

Materials and methods

18S rDNA sequences acquired from Genbank and EMBL were augmented by
additional sequences generated in my lab. Sequences were generated using the
primers and methodology as described in Whiting et al. (1997), except that the
entire region of 18S was amplified and sequenced. Only sequences of 1kB or
greater were used in this study, with only one sequence selected per
holometabolan insect family, with the exception of two relatively diverse families, the Carabidae and Tipulidae. Emphasis was placed on sampling
Holometabola as thoroughly as possible, with an adequate number of outgroup
taxa, primarily representing the Paraneoptera, the sister group to the
Holometabola (Kristensen, 1991; Whiting et al., 1997). This sampling resulted in 122 sequences, 100 from Holometabola and 22 from outgroup taxa.
While other holometabolous insect families have been sequenced for 18S
rDNA (e.g., Farrell, 1998), only 92 sequences were available in the public
domain as of November, 1999; the other 30 sequences were generated in this
study (Tab. 1). Roughly half of these sequences consist of the entire 18S rDNA
region and the other half of just over 1 Kb of sequence data.

Sequences were assembled in Sequencher™ 3.1.1 (Genecodes, 1999) and a
gross alignment was performed by manually aligning the conserved domains
across the taxa. Each conserved domain, and variable regions between the
domains, were removed in sections and entered into the computer program
POY (Gladstein and Wheeler, 2000a) to undergo more exhaustive alignment.
POY was implemented using gap cost = 2, change cost = 1, with TBR branch
swapping on 100 alignments, with the option "implied alignment" implemented. While POY is designed to construct a topology while simultaneously performing alignment (Wheeler, 1999a), the implied alignment option outputs a
multiple alignment which is more optimal than those typically found by other
alignment algorithms such as MALIGN (Wheeler and Gladstein, 1995) or
Clustal W (Thompson et al., 1996; W. Wheeler, pers. comm., 2000). In the
case of multiple equally parsimonious alignments, the first alignment was arbitrarily selected for subsequent analyses.

Variable regions which appeared ambiguously aligned across the insect
orders, but relatively conserved within each order, were aligned independently within each holometabolous insect order using POY with the parameters as
described above. Alignments were constructed in blocks for each
holometabolous insect order. Mecoptera and Siphonaptera were treated as a

Table 1. List of taxa used in the analysis

Ephemeroptera	*Ephemerella* sp.	U65107
Odonata	*Calopteryx maculata*	U65108
Odonata	*Libellula pulchella*	U65109
Embioptera	*Oligotoma saundersii*	U65117
Blattodea	*Blaberus* sp.	U65112
Mantodea	*Mantis religiosa*	U65113
Psocodea	*Cerastipsocus venosus*	U65118
Hemiptera		
Cicadidae	*Okanagana utahensis*	U06478
Cercopidae	*Prosapia plagiata*	U16264
Cicadellidae	*Graphocephalus atropunctata*	U15213
Membracidae	*Spissistilus festinus*	U06477
Delphacidae	*Prokelisia marginata*	U09207
Cixidae	*Olarius hesperinus*	U15215
Dictyopharidae	*Scolops fumida*	U15216
Issidae	*Hysteropterum severini*	U15214
Flatidae	*Siphanta acuta*	U06481
Peloridiidae	*Hemiowoodwardia wilsoni*	AF131198
Gerridae	*Aquarius remigis*	U15691
Saldidae	*Saldula pallipes*	U65121
Notonectidae	*Buenoa* sp.	U65120
Lygaeidae	*Lygus lineolaris*	U65122
Pentatomidae	*Rhaphigaster nebulosa*	X89495
Coleoptera		
Hydroscaphidae	*Hydroscapha natans*	AF012525
Trachypachidae	*Trachypachus gibbsii*	AF002808
Carabidae	*Cicindela s. sedecimpunctata*	AF012518
Carabidae	*Metrius contractus*	AF012515
Carabidae	*Clinidium calcaratum*	AF012521
Carabidae	*Omophron obliteratum*	AF012513
Carabidae	*Omus californicus*	AF012519
Noteridae	*Suphis inflatus*	AF012523
Dytiscidae	*Copelatus chevrolati renovatus*	AF012524
Staphylinidae	*Xanthopyga cacti*	AF002810
Clambidae	*Clambus arnetti*	AF012526
Scarabaeidae	*Dynastes granti*	AF002809
Elateridae	*Octinodes* sp.	U65128
Lampyridae	*Photuris pennsylvanica*	U65129
Rhipiphoridae	*Rhipiphorus fasciatus*	U65130
Meloidea	*Meloe proscarabaeus*	X77786
Cerambycidae	*Tetraopes tetropthalmus*	U65131
Megaloptera		
Sialidae	*Sialis* sp.	X89497
Raphidioptera		
Raphidiidae	*Agulla adnixa*	AF286301
Inoceliidae	*Nehga inflata*	AF286272

(continued on next page)

Table 1. (continued)

Neuroptera		
Ithonidae	*Oliarces clara*	AF012527
Berothidae	*Loloymia texana*	U65134
Mantispidae	*Mantispa pulchella*	U65135
Hemerobiidae	*Hemerobius stigmata*	U65136
Chrysopidae	*Anisochrysa carnea*	X89482
Myrmeleontidae	*Myrmeleon immaculatus*	U65137
Hymenoptera		
Cephidae	*Hartigia cressonii*	L10173
Orussidae	*Orussus thoracicus*	L10174
Tenthredinidae	*Hemitaxonus* sp.	U65150
Trigonalyidae	*Bareogonalos canadensis*	L10176
Ichneumonidae	*Ophion* sp.	U65151
Braconidae	*Trioxys pallidus*	AJ009351
Evaniidae	*Evania appendigaster*	L10175
Pteromalidae	*Mesopolobus* sp.	L10177
Bethylidae	*Epyris sepulchralis*	L10180
Chrysididae	*Caenochrysis doriae*	L10179
Pompilidae	*Priocnemus oregana*	L10181
Mutillidae	*Dasymutilla gloriosa*	U65152
Vespidae	*Apoica* sp.	U65153
Vespidae	*Monobia quadridens*	U65154
Formicidae	*Doronomyrmex kutteri*	X73274
Lepidoptera		
Micropterigidae	*Micropterix calthella*	AF136883
		AF136863
Agathiphagidae	*Agathiphaga queenslandensis*	AF136884
		AF136864
Heterobathmiidae	*Heterobathmia pseuderiocrania*	AF136887
		AF136867
Eriocraniidae	*Eriocrania semipurpurella*	AF136886
		AF136866
Hepialidae	*Sthenopis quadriguttatus*	AF136891
		AF136871
Prodoxidae	*Tegeticula yuccasella*	AF136889
		AF136869
Psychidae	*Thyridopteryx ephemeraeformis*	AF136894
		AF136874
Tineidae	*Tineola bisselliella*	AF136893
		AF136873
Pyralidae	*Galleria mellonella*	AF286298
Papilionidae	*Papilio troilus*	AF286299
Saturniidae	*Hemileuca* sp.	AF286273
Lymantriidae	*Lymantria dispar*	AF136892
		AF136872
Noctuidae	*Ascalapha odorata*	U65140

(continued on next page)

Table 1. (continued)

Trichoptera		
Philopotamidae	*Wormaldia moesta*	AF136881
		AF136861
Brachycentridae	*Brachycentrus nigrosoma*	AF136880
		AF136860
Limnephilidae	*Pycnopsyche lepida*	AF286292
Limnephilidae	*Hydropsyche* sp.	AF286291
Leptoceridae	*Oecetis avara*	AF286300
Siphonaptera		
Pulicidae	*Tunga monositus*	AF286279
Rhopalopsyllidae	*Polygenis pradoi*	AF286277
Hystricopsyllidae	*Hystrichopsylla t. talpae*	AF286281
Coptopsyllidae	*Coptopsylla africana*	AF286275
Pygiopsyllidae	*Acanthopsylla r. rothschildi*	AF286283
Stephanocircidae	*Craneopsylla minerva wolffheuglia*	AF286286
Ctenophthalimidae	*Megarthroglossus divisus*	AF286276
Ischnopsyllidae	*Myodopsylla palposa*	AF286282
Leptopsyllidae	*Frontopsylla nakagawai*	AF286280
Ceratophyllidae	*Orchopeas sexdentatus*	AF286274
Mecoptera		
Boreidae	*Boreus coloradensis*	AF286285
Boreidae	*Caurinus dectes*	AF286288
Meropeidae	*Merope tuber*	AF286287
Bittacidae	*Bittacus strigosus*	AF286290
Apteropanorpidae	*Apteropanorpa evansi*	AF286284
Choristidae	*Chorista australis*	AF286289
Panorpodidae	*Brachypanorpa carolinensis*	AF286296
Panorpidae	*Panorpa takenouchii*	AF286278
Diptera		
Tipulidae	*Tanyptera dorsalis*	AF286295
Tipulidae	*Epiphragma fasciapenne*	AF286294
Tipulidae	*Dolichopeza subalbipes*	AF286297
Tipulidae	*Nephrotoma altissima*	U48379
Chaoboridae	*Corethrella wirthi*	U49736
Dixidae	*Dixella cornuta*	U48381
Chironomidae	*Ablabesmyia rhamphe*	U48384
Simuliidae	*Simulium vittatum*	U48383
Psychodidae	*Lutzomyia shannoni*	U48382
Tabanidae	*Chrysops niger*	AF073889
Asilidae	*Laphria* sp.	AF286293
Bombyliidae	*Mythicomyia atra*	U65158
Tephritidae	*Ceratitis capitata*	AF096450
Drosophilidae	*Drosophila melanogaster*	M21017
Hippoboscidae	*Ornithoica vicina*	AF073888
Stylopidae	*Stylops melittae*	X89440
Stylopidae	*Xenos pecki*	U65164
Stylopidae	*Crawfordia* n.sp.	U65163

single order because prior evidence suggests that the Mecoptera is paraphyletic with respect to Siphonaptera (Bilinski et al., 1998; Schlein, 1980; Whiting et al., 1997). These variable regions were excluded from the outgroups because resolution among the outgroups is not the focus of this study. The alignment blocks were assembled together by scoring the taxa not included in the alignment for a particular order with missing values, and assembling the blocks into a single alignment. This produces a blocked alignment for the variable regions, and each of these blocks were spliced together into a single alignment to form one variable blocked alignment (Fig. 1). Only well-supported monophyletic groups, as inferred from the conserved portions of the alignment and/or morphology, were designated as blocks in this alignment. The variable blocked regions and conserved regions were then assembled into a single alignment, with the conserved regions flanking the variable, blocked regions. The alignment can be found at http://dnasc.byu.edu/~Whitinglab/data_sets.htm.

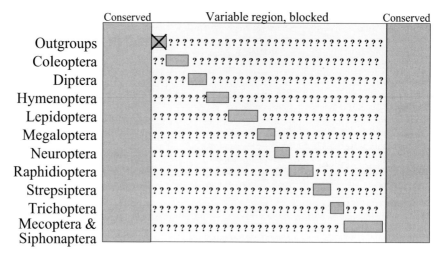

Figure 1. Scheme of alignment for blocked, variable regions. Regions that are ambiguously aligned between orders, but unambiguously aligned within orders, were aligned as blocks for each holometabolous insect order (represented as light boxes). Taxa outside of the block regions were coded with missing data. These regions were spliced together to create a step-like formation in the total alignment, where the variable block regions are flanked by conserved regions. The blocks were combined for Mecoptera and Siphonaptera and excluded for the outgroups.

Trees were reconstructed under parsimony with gaps treated as missing data. The program NONA (Goloboff, 1994) was employed to search the tree space, with 100 random addition sequences, holding 20 trees for each swapping cycle, and then performing TBR branch swapping on the set of most optimal trees found during random addition. Bremer supports (Bremer, 1994) were calculated in NONA by saving trees up to 5 steps away from the most parsimonious solution. Bootstrapping was performed in PAUP (Swofford, 2000) using a full heuristic search with 1000 replicates. Trees were reconstructed with the blocked, variable regions included and excluded, and the order Coleoptera was either constrained or unconstrained as a monophyletic group.

Results

Alignment of 18S rDNA using the methods described above resulted in 10 conserved regions (regions aligned across all taxa) and three variable regions which were blocked out. Two highly autapomorphic strepsipteran insert regions were excluded from the analysis. Phylogenetic analysis of the data with variable, blocked regions included resulted in 60 trees of length 7120, the strict consensus of which is presented in Figure 2. This topology supports the monophyly of Holometabola, although with low bootstrap and Bremer support values. The hemipteran suborder Auchenorrhyncha is paraphyletic and the Heteroptera is monophyletic, in accord with the results of other 18S studies (Campbell et al., 1995; Von Dohlen and Moran, 1995). Relationships among the outgroups are largely unresolved, though the monophyly of Neoptera is

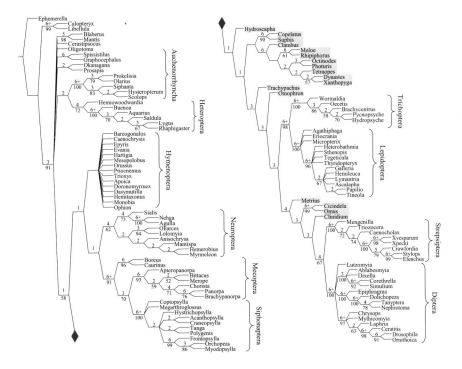

Figure 2. Strict consensus of 60 most parsimonious trees (L = 7120, CI = 38, RI = 49) from the 18S rDNA alignment with blocked variable regions included. Coleoptera is paraphyletic and the beetle taxa are highlighted. Numbers above nodes are Bremer support; numbers below nodes are bootstrap values.

supported. Except for Coleoptera and Mecoptera, all other holometabolous insect orders are supported as monophyletic. Coleoptera is grossly paraphyletic with the Polyphaga forming one clade, Dytiscidae + Noteridae another. The family *Carabidae* is polyphyletic, and *Hydroscapha* (Myxophaga) and *Trachypachus* (Adephaga) are placed in two other positions on the tree. Maddison et al. (1999) suggested that the taxa *Omophron, Metrius, Cicindela, Omus* and *Clinidium* had long branches for 18S and thus excluded them in some analyses. Exclusion of these taxa from this data set still yields a paraphyletic Coleoptera with the Polyphaga and Adephaga placed in different portions of the tree.

Mecoptera is paraphyletic with respect to Siphonaptera in these results. The family Boreidae is the most basal taxon of this lineage, and fleas form a sister group relationship with the remainder of the Mecoptera. With a more thorough sampling of mecopteran and flea taxa, 18S supports the Siphonaptera as sister group to Boreidae (Whiting, in prep). The tribe Panorpiini ((Choristidae + (Panorpodidae + Panorpidae)) is supported, but this result disagrees with Willmann (1987) in the placement of Meropeidae as sister group to Bittacidae. Phylogenetic relationships among fleas are poorly known (Lewis and Lewis, 1985), though this analysis does support the monophyly of Ceratophylloidea (Leptopsyllidae + Ceratophyllidae + Ischnopsyllidae). Amphiesmenoptera is considered the best-supported ordinal assemblage among insects, supported by well over 20 morphological characters (Kristensen, 1991, 1995), and was recovered in this analysis with high Bremer and bootstrap values. The monophyly of Lepidoptera was also well supported, though Glossata is paraphyletic with *Eriocrania* placed with the mandibulate moths as the most basal taxon. The majority of lepidopteran taxa sampled have only been sequenced for about 1 kb of 18S, so inferred relationships may change with the addition of missing sequence data. Trichoptera is also well supported as a monophyletic order

The controversial group Halteria (Strepsiptera + Diptera) was supported in this analysis. The group was first proposed by Whiting and Wheeler (1994), and the grouping of these taxa was subsequently challenged to be the result of long-branch attraction (Carmean and Crespi, 1995; Felsenstein, 1978; Huelsenbeck, 1997), though this relationship is most congruent with both morphological and molecular data (Whiting, 1998a; Whiting, 1998b). Reanalysis of the original Whiting et al. (1997) data by Huelsenbeck (1998), with likelihood methods that account for rate heterogeneity, could neither support nor refute this hypothesized sister-group relationship. The current analysis is unique in that it contains what is presumed to be the most primitive strepsipteran genus *Mengenilla*, which is placed as the most basal taxon in this analysis. Strepsiptera and Diptera are both well supported as monophyletic groups. Within Diptera, Brachycera, Cyclorrhapha and Tipulidae are monophyletic; Nematocera is at the base of Diptera, though relationships within Nematocera are partially unresolved.

Neuropterida (Megaloptera + Raphidioptera + Neuroptera) is supported with the Megaloptera as sister group to Raphidioptera, as suggested by morphology (Kristensen, 1991; Whiting et al., 1997). While Hymenoptera is supported as a monophyletic group, relationships within this order are entirely unresolved. In terms of overall relationships, this analysis divides the Holometabola into two major clades: (Hymenoptera + (Neuropterida + Mecoptera/Siphonaptera)) and Amphiesmenoptera + Halteria with the Coleoptera scattered throughout the second group. The Antliophora (Diptera + Strepsiptera + Mecoptera + Siphonaptera) is paraphyletic.

Constraining the monophyly of Coleoptera generates 160 trees, the strict consensus of which supports a topology largely congruent with the unconstrained tree for the non-beetle taxa (Fig. 3). The Coleoptera are placed as sister group to Amphiesmenoptera + Halteria, though this is rather poorly supported in terms of bootstrap and Bremer values. It requires 63 additional steps to constrain Coleoptera as monophyletic. The results of this constrained analysis support a monophyletic Polyphaga (as in the unconstrained tree), but the Adephaga are unresolved, as are the relationships among the suborders. When the variable blocked alignment regions are excluded from the analysis, the same general topology of interordinal relationships is produced, except that the

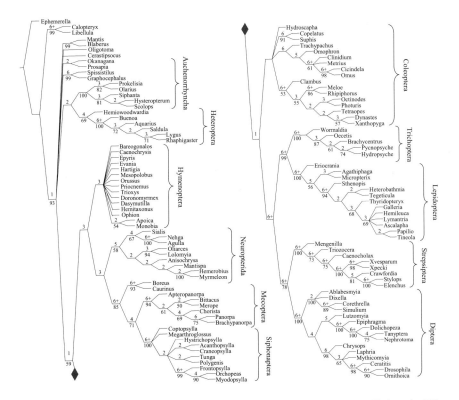

Figure 3. Strict consensus of 160 most parsimonious trees (L = 7183, CI = 39, RI = 55) from the 18S rDNA alignment with blocked variable regions included and the Coleoptera forced as monophyletic. Numbers above nodes are Bremer support; numbers below nodes are bootstrap values.

Hymenoptera is not supported as a monophyletic group. Removing these blocks reduces resolution within the Lepidoptera, Trichoptera, Neuroptera and Polyphaga, and changes relationships within the Diptera (Nematocera is paraphyletic), Strepsiptera (*Triozocera* + *Crawfordia* is the most apical group) and Mecoptera (*Bittacus* + *Merope* are the most basal mecopteran taxa, excluding Boreidae). Generally these relationships are less congruent with morphological hypotheses than the tree generated from the alignment with the blocked regions intact.

Discussion

The results of these analyses are similar to those found when 65 ingroup and 20 outgroup taxa were analyzed for 1 kb of 18S rDNA (Whiting et al., 1997), namely, that 18S rDNA sequence data do a reasonably good job of supporting the monophyly of each holometabolous insect order (with the exception of Coleoptera, discussed below), but that the data do only an adequate job of supporting interordinal relationships. Overall, support for interordinal relationships is rather weak, with the exception of Amphiesmenoptera and Raphidioptera + Megaloptera which are supported by bootstrap values of 98+ and Bremer supports of 6+. These results also suggest that Siphonaptera is nested within Mecoptera, though the exact position of the fleas relative to the Mecoptera is not well supported in this analysis. In most cases, the 18S rDNA data elucidate the same basic pattern of relationships within each order as supported by other data (e.g., in Lepidoptera, Strepsiptera, Diptera, Siphonaptera and Mecoptera), though in one case the results are entirely unresolved (Hymenoptera) or altogether confusing (Coleoptera).

These results highlight some of the difficulties associated with inferring higher-level phylogenetic relationships from molecular data. While it is generally recognized that taxon sampling plays a critical role in phylogeny estimation, particularly at the higher levels (reviewed in Hillis, 1998), the difficulty of adequately sampling the Holometabola has received little attention. The Holometabola encompasses what is the most speciose and arguably the most diverse group of organisms on earth, and it is a challenge to adequately sample the diversity of this clade. In this analysis, I used 102 ingroup sequences to represent the Holometabola, and while this is a considerable improvement over earlier studies, it must be acknowledged that the current sampling uses one sequence to represent the diversity of about 85,000 species (Wilson, 1988). Thus one might anticipate that inadequate taxon sampling may lead to unusual results in the phylogeny, and this may be the case in the topologies generated above.

The importance of thorough taxon sampling to phylogenetic estimation is perhaps best illustrated by the status of Coleoptera as based on 18S rDNA studies. In the first such study, Coleoptera was represented by a single sequence, making it impossible to test for beetle monophyly (Carmean et al.,

1992). In a second study, Coleoptera was represented by two sequences from a single polyphagan superfamily, and resulted in the Coleoptera forming a monophyletic group (Chalwatzis et al., 1996). It now appears that this presumably correct answer was an artifactual result of including only two exemplars of a single superfamily. Indeed, this analysis was a relatively weak test of beetle monophyly because the diversity of the order was not characterized as broadly as it might have been with one exemplar from each of the major suborders. Two 18S rDNA studies have done an admirable job of representing the Adephaga and Polyphaga, though neither study specifically tested for coleopteran monophyly, and they lacked the proper outgroup selection to do so. In the study of Adephaga, the tree was rooted only to the Neuropterida (represented by five sequences) (Maddison et al., 1999), on the assumption that this sister group relationship was well supported by morphology, though it is based on a single character suite which may or may not be reliable (Mickoleit, 1973; Whiting et al., 1997). In the study of Polyphaga, the tree was rooted to other beetle taxa from within the Polyphaga (Farrell, 1998), so beetle monophyly could not be tested. However, as demonstrated above (in agreement with Whiting et al., 1997), when a wide range of beetle and outgroup taxa are included, the disturbing result is a paraphyletic Coleoptera, generally with the Polyphaga forming one clade and the Adephaga forming another, but each placed apart on the topology. While I agree with Hennig (1981) that "the Coleoptera are as well founded a monophyletic group that we could ever hope to find" (p. 300), the point is that the current 18S rDNA data do not actually support a monophyletic Coleoptera, though under some sampling strategies and in some analyses they have appeared to do so. This lack of support may be due in large part to the difficulty of adequately representing coleopteran diversity in a molecular phylogenetic study, at least with 18S rDNA. It is tempting to think that the inclusion of additional taxa would remedy this problem, and it may do just that, but it is also possible that the Coleoptera are so diverse that it does not matter how many taxa are sampled, the group will always be paraphyletic when relationships are inferred from 18S rDNA. Notice, however, that other insect orders which are also very large (Lepidoptera and Hymenoptera), are supported as monophyletic with the current 18S rDNA data, so one cannot simply conclude that large and diverse insect groups cannot be supported as monophyletic with 18S rDNA data.

Unusual phylogenetic results (such as a paraphyletic Coleoptera) are often attributed to the failure of parsimony to account for rate heterogeneity in sequence data, and it is assumed that if the "long-branched" sequences are removed from the analysis (e.g., Maddison et al., 1999), or a different method of phylogenetic estimation is used for tree reconstruction (e.g., maximum likelihood under models that include parameters for rate heterogeneity; Huelsenbeck, 1998), then these unusual results will disappear. However, the first approach can lead to the elimination of a large percentage of the observed data for a group of organisms in order to "correct" for parsimony's presumed failure (~20% in the case of Adephaga; Maddison et al., 1999). The second approach is currently unable to search effectively data sets that include a thorough sampling of taxa and genes, so data must be eliminated prior to analysis. Even so, this approach may not resolve all "unusual results" as claimed by proponents, and may actually introduce new "unusual results" as was the case for Diptera + Strepsiptera (Huelsenbeck, 1997, 1998; Whiting, 1998a, 1998b). In either approach, the proposed solution requires the elimination of a large portion of data which is antithetical to what we know about the sensitivity of higher-level phylogeny to taxon sampling; we should include more data from more taxa, not less. Moreover, one should question the scientific rigor of eliminating observations simply because they generate results which do not conform to preconceived notions of phylogeny. Wherever the solution may lie, elimination of data seems a treacherous path to pursue and these solutions are not as simple as they are often portrayed.

A second major problem in higher-level phylogeny estimation is the difficulty of aligning the sequence data from divergent groups. It is very common for investigators to align sequences entirely by hand, and while this may be adequate for protein coding genes which have relatively conserved amino acid sequences, it is more difficult with non-protein coding regions, such as the ribosomal genes examined in this study. Moreover, alignment by hand does not allow any sort of optimality criterion to be applied to the data during the alignment process. Since the majority of systematists would argue that an optimality criterion (*sensu* Swofford et al., 1996) should be used in reconstructing a tree from the aligned data, it is ironic that the majority of current practitioners do not use an optimality criterion in computing an alignment. But alignments and phylogenetic trees are both simply statements of character homology, and one could argue that the same methodology used in tree reconstruction must be applied to sequence alignment in order to be methodologically consistent (Wheeler, 1999a). Current algorithms are computationally limited by the same difficulties found in phylogeny reconstruction, i.e., a large landscape of possible alignments to compute, from which a most optimal alignment must be selected, which grows exponentially as additional taxa are added. Hence one must grapple with the many difficulties of heuristic searches during alignment as during tree reconstruction. It is very common for practitioners to take the output from an alignment algorithm and correct the alignment by hand. This is an acceptable practice only if the revised alignment is shown to be more optimal under the criterion used to generate that initial alignment, i.e., a more optimal score should be computed for that revised alignment; otherwise this becomes an computer-assisted alignment void of any optimality criterion. Even if a most optimal alignment can be computed, it is likely that there are multiple most optimal alignments, each of which may generate a different phylogenetic hypothesis. In this study, the first alignment was selected, but this was entirely arbitrary and the other optimal alignments may (or may not) produce different phylogenetic conclusions. Moreover, it is likely that I did not generate the entire set of optimal alignments, so the effects of equally parsimonious alignments on these phylogenetic conclusions will require further investigation.

Alignment ambiguity is particularly pronounced with the ribosomal genes and it is common to exclude regions that are considered "unalignable" from a phylogenetic analysis. By doing this, the investigator attempts to eliminate regions of uncertain positional homology which may bias the phylogenetic results, hoping, in essence, to remove homoplasy from the data set. This is touted as being the most conservative approach to sequence alignment (e.g., Chalwatzis et al., 1996; Farrell, 1998; Maddison et al., 1999; Yeates and Wiegmann, 1999). However, the elimination of these data is typically based on subjective criteria not easily replicated from worker to worker. Moreover, the problem is compounded by the fact that alignment ambiguity is related to taxon sampling; more closely related taxa generally have less alignment ambiguity than more distantly related taxa. Adding more taxa to a study may result in resolving ambiguity in some alignment regions, since these new taxa may have intermediate states that help clarify homology statements across the alignment. Conversely, adding highly autapomorphic sequence to an alignment can result in ambiguity where none existed before, but whether a sequence is autapomorphic or not is also related to taxon sampling.

While it may be difficult to align sequences across all taxa in an analysis, within a subset of taxa used to represent a higher group, particularly in studies that do a thorough job of sampling taxa, alignment may be relatively unambiguous. For example, there is little ambiguity in aligning the entire 18S rDNA across all families of Mecoptera, but there is considerable ambiguity in aligning some regions of mecopteran 18S sequence with dipteran sequence. By excluding these alignment regions from the analysis, one is excluding potentially phylogenetically informative characters which may provide greater resolution to taxa comprising these higher groups. But by placing these variable regions into alignment blocks as described above, the variable portion of the alignment can designate relationships at the tips of the trees while eliminating the ambiguity in the alignment which may bias intergroup relationships. This is exactly the result as demonstrated above. In essence, these "unalignable regions" are treated as inapplicable data between groups but applicable within a group. For instance, imagine two major groups represented by multiple taxa. In the first group there is a multi-nucleotide insert that shows structure such that the sequence of the insert resolves all terminals in that group; the second group lacks the insert. The presence of the insert is a character supporting the monophyly of the first group, but the actual sequence of the insert, those regions which provide structure for the topology of all the terminals in that group, are inapplicable as compared to the sequences in the second group which lack the insert. To remove the insert sequence from any consideration in phylogenetic analysis is to lose characters that would provide resolution to the first group, simply because they are inapplicable to the second group. Now imagine a situation where the second group also has an insert which resolves all of its terminals, but that insert is so different from that of the first group that it cannot be aligned with any degree of confidence between the groups; this is the case we see among the insect orders. To eliminate these regions from the analysis is to throw away important phylogenetic information that is applicable for deciphering relationships within the orders, but may be inapplicable between the orders.

One possible negative effect of treating sequence blocks as inapplicable data is the introduction of large amounts of missing data to an analysis, which can cause the taxa to attach to multiple positions in the tree and collapse the consensus (Strong and Lipscomb, 1999). However, at least in the case of these18S rDNA data, this does not seem to be a problem because the structure in the conserved, non-blocked regions of the alignment prevents the taxa with missing data from randomly shuffling positions on the tree. It is important that the set of sequences designated as blocks be derived from well-supported monophyletic groups, in order to avoid the potential of biasing the results towards supporting monophyly of the sequences that happen to be placed together in a block. However, placing sequences together in a block does not necessarily constrain monophyly of the taxa in that block (as is the case for Coleoptera), because of the way that missing data are optimized on topologies. For example, if the conserved regions of the alignment suggest that one particular beetle taxon is placed as sister group to Amphiesmenoptera, then the missing data for that taxon will be optimized to the amphiesmenopteran states. There is still a certain amount of subjectivity in deciding when sequence alignment among taxa is ambiguous enough to break into subgroup alignment blocks, but this procedure is less subjective than the sheer elimination of these regions as a whole because the data are still allowed to influence the analytical results. In situations where monophyly is well established, as is the case with these insect orders, this procedure seems to better preserve valuable phylogenetic information than the simple elimination of variable regions.

One of the frustrations of current insect ordinal systematics is that the information recovered from molecular data is more redundant with the information from other sources than we would perhaps prefer. While on the one hand it is comforting that the molecular data recover the monophyly of most insect orders, on the other hand these monophyletic groups were already well supported. It is really the interordinal relationships that we are interested in, and they seem to be the most difficult to extract from the current molecular data with any degree of confidence. Perhaps the greatest problem in this study, and that of many other higher-level phylogenetic studies, is the reliance on a single marker for phylogenetic inference. Systematics seems to have a constant vibrato of the superiority of one character system over another, but if the history of systematics has taught us only one thing, it is that single character systems are nearly guaranteed to fail, at least in some portion of the topology. This is true whether one uses morphology, molecules, or developmental data to infer phylogeny. We should not be surprised to find that the 18S rDNA data does a good job on one portion of the topology but is rather ill-behaved on other portions; why should it perform differently than any other character system? The future of insect molecular systematics lies not only in increasing the taxon size for a particular marker, but increasing the range of markers used in phylogenetic inference. The combination of careful taxon sampling and careful analyses will undoubtedly lead to greater insights into the evolution of the most diverse group of organisms on the earth, the Holometabola.

Acknowledgements
I thank Paige Humphreys and Alison Whiting for assistance in generating the sequence data, and Matthew Terry, Taylor Maxwell and Jason Cryan for comments on the manuscript. This work was supported by NSF grants DEB-9615269, DEB-9806349, and NSF CAREER award DEB-9983195.

Molecular Systematics and Evolution: Theory and Practice
ed. by R. DeSalle, G. Giribet and W. Wheeler
© 2002 Birkhäuser Verlag/Switzerland

Relationships among metazoan phyla as inferred from 18S rRNA sequence data: a methodological approach

Gonzalo Giribet

Department of Organismic and Evolutionary Biology, Harvard University, Museum of Comparative Zoology, Cambridge, MA 02138, USA

Summary. The relationships among the phyla of Metazoa have been investigated by several authors. Different genes have been applied to this problem, but only the ribosomal gene 18S rRNA has been investigated for enough phyla so as to attempt an answer to the question of how the current living forms are related to each other (only one phylum, the Loricifera, is missing). In this chapter, I propose an alternative way to analyze the data obtained from ribosomal genes, or other non-coding genes that show sequence length variation.

Introduction

Early phylogenetic studies of animal relationships used 18S rRNA sequence data of a few metazoan groups (Field et al., 1988; Lake, 1989; Raff et al., 1989; Lake, 1990), and since then, more and more sequences have been added to the small nuclear ribosomal data set. Later on, more elaborated studies, accounting for larger diversity of metazoan phyla of the entire, or part of the metazoan tree were attempted (Turbeville et al., 1991; Turbeville et al., 1992; Wainright et al., 1993; Philippe et al., 1994; Turbeville et al., 1994; Halanych et al., 1995; Winnepenninckx et al., 1995; Garey et al., 1995; Halanych, 1996; Halanych et al., 1996; Garey et al., 1996; Cavalier-Smith et al., 1996; Aguinaldo et al., 1997; Carranza et al., 1997; Eernisse, 1998; Cohen et al., 1998; Littlewood et al., 1998; Winnepenninckx et al., 1998; Aguinaldo and Lake, 1998; Kim et al., 1999; Giribet and Wheeler, 1999a). The latest of these studies began to include representatives of almost all animal phyla, although some lacunas of taxon sampling were still present. More recently, a few authors started combining the vast available 18S rRNA data with morphological data, in a simultaneous analysis framework (Zrzavý et al., 1998; Giribet, 1999; Giribet et al., 2000).

Besides the 18S rRNA locus, a few other molecular markers have been explored with the aim of adding new genetic data to the current ribosomal database. Specifically, these molecular markers are: elongation factor-1 alpha (EF-1α) (Kojima et al., 1993; Kobayashi et al., 1994; Friedlander et al., 1994; Kobayashi et al., 1995; Kobayashi et al., 1996; Regier and Shultz, 1997; McHugh, 1997; Kojima, 1998; Regier and Shultz, 1998), heat shock protein 70

(Hsp70) (Borchiellini et al., 1998; Müller et al., 1998), cytochrome *c* oxidase I (COI) (Folmer et al., 1994), dopa decarboxylase (DDC) (Friedlander et al., 1994), Phosphoenolpyruvate carboxykinase (PEPCK) (Friedlander et al., 1994), the nonrepeating portion of the largest subunit of RNA polymerase II (POL II) (Friedlander et al., 1994; Regier and Shultz, 1997), elongation factor-2 (EF-2) (Friedlander et al., 1994), histone H3 (H3) (Colgan et al., 1998; Brown et al., 1999) and U2 snRNA (Colgan et al., 1998; Brown et al., 1999). However, these markers have been poorly sampled in the context of studying metazoan evolution (only samples of a few phyla are available for each one of them), and to date, no other systems are comparable to the available data for 18S rRNA.

A few authors adopted multigene approaches (Wray et al., 1996; Nikoh et al., 1997; Ayala and Rzhetsky, 1998; Gu, 1998) by sacrificing taxon sampling, although these studies were more interested in dating certain evolutionary events than in metazoan phylogeny.

In summary, the only molecular data set currently available for the full diversity of metazoan taxa is the 18S rRNA data set. Despite the criticisms raised against 18S rRNA usefulness (Philippe et al., 1994; Abouheif et al., 1998), and although I agree with these authors in that more genes are needed, proper analyses and taxon representation of the 18S rRNA data set are still needed.

Analyzing ribosomal genes

From a methodological point of view, the analysis of ribosomal genes applied to deep divergences is not an easy task. Sequence length variation occurs for almost every pair of sequences compared; sometimes the variation is extremely large. The fact that hundreds of sequences can now be included in an analysis makes the problem of length variation even worse. This has led some investigators to explicitly disregard ribosomal genes, and in particular the 18S rRNA locus, to study deep metazoan divergences (i.e., Ayala et al., 1998).

Other authors believe that there is not enough information in ribosomal genes to explain evolutionary events that occurred within the 40 million years that comprises the episode known as the "Cambrian explosion" (e.g., Philippe et al., 1994), or even that the 18S rRNA gene is an "unsuitable candidate" for reconstructing metazoan phylogeny, although the polytomies currently observed (for the triploblastic taxa) are not reliable evidence for inferring the existence of a Cambrian explosion (Abouheif et al., 1998).

Three salient points are fundamental in the study of Philippe et al. (1994). First, rapid evolving sites were eliminated, which might be important or fundamental to resolve the intervals of time that they affirm that cannot be resolved. Second, a molecular clock is assumed for 18S for metazoan divergences, which might be one of the most unrealistic cases where a molecular clock could be applied, if any. Third, although 69 18S rRNA sequences were used (corresponding to 15 animal phyla), the taxon sampling was poor in terms of numbers of phyla. At best, the experimental design of Philippe et al. (1994) was not a good scenario to test whether the 18S rRNA locus can resolve the "Cambrian explosion" or not. To do so, ideally we should include all members derived from such an evolutionary event (which cannot be done due to extinction), and all the information contained in the molecule, which was not done since "a major source of artifact" was eliminated.

It seems logical though, that any single molecule cannot be used to reliably solve such a complicated phylogenetic problem. This is related to the ratio between the number of characters per taxon. It has been estimated that in a tree without homoplasy, three characters are required to resolve a node with a bootstrap value of 95% (Felsenstein, 1985a). This means that if we wanted to resolve the phylogeny of all "orders" of metazoans (about 230 extant lineages) and wanted to represent each one of them by two sequences, we would need about 1400 unambiguous characters, if all them were distributed in groups of three for each node. This is obviously far from the reality, not only for the 18S rRNA gene, but for any single gene system.

The other notion that has been discussed in the context of metazoan phylogeny is whether an investigator should use a few "representative" sequences (i.e., Aguinaldo et al., 1997; Ruiz-Trillo et al., 1999), or a large selection of taxa (i.e., Giribet and Ribera, 1998; Giribet et al., 2000). All these strategies seemed to agree in many aspects of metazoan phylogeny, although there are important differences. An exemplar taxon that illustrates this issue is the phylum Nematoda. Aguinaldo et al. (1997) placed the nematodes together with the arthropods by eliminating all nematode sequences but the least divergent one. But the same result was found by Giribet and Ribera (1998) even when the most divergent nematodes were included. That the addition of taxa increases phylogenetic accuracy was also noticed by Wheeler (1992), Hillis (1996) and Graybeal (1998), although under very particular circumstances subsampling taxa does not constitute a problem in phylogenetic reconstruction (Poe, 1998), or the addition of taxa can even worsen accuracy (Poe and Swofford, 1999).

The problem of metazoan phylogeny is that subsamples of taxa are insufficient to explain the evolution of the group, because the crown group is too diverse, more so if considering the large amount of extinct lineages. Then, irrespective of whether the addition of taxa improves or decreases accuracy, the taxa must be sampled if a global picture is to be achieved. Thus, strategies are required that maximize the use of information of a large number of sequences, and allow for the comparison of hundreds of sequences in a repeatable way.

Alternatives to the strategy of Philippe et al. (1994) are phylogenetic methods of DNA analysis that do not require multiple sequence alignments. These methods can account for higher levels of sequence length variation, and thus avoid entirely (or almost entirely) the necessity of data removal. Such is the case of the "direct optimization" (Wheeler, 1996) that allows using the entire molecule without performing multiple sequence alignments, and thus allows analyzing larger numbers of terminals than multiple sequence alignments do. Therefore, I present here an analysis of 145 18S rDNA sequences using direct

---

Box 1. Direct optimization

This method uses base-to-base correspondences to compare pairs of DNA sequences, as multiple sequence alignments do. A binary guide-tree relating all terminals is generated (i.e., randomly) and the sequences are optimized in the tree in a down-pass procedure, taking note of the transformations (indels and substitutions) required during every step of the optimization process. Other guide-trees are generated (i.e., using TBR branch-swapping) and examined. The best trees are those that require a minimum number of steps (which could be weighted). Gaps are however not interpreted as patterns, but just as processes, unlike in "fixed alignments".

---

optimization. Comparisons of the results with those of multiple sequence alignments are not presented due to the impossibility of performing multiple alignments for such a large number of taxa.

Secondary structure and DNA sequence comparisons

The use of secondary structure in ribosomal genes to refine alignments (or in a broader sense, to refine homology hypotheses) is fairly new (e.g., Kjer et al., 1994; Kjer, 1995). More recently, this strategy has been applied in different phylogenetic studies using both multiple sequence alignments (Titus and Frost, 1996; Giribet and Wheeler, 1999b) and direct optimization (see below).

Irrespective of the method of DNA sequence comparison that we decide to use (multiple alignments, direct optimization, etc.), ribosomal genes can be too different in length as to perform accurate estimates of phylogeny by using the entire molecule as a single homologous piece. A strategy that I have applied to the study of ribosomal DNA sequences entails splitting the molecules into the smallest possible unambiguous fragments, and submitting these fragments to the next step of the phylogenetic analysis (alignment or optimization) (Giribet, 1999, 2001; Edgecombe et al., 1999; Giribet et al., 2000; Giribet and Ribera, 2000). The criterion for splitting is based on regions contained within known primer sequences and information from secondary structure models that define unambiguously homologous regions, which can be identified in all the terminals included in the analysis. This allows for splitting the 18S rRNA locus in up to about 50 regions (36 bp in length in average), which can be given as linked regions to the phylogenetic analysis program. As with the similar strategies mentioned earlier (Kjer, 1995; Titus and Frost, 1996), the idea is to delimit smaller regions to be submitted to the analysis step. This should in theory ensure better hypotheses of homology on the one hand, and decrease computation time on the other.

This strategy also allows the investigator to identify, *a priori*, conserved and non-conserved regions of the molecule, and thus to study the effect of such

---

Box 2. Sensitivity analysis and character congruence

The notion of *sensitivity analysis* was introduced into phylogenetic analysis by Wheeler (1995) due to the fact that different parameters as applied at the alignment step might yield different alignments, and thus different phylogenetic hypotheses (see also Fitch and Smith, 1983). A sensitivity analysis implies simply exploring different parameters when comparing DNA sequences (e.g., different indel costs and different substitution costs). Then the different hypotheses (generally one per parameter set utilized) can be presented.

  The concept of *character congruence* as an optimality criterion was also introduced by Wheeler (1995) into phylogenetic analysis. Character congruence among different partitions (as measured by the ILD test) can be used as an external criterion to choose among the multiple hypotheses obtained after performing a sensitivity analysis. The parameter set that minimizes incongruence among partitions is chosen as the "best" parameter set, and the hypothesis obtained under such a parameter set is chosen as the "preferred" hypothesis. Other criteria for choosing the "best" parameter set might be applied. Others might prefer not to choose a parameter set, but just show all the results obtained for all the parameters examined, and search for common structure among the different trees.

---

areas in the phylogenetic reconstruction. It also allows the identification of taxa containing certain hypervariable regions, so that decisions can be taken about the inclusion or exclusion of these regions. An example of this coding strategy is illustrated with an analysis of triploblastic taxa (see Giribet, 1999; Giribet et al., 2000).

An example of 18S rRNA analysis for triploblastic animals

As mentioned earlier, the first level of homology adopted here is fragments of DNA that are delimited according to secondary structure features, followed by a dynamic base-to-base correspondence. This was achieved by dividing the 145 18S rRNA sequences (see Appendix 1) into the smallest possible unambiguously recognizable homologous regions. In total, the 18S rRNA molecule was divided into 47 regions (excluding the external primers 1F and 9R) for each terminal taxon. The 47 input files contained the unaligned sequences of all terminal taxa. All these 47 sequence files, parameter files and batch files are available from http://www.mcz.harvard.edu/Departments/InvertZoo/Giribet/data/

  Sequence data were analyzed using the direct optimization method described by Wheeler (1996) and implemented in the computer program POY (Gladstein and Wheeler, 1997).

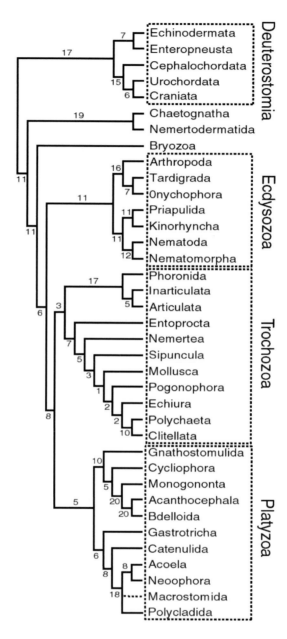

Figure 1. Summary tree of metazoan relationships when combining the 18S rDNA sequence data and the morphological data set of Zrzavý et al. (1998) (see Giribet et al., 2000 for details). The dashed line in Macrostomida indicates nonmonophyly. Numbers in branches indicate Bremer support values.

A sensitivity analysis (*sensu* Wheeler, 1995) of several parameter combinations was undertaken to assess the effects of different parameter combinations

Table 1. An example of one of the variable regions excluded from one set of analyses showing the large amount of sequence divergence among taxa

---

Enoplus
```
1 GGCAGCAAAT TTTGTTGTTT GTTGAA
```
Ephemera
```
1 GGCGCCGAGA TCCTTGTGCT CTCGGCGCTT ACT
```
Euperipatoides
```
1 GCCCGGAGCG GTTTTTGATC TTTGCCGTCC GTCCTTTGTT TTTTCCCTTT TTCCTGTCCT
1 CGGCCGACCG GTTCGGCGGG TACGAGGGAA GGCGGGGGCG GAGGGGTCAT CCGCAGGCCG
1 TCCGCGGTCT GCGGGGAAAC CTTTTCGTCA CTTTACCTTC TCGCCGTCTT TGCCGTTTCC
1 CGAGCGGTCG AGACGGAAGG GGAGCTTAGG CGGAAGCATA ACCGAAGCCG CCGGCTTTTA
1 CAGGAAG
```
Geocentrophora
```
1 GTACACTGGT CAAACATGCC GGTGCATA
```
Glossiphonia
```
1 GTACAGCCGC TGTCAGCCGC AACGTCTTCG GGCGTTCGGA CGTCTGGAAA CAGCGCTGCC
1 GGTGCAGA
```
Glossobalanus
```
1 GTTACGCGAC CCACCGGGTC GGCGTCCAA
```
Glycera
```
1 GTTCGCCGAT TCTATGTCGG TGCCAA
```
Gnathostomula paradoxa
```
1 GTGACTGGCT GCCTTGTGCA GTGCAGTTG
```
Gnathostomula sp.
```
1 GCCGCCTGCT GGCTCTTGCC AGTTGGGTGA AATTGG
```
Gordius
```
1 GAACGTCGAT CATTTTCGTC GGCGAAG
```
Haplognathia
```
1 GCACTTTGAT GGCTCTGCCG TCATATGGCG
```
Heterodon
```
1 GTTATGCGAC CCCCNAGCGG TCGGNNTCCA A
```
Hirudo
```
1 GTACAGCCGC TGTCAGCCGC AACGTCTTCG GGCGTTCGGA CGTCTGGAAA CAGCGCTGCC
1 GGTGCAGA
```

---

on phylogenetic conclusions. A parameter space of two analytical variables was examined: insertion-deletion ratio, and transversion-transition ratio (as in Wheeler, 1995). When the transversion-transition ratio was set at a value other than one, the insertion-deletion cost was set according to the cost of transversions (i.e., tv:ts = 2; gap:change = 2; gaps cost twice as much as transversions, and transversions cost twice as much as transitions). In total, seven combinations of parameters were employed in the analysis (gap:tv:ts = 111, 121, 141, 211, 221, 241, 411). This is considered a way to explore the data and to discern between well-supported relationships (those supported throughout a wide range of parameters) and poorly supported relationships (those that only appear with very particular parameter sets).

The molecular analyses were performed for the complete data set (all 47 fragments), as well as for a reduced data set excluding 5 of the 47 regions (E10-1, E10-2, E21-1-2, 41 and 47) that showed large variations in sequence length among the sampled taxa (Tab. 1).

The results of the combined analysis of 18S rRNA and morphology have been presented elsewhere (Giribet, 1999; Giribet et al., 2000; see Fig. 1). My intention is to discuss in more detail the molecular section of those analyses, in particular the differences among trees obtained when the entire molecule is analyzed *versus* when the hypervariable regions are excluded. This is illustrated in Figures 2 and 3, which correspond to the same parameter set analyzed for the entire molecule (Fig. 2), or excluding the five hypervariable regions (Fig. 3). Taxon names have been assigned a specific coding, according to the optimal tree obtained for the combined analysis (Giribet, 1999; Giribet et al., 2000). When these two trees are compared, the one that was analyzed excluding the variable regions is more congruent with the morphological hypothesis, indicating that the use and examination of secondary structure features may make a good criterion for deciding whether part of the data can be discarded *a priori*.

It is also important to notice that different parameters lead to different phylogenetic hypotheses (e.g., compare trees from Figs 3–5). Thus it is important to stress the necessity of examining the hypotheses under these different parameters, and to use an external criterion, such as character congruence (as in W.C. Wheeler, 1995) to choose the best hypothesis (Tab. 2).

Table 2. Tree length for the individual partitions

IndelC	Tv/Ti	molecC	18S	Morph	Total	ILD
1	1	1	11,336	2,185	13,628	0.00785
1	2	2	16,892	4,370	21,563	0.01396
1	4	4	27,848	8,740	36,952	0.00985
2	1	2	12,799	4,370	17,329	0.00923
2	2	4	19,691	8,740	28,769	0.01175
2	4	8	33,013	17,480	51,312	0.01596
4	1	4	15,213	8,740	24,373	0.01723

18S=18S rDNA data set; Morph = morphological data set weighted as the highest molecular cost [molecC]) and for the combined data set (Total) (from Giribet, 1999; Giribet et al., 2000). The congruence among partitions is measured using the ILD metrics (Mickevich and Farris, 1981). IndelC = insertion-deletion cost ratio; Tv/Ti = transversion-transition ratio; molecC = highest molecular cost (is calculated by multiplying IndelC × Tv/Ti).

Conclusions

The results presented show that the general patterns reflected in the evolution of the 18S rRNA loci are highly congruent with the morphological data and

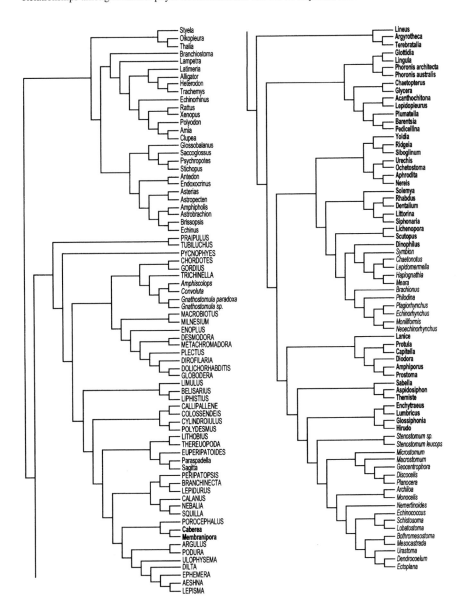

Figure 2. 18S rDNA tree for optimal parameter set 111 (gap = change; tv = ts) at 20,495 steps, when the complete gene sequence is used. Taxon coding as follows: regular font for deuterostomes and Chaetognatha, capitals for ecdysozoans, italics for platyzoans, and bold for trochozoans.

with the current views of metazoan taxonomy. Only a few regions of the 18S rRNA loci are too variable when compared among distant taxa, but an *a priori* identification of such regions and their exclusion from analyses (as a morphol-

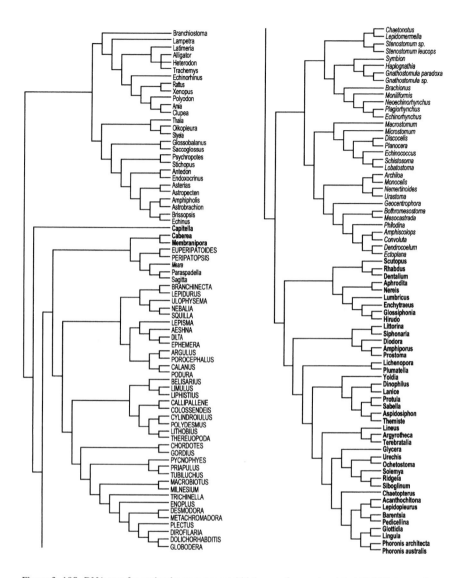

Figure 3. 18S rDNA tree for optimal parameter set 111 (gap = change; tv = ts) at 11,336 steps, when the most heterogeneous regions are removed from the analysis. Taxon coding as in Figure 2.

ogist tries to sort putative homologies) certainly increase the "quality" of the phylogenetic results.

According to the results obtained here, I propose that it is reasonable to use molecular data sets based on 18S rRNA to study metazoan relationships despite the criticism that other researchers try to promote based on depauperate taxon sampling and manual alignments. However, the addition of new

Figure 4. 18S rDNA tree for suboptimal parameter set 211 (gap = change; tv = ts) at 12,799 weighted steps, when the most heterogeneous regions are removed from the analysis. Taxon coding as in Figure 2.

sources of phylogenetic data will always be welcome when studying phylogenetic relationships of groups as large as the Metazoa.

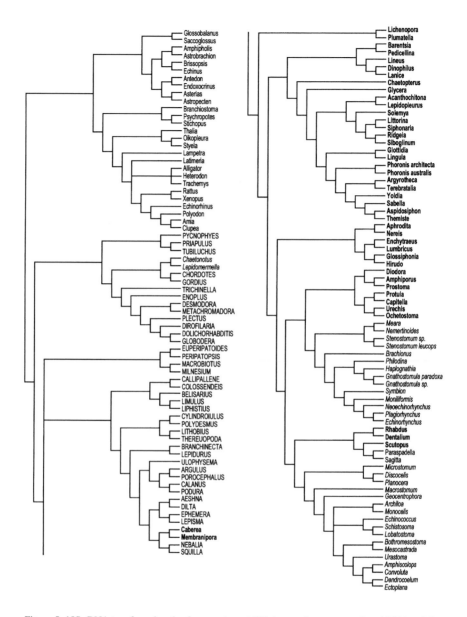

Figure 5. 18S rDNA tree for suboptimal parameter set 121 (gap = change; tv = ts) at 16,892 weighted steps, when the most heterogeneous regions are removed from the analysis. Taxon coding as in Figure 2.

Acknowledgements
I want to thank Rob DeSalle and an anonymous reviewer for comments that improved this chapter. Ward Wheeler is acknowledged for his generosity in computing time.

Appendix 1. Taxon sampling used in the molecular analyses and GenBank accession codes

Annelida – Polychaeta (9 spp.)		
Order Phyllodocida	*Nereis virens*	Z83754
	Aphrodita aculeata	Z83749
	Glycera americana	U19519
Order Spionida	*Chaetopterus variopedatus*	U67324
Order Capitellida	*Capitella capitata*	U67323
Order Terebellida	*Lanice conchilega*	X79873
Order Sabellida	*Sabella pavonina*	U67144
	Protula sp.	U67142
Order Dinophilida	*Dinophilus gyrociliatus*	AF119074
Annelida – Clitellata (4 spp.)		
Order Prosothecata	*Enchytraeus* sp.	Z83750
Order Opisthopora	*Lumbricus rubellus*	Z83753
Order Arhynchobdellida	*Hirudo medicinalis*	Z83752
Order Rhynchobdellida	*Glossiphonia* sp.	Z83751
Mollusca (10 spp.)		
Class Caudofoveata	*Scutopus ventrolineatus*	X91977
Class Polyplacophora	*Lepidopleurus cajetanus*	AF120502
	Acanthochitona crinita	AF120503
Class Gastropoda	*Diodora graeca*	AF120513
	Littorina obtusata	X94274
	Siphonaria pectinata	X94274
Class Scaphopoda	*Dentalium pilsbryi*	AF120522
	Rhabdus rectius	AF120523
Class Bivalvia	*Solemya velum*	AF120524
	Yoldia limatula	AF120528
Sipuncula (2 spp.)		
Class Phascolosomida	*Aspidosiphon misakiensis*	AF119090
Class Sipunculida	*Themiste alutacea*	AF119075
Echiura (2 spp.)		
Order Echiuroinea	*Ochetostoma erythrogrammon*	X79875
Order Xenopneusta	*Urechis caupo*	AF119076
Pogonophora (2 spp.)		
Class Perviata	*Siboglinum fiordicum*	X79876
Class Obturata	*Ridgeia piscesae*	X79877
Nemertea (3 spp.)		
Class Anopla	*Lineus* sp.	X79878
Class Enopla	*Prostoma eilhardi*	U29494
	Amphiporus sp.	AF119077
Brachiopoda (4 spp.)		
Class Inarticulata	*Glottidia pyramidata*	U12647
	Lingula lingua	X81631

(continued on next page)

Appendix 1. (continued)

Class Articulata	*Terebratalia transversa*	U12650
	Argyrotheca cordata	AF119078
Phoronida (2 spp.)		
	Phoronis architecta	U36271
	Phoronis australis	AF119079
Bryozoa (4 spp.)		
Class Stenolaemata	*Lichenopora* sp.	AF119080
Class Gymnolaemata	*Membranipora* sp.	AF119081
	Caberea boryi	AF119082
Class Phylactolaemata	*Plumatella repens*	U12649
Entoprocta (2 spp.)		
	Pedicellina cernua	U36273
	Barentsia hildegardae	AJ001734
Cycliophora (1 sp.)		
	Symbion pandora	Y14811
Rotifera (2 spp.)		
Class Bdelloidea	*Philodina acuticornis*	U41281
Class Monogononta	*Brachionus plicatilis*	U29235
Acanthocephala (4 spp.)		
Class Palaeoacanthocephala	*Plagiorhynchus cylindraceus*	AF001839
	Echinorhynchus gadi	U88335
Class Archiacanthocephala	*Moniliformis moniliformis*	Z19562
Class Eoacanthocephala	*Neoechinorhynchus pseudemydis*	U41400
Gastrotricha (2 spp.)		
Order Chaetonotida	*Chaetonotus* sp.	AJ001735
	Lepidodermella squammata	U29198
Gnathostomulida (3 spp.)		
Order Filospermoidea	*Haplognathia* sp.	AF119084
Order Bursovaginoidea	*Gnathostomula paradoxa*	Z81325
	Gnathostomula sp.	AF119083
Plathelminthes – Nemertodermatida (2 spp.)		
	Meara stichopi	AF119085
	Nemertinoides elongatus	U70084
Plathelminthes – Acoela (2 spp.)		
	Convoluta naikaiensis	D83381
	Amphiscolops sp.	D85099
Plathelminthes – Catenulida (2 spp.)		
	Stenostomum leucops	U70085
	Stenostomum sp.	U95947

(continued on next page)

Appendix 1. (continued)

Plathelminthes – Rhabditophora (15 spp.)		
Order Macrostomida	*Macrostomum tuba*	U70081
	Microstomum lineare	U70083
Order Polycladida	*Discocelis tigrina*	U70079
	Planocera multitentaculata	D17562
Order Lecithoepitheliata	*Geocentrophora* sp.	U70080
Order Proseriata	*Archiloa rivularis*	U70077
	Monocelis lineata	U45961
Order Rhabdocoela	*Bothromesostoma* sp.	D85098
	Mesocastrada sp.	U70082
Order Prolecitophora	*Urastoma* sp.	U70086
Order Tricladida	*Dendrocoelum lacteum*	M58346
	Ectoplana limuli	D85088
Cestoda	*Echinococcus granulosus*	U27015
Trematoda	*Schistosoma mansoni*	M62652
	Lobatostoma manteri	L16911
Priapulida (2 spp.)		
	Priapulus caudatus	D85088
	Tubiluchus corallicola	AF119086
Kinorhyncha (1 sp.)		
Order Homalorhagida	*Pycnophyes kielensis*	U67997
Nematomorpha (2 spp.)		
Class Gordioida	*Chordotes morgani*	AF036639
	Gordius aquaticus	X80233
Nematoda (8 spp.)		
Order Araeolaimida	*Plectus aquatilis*	AF036602
Order Desmodorida	*Desmodora ovigera*	Y16913
Order Chromadorida	*Metachromadora* sp.	AF036595
Order Enoplida	*Enoplus brevis*	U88336
Order Trichocephalida	*Trichinella spiralis*	U60231
Order Rhabditida	*Dolichorhabditis* sp.	AF036591
Order Tylenchida	*Globodera pallida*	AF036592
Order Spirurida	*Dirofilaria immitis*	AF036638
Onycophora (2 spp.)		
	Peripatopsis capensis	AF119087
	Euperipatoides leukarti	U49910
Tardigrada (2 spp.)		
Class Eutardigrada	*Macrobiotus hufelandi*	X81442
	Milnesium tardigradum	U49909
Arthropoda (22 spp.)		
Class Pycnogonida	*Colossendeis* sp.	AF005440
	Callipallene sp.	AF005439

(continued on next page)

Appendix 1. (continued)

Class Chelicerata	*Limulus polyphemus*	U91490
	Belisarius xambeui	U91491
	Liphistius bicoloripes	AF007104
Class Branchiopoda	*Branchinecta packardi*	L26512
	Lepidurus packardi	L34048
Class Maxillopoda	*Argulus nobilis*	M27187
	Ulophysema oeresundense	L26521
	Calanus pacificus	L81939
	Porocephalus crotali	M29931
Class Malacostraca	*Nebalia* sp.	L81945
	Squilla empusa	L81946
Class Myriapoda	*Thereuopoda clunifera*	AF119088
	Lithobius variegatus	AF000773
	Polydesmus coriaceus	AF005449
	Cylindroiulus punctatus	AF005448
Class Hexapoda	*Podura aquatica*	AF005452
	Dilta littoralis	AF005457
	Lepisma sp.	AF005458
	Aeschna cyanea	X89481
	Ephemera sp.	X89489

Enteropneusta (2 spp.)

	Glossobalanus minutus	AF119089
	Saccoglossus kowalevskii	L28054

Echinodermata (10 spp.)

Class Crinoidea	*Antedon serrata*	D14357
	Endoxocrinus parrae	Z80951
Class Holothuroidea	*Stichopus japonicus*	D14364
	Psychropotes longicauda	Z80956
Class Echinoidea	*Echinus esculentus*	Z37125
	Brissopsis lyrifera	Z37119
Class Ophiuroidea	*Amphipholis squamata*	X97156
	Astrobrachion constrictum	Z80948
Class Asteroidea	*Asterias amurensis*	D14358
	Astropecten irregularis	Z80949

Urochordata (3 spp.)

Class Apendicularia	*Oikopleura* sp.	D14360
Class Thaliacea	*Thalia democratica*	D14366
Class Ascidiacea	*Styela plicata*	M97577

Cephalochordata (1 sp.)

	Branchiostoma floridae	M97571

Craniata (11 spp.)

Cephalaspidomorphi	*Lampetra aepyptera*	M97573
Chondrichthyes	*Echinorhinus cookei*	M91181
Actinopterygii	*Amia calva*	X98836
	Polyodon spathula	X98838
	Clupea harengus	X98845

(continued on next page)

Appendix 1. (continued)

Coelacanthiformes	*Latimeria chalumnae*	L11288
Amphibia	*Xenopus laevis*	X04025
Testudines	*Trachemys scripta*	M59398
Lepidosauria	*Heterodon platyrhinos*	M59392
Archosauria	*Alligator mississippiensis*	M59383
Eutheria	*Rattus norvegicus*	X01117
Chaetognatha (2 spp.)		
Order Phragmophora	*Paraspadella gotoi*	D14362
Order Aphragmophora	*Sagitta elegans*	Z19551

Part 2
Current problems in
molecular systematics

Introduction to part 2

Current problems in molecular systematics

The second section examines seven topics that have arisen either as controversies or as new inroads into systematic and evolutionary analysis over the past few years. Alignment is one of the first analytical steps in any evolutionary study that utilizes molecular sequences and Giribet, Wheeler and Muona examine the pitfalls of this initial step. Analyzing the aligned data with large numbers of taxa and characters to produce a phylogenetic tree is a complex and difficult endeavor. The next three chapters in this section examine the process of tree construction by looking at parallel computing (Janies and Wheeler), statistical aspects of maximum parsimony (Steele) and issues related to comparing phylogenies produced by methods based on philosophically different premises such as maximum likelihood, maximum parsimony and distance methods (Giribet, DeSalle and Wheeler). The final three chapters examine three specific problems in classical systematics and evolution studies that have arisen as a direct consequence of molecular data. The first problem concerns our view of species and the levels above and below the species boundary. Goldstein and Brower examine the role of molecular characters in constructing cladograms at and around the species boundary and what these cladograms might or might not mean. Because molecular data have often been juxtaposed with classical morphological data, an assessment of the role of both in interpreting systematic patterns is important. Baker and Gatesy ask the question "Is morphology still relevant?" in their chapter. In addition to answering that morphology is relevant, they also outline approaches for assessing the relative importance of different data partitions in systematic analysis. In the same vein, new developmental data have been used to make statements about the origin of evolutionary novelties. Bang, Schultz and DeSalle conclude this section of the book by examining the systematic ramifications of using developmental data to make evolutionary inferences.

R. DeSalle, G. Giribet and W. Wheeler

Molecular Systematics and Evolution: Theory and Practice
ed. by R. DeSalle, G. Giribet and W. Wheeler
© 2002 Birkhäuser Verlag/Switzerland

DNA multiple sequence alignments

Gonzalo Giribet[1], Ward C. Wheeler[2] and Jyrki Muona[3]

[1] *Department of Organismic and Evolutionary Biology, Museum of Comparative Zoology, Harvard University, Cambridge MA 02138, USA*
[2] *Division of Invertebrate Zoology, American Museum of Natural History, New York, NY 10024, USA*
[3] *Zoological Museum, Division of Entomology, Finnish Museum of Natural History, University of Helsinki, FIN-00014 Helsinki, Finland*

Summary. In this chapter we examine the procedure of multiple sequence alignment. We first examine the heuristic procedures commonly used in multiple sequence alignment. Next we examine sources of ambiguity involved in the alignment procedure. We suggest that several alignment parameters be employed to examine alignment sensitivity. We end by presenting an experiment with humans showing the ambiguity involved in manual alignment.

Introduction

Multiple sequence alignment is a procedure to turn unequal length sequences into equal length character strings via the insertion of gaps. These gaps are mere placeholders which indicate that an insertion or deletion has occurred somewhere after the compared sequences diverged from a common ancestor, resulting in a lack of homologous nucleotides at that position for that taxon.

Despite the existence of new methods for phylogenetic analysis that entirely avoid alignments, the issue of using multiple sequence alignments (fixed alignments) as a source for the primary homology statements for phylogenetic analysis is still important for certain areas of knowledge. An investigator may choose a fixed alignment *versus* a dynamic alignment, and base-to-base correspondences *versus* fragment-to-fragment correspondences for several reasons. For example, below the population level, or to study molecular evolution (of both DNA and proteins), certain methods that are commonly applied require the use of fixed alignments (see Wheeler 2002).

Three issues appear important to us in this respect. First, how multiple sequence alignments are generated algorithmically (and the inherent problematica of alignments). Second, how the available software performs alignments (implementation). Third, how parameter sensitivity enters into exploring phylogenetic hypotheses at the alignment level. This last issue also applies to all other methods of sequence comparison (i.e., optimization of DNA fragments).

Background

The first step in any phylogenetic analysis involves some sort of pairwise comparisons of DNA data (or amino acids; but from here on, we will refer to DNA data analyses). Two families of comparison methods are available: local comparisons, meant to search for homologous domains among sequences, such as the BLAST family of programs (e.g., Altschul et al., 1997), and global comparisons, the type applied to phylogenetic inference, where the entirety of two putatively homologous strings of DNA is compared to assign base-to-base correspondences.

The fundamental method of pairwise sequence alignment was first described by Needleman and Wunsch (1970), and extended to multiple dimensions by Sankoff and Cedergren (1983). The Needleman and Wunsch algorithm calculates the minimum edit distance between two DNA sequences, which is the minimum number of transformations required to go from one sequence to another. In its simplest incarnation, two parameters need to be specified, the gap penalty (or indel cost: the cost assigned to insertion or deletion events), and the change cost (the cost assigned to go from one base to any other). This change cost can be categorized in many different ways, assigning independent costs for every particular type of transformation, or assigning costs to certain categories (i.e., transversions, transitions, etc.). These costs need to be explicit in any algorithmic comparison, and have some lower boundaries delimited by the triangle inequality (Wheeler, 1993). The Needleman and Wunsch algorithm can be expressed as a minimization process, but other optimization procedures for DNA sequence comparisons might be used as well (e.g., maximization of base matches). The specific mechanics of the Needleman and Wunsch algorithm have been reviewed elsewhere (Wheeler, 1994), and we are not going to review the process in detail, but just note certain relevant aspects.

In order to align two sequences of length $(N-1)$ and $(M-1)$, a matrix of $N \times M$ cells is created, and the minimum cost path through this matrix (given specific parameter costs) is calculated. The matrix is traversed in such a fashion that only the adjacent three cells (usually the cells above, to the left, and diagonally up to the left) are examined to determine the cost of each cell and the most efficient path to it (Needleman and Wunsch, 1970). This means that for each of the $N \times M$ cells, three cells are involved in the calculation of each other internal cell. While this is manageable for two sequences (the cost of computation being roughly proportional to the product of the sequence lengths), and significant shortcuts are known, extensions to phylogenetically interesting numbers of sequences are extremely computationally intensive. The alignment matrix for n sequences would have n axes, and each cell would require knowledge of $2^n - 1$ other cells. Furthermore, while the cost of spanning two sequences is simply the summed difference, when four or more sequences are involved, some tree search or prior knowledge is required to determine the alignment and its overall cost (Sankoff and Cedergren, 1983).

These complicating factors have made true multiple alignment unachievable for anything but the smallest number of taxa. Real data sets require, at least, heuristic solutions (Wheeler, 2000b). In fact Slowinski (1998) showed that there are 1.05×10^{18} different alignments for five DNA sequences of five nucleotides each, and thus recommends not even attempting to perform multiple sequence alignments, since any optimality criterion is "virtually guaranteed to fail".

The heuristic strategy followed in multiple sequence alignment procedures is quite simple. Since aligning two sequences is easy, the procedure adds sequences via a "guide tree". All programs for multiple sequence alignment in common use today follow this idea, but differ in how they get the binary "guide" tree, and how they add the pairwise results together to generate the complete alignment.

Three implementations of heuristic multiple sequence alignment algorithms that are in some use today are the CLUSTAL family (Higgins and Sharp, 1988, 1989; Higgins et al., 1992, 1996; Higgins, 1994; Thompson et al., 1994, 1997; Jeanmougin et al., 1998), TREEALIGN (Hein, 1989, 1990), and MALIGN (Wheeler and Gladstein, 1994, 1995). These three programs rely on guide trees to accrete pairwise alignment. In the case of CLUSTAL and TREEALIGN, a distance tree is calculated from all the pairwise sequence similarity scores, and this distance tree becomes the guide tree. In the case of CLUSTAL, this is a Fitch-Margoliash tree; TREEALIGN uses a method developed by Hein (1989, 1990). At the nodes (vertices) of the guide trees, consensus (CLUSTAL) or quasi-optimized (TREEALIGN) single sequences are created from the aligned pair, which is then submitted to another pairwise alignment further down the tree. When the root of the guide tree is reached, the various gaps inserted on the way down are placed into the sequences at the tips creating sequences of equal length—the multiple alignment.

MALIGN also uses guide trees, but differs from the other programs in that it examines multiple guide trees. These guide trees are generated through standard tree search procedures of tree building and branch swapping. Furthermore, no individual sequences are created at the internal vertices, but the partial alignment of sequences descending from that node are carried along and aligned in a modified pairwise manner. During the search procedure, a complete multiple alignment is generated for each candidate guide tree, and a heuristic phylogenetic search is performed on the multiple alignment. The entire procedure involves two levels of heuristics, one to generate the alignment and another to perform tree searches on each one of the alignments. The alignment (or alignments, if multiple solutions are found) which produces the most parsimonious phylogenetic result (i.e., lowest cost) is chosen as the "best" multiple alignment. As a result of this search procedure, MALIGN will frequently examine many thousands or millions of candidate alignments (usually n^3 for n sequences). Not surprisingly, CLUSTAL and TREEALIGN frequently generate results more rapidly than MALIGN.

Furthermore, sequence comparison can well include evolutionary models and be based on statistical approaches. Maximum likelihood methods for alignment of DNA sequences have been proposed (Thorne et al., 1991; Thorne and Churchill, 1995), although these methods have not yet been applied to phylogenetically interesting data sets.

In summary, irrespective of which program or method is used, multiple sequence alignment is a computationally expensive technique, and only heuristic solutions can be achieved.

Sources of ambiguity

That alignments originated from different sources might result in alternative phylogenetic hypotheses is logical, and has also been demonstrated empirically (e.g., Wägele and Stanjek, 1995; Winnepenninckx and Backeljau, 1996). In a recent review on DNA sequence alignments, Wheeler (1994) enumerated three sources of ambiguity in multiple sequence alignments (sources of non-unique alignments). An extra source of difficulty, as mentioned above, is the necessity of heuristics in solving alignment problems. Three sources of ambiguity are:

Parameter variation

That different parameters can result in different alignments, and consequently in alternative phylogenetic hypotheses, is a well-known phenomenon, first described by Fitch and Smith (1983; see also Waterman et al., 1992; Wheeler, 1995; Morrison and Ellis, 1997; Cerchio and Tucker, 1998; Giribet and Wheeler, 1999b). Since there is *a priori* no way to determine directly the appropriate gap or change values, more or less arbitrary decisions must be made when choosing a particular cost regime (Giribet and Wheeler, 1999b). An obvious solution to this problem is to examine a wide space of parameters. For example, using a large regime of gap and change costs would show which areas of the alignment are conserved, and which are more parameter-dependent. What the investigator does with this information is another matter.

Describing the enormous parameter space that can be explored by multiple sequence alignments, Higgins et al. (1996) stated that:

> "We justify this by asking the user to treat CLUSTAL W as a data exploration tool rather than as a definitive analysis method. It is not sensible to automatically derive multiple alignments and to trust particular algorithms as being capable of always getting the correct answer".

Many investigators remove "gappy" areas (whether obtained automatically or manually), appealing to the idea that these areas do not reflect true homologies, or that the pattern of homology cannot be recognized. This could lead to extremes in which all informative data are removed, especially if hundreds of sequences are examined, even if they were coding genes (never underestimate the power of mutation!). Furthermore, many times this is done because the alignments have been generated manually, or by using bogus algorithms. Other more objective alternatives have been proposed, such as Cull or Elision (Gatesy et al., 1993; DeSalle et al., 1994; Wheeler et al., 1995). However, Cull could also end up with all the information removed from the alignment. Elision is neater in the sense that it acts as a weighting function, downweighting all these positions with ambiguous alignments.

A third solution was proposed by W.C. Wheeler (1994, 1995), which is the use of congruence with other sources of information to decide which alignment best explains evolution of all sources of phylogenetic evidence. Character congruence (Mickevich and Farris, 1981; Farris et al., 1995) or topological congruence (Wheeler, 1999a) are our preferred criteria. In these cases, no information is discarded or downweighted, which accounts for analyses that accommodate a wider scheme of phylogenetic variation. This does not mean that the alternative alignments should not be explored to test for phylogenetic stability to parameter choice. Obviously, more parameters can always be analyzed. Another critique of the use of parameters is that the scheme of parameters is applied uniformly to all the positions in the analyses. But as the phylogenetic data come today, it seems the best way to account for the first source of ambiguity in multiple sequence alignment.

Multiple order-dependent solutions

When multiple alignments are created, whether by exact or by heuristic means, the notion of alignment order comes into play. Heuristic multiple alignment solutions are built typically from a series of pairwise alignments. Initially two sequences are aligned and this result aligned to a third sequence, maintaining the relative alignment between the first two ("once a gap, always a gap"; Feng and Doolittle, 1987, 1990) and so on. This procedure is obviously order-dependent. A different addition order might well yield a different alignment, even when the exact same parameters are chosen. So, not only can different parameter sets result in different alignments, but also the same parameter sets might yield different alignments if a different guide tree is used. This was considered by Wheeler (1994) to be analogous to the existence of multiple optimal trees in a standard parsimony phylogenetic analysis.

Multiple path-dependent solutions

The third source of alignment ambiguity is path variation. Path variation occurs when the alignment algorithm can follow multiple paths through the alignment space, yielding again multiple solutions (even for the same parameter space and for the same guide tree). Path variation occurs when the alignment can either insert a gap or match the bases with equal cost. Every time that this happens, the number of optimal solutions multiplies, and if this happens repeatedly, the result is a large number of equally costly, but different alignments.

Alignments are just hypotheses of homology

Alignments are not "given static hypotheses of homology", or any phenomenon that we can observe in nature. This is a truth that is hard to accept for many investigators. The same applies to particular base transformations, insertions, deletions, etc. The path from one sequence to another, connected by a common ancestor, may suggest such a phenomenon, but alignments of multiple taxa are missing way too many of these events, too many terminals, and too many ancestors (nodes). This implies that any information that we report in the form of an alignment is the most accurate estimate of these unobserved processes, and thus we should not be afraid to explore the solutions suggested by alternative alignments. Otherwise we would be fooling ourselves by believing that we got "the" alignment.

As an example, there are several possible alignments for the following two sequences (1) AATCGCG and (2) AACCCGG. Four of these possibilities are shown here:

(a)	AATCGCG		(c)	AATCGCG–
	AACCCGG			AA–CCCGG
(b)	AATCGCG–		(d)	AATCGC–G–
	AACC–CGG			AA–C–CCGG

Depending on the parameter set adopted, some alignments will be "better" (shorter) than others. For example, if we consider all transformations as equal and assign them a cost of 1 (gap cost = 1; tv cost = 1; ts cost = 1), alignment (a) requires three transformations (a total cost of 3), as it does alignment (b) and (c), while alignment (d) requires four transformations (a total cost of 4). Thus, applying this model, alignments (a), (b) and (c) are equally supported. If we apply a second model with gap costs weighted twice as much as base transformations (gap cost = 2; tv cost = 1; ts cost = 1), then alignment (a) requires no gaps, 2 transversions and 1 transition (3 base transformations; total cost of 3). Alignment (b) requires two indel events and one transition (total cost of 5), alignment (c) requires two indel events and one transversion (total cost of 5), and alignment (d) requires 4 indel events and no base transformations (total cost of 8). Yet other models could be applied, resulting in favored alignments (a) and (b) (gap cost = 2; tv cost = 2; ts cost = 1); (b) (gap cost = 1; tv cost = 2; ts cost = 1), etc. (Tab. 1).

Table 1. Total cost of alignments (a), (b), (c) and (d) at different parameter values

gap	tv	ts	alignment	total
1	1	1	(a)	3
			(b)	3
			(c)	3
			(d)	4
2	1	1	(a)	3
			(b)	5
			(c)	5
			(d)	8
2	2	1	(a)	5
			(b)	5
			(c)	6
			(d)	8
1	2	1	(a)	5
			(b)	3
			(c)	4
			(d)	4

What we are trying to illustrate with this example is the notion that the decision on which is the "best" alignment is not trivial, and certainly decisions made "by eye" would probably choose alignment (a) *versus* the alternative ones, although (b) and (c) might be as good or even better under a wide range of parameters. This, we guess, stresses the necessity of being explicit and repeatable, two conditions only mutually satisfied by automatic alignments. The lack of sufficiently good algorithms to perform multiple alignments should not be taken as a critique of a philosophically superior method.

Due to the existence of sources of ambiguity in multiple sequence alignments, different alignments based on different parameter sets should be explored. With these multiple hypotheses of positional homology, phylogenetic analyses increase in complexity, but decrease in the degree of arbitrariness.

Experimenting with humans

In order to evaluate "manual alignments," eight student investigators were given a set of sequences, the "original data", and were asked to align them. The original data consisted of ten sequences, most starting with the motif "AAGAAGAAT", and all of them ending with the motif "TTTATTTTGA". The students knew what homology meant and were supposed to align the sequences so that they would be equally long and have the "highest possible" base-to-base concordance, any way they could.

The same ten sequences were submitted to ClustalW and Malign for multiple sequence alignments with parameters set at gap cost = 10; change cost = 1; gap extension penalty options off (each gap was given the same cost value, as if they were independent). The sequences were also optimized in POY

Table 2. Tree length of the alignment of 10 sequences

One	388
Two	362 (some bases deleted)
Three	374 (some bases deleted)
Four	367 (some bases deleted)
Five	388
Six	367 (some bases deleted)
Seven	362 (some bases deleted)
Eight	357 (some bases deleted)
CLUSTAL	395
MALIGN	386
POY	379

One to Eight indicate manual alignments generated by 8 students. MALIGN, CLUSTAL and POY indicate the alignments obtained with the respective programs.

(Gladstein and Wheeler, 1997) and the "implied alignment" corresponding to the topology optimized was compared to the other alignments. Manual alignments and computer-generated alignments were evaluated using the parsimony program NONA v 2.0 (Goloboff, 1994), counting gaps as a character state (gap cost arbitrarily set at 1) using tbr branch swapping (h1000;h/10;mult*100;).

For the manual alignments, in general a few gaps were added, but in most cases, a few bases were removed as well, making tree-length comparisons impossible. Of course removing bases was incorrect, but these were just examples. Tree lengths for all the alignments are given in Table 2. What we can observe from this simple experiment is that manual alignments are unpredictable. In addition, they incorporate a high degree of subjectivity and are error-prone.

Computer-generated alignments dependent on guide trees are more parsimonious if several guide trees are examined (MALIGN *versus* CLUSTAL), although shorter alignments might exist (implied alignment from POY).

Multiple sequencing alignment is a complicated process that requires considerably large amounts of computation. Methods using a single guide tree can be improved by doing multiple runs with different starting points, giving several randomly generated guide trees. However, this is tedious and this is why programs such as MALIGN, which examines multiple guide trees (using multiple random addition), are superior. Even when multiple guide trees are examined, there is no guarantee, as in any other heuristic procedure, that the optimal (shortest) alignment will be found. Shorter alignments can be found much faster by outputting the implied alignment with a tree generated via DNA direct optimization (Wheeler, 1996).

Acknowledgements
We want to thank Rob DeSalle and John Gatesy for comments that improved this chapter.

Molecular Systematics and Evolution: Theory and Practice
ed. by R. DeSalle, G. Giribet and W. Wheeler
© 2002 Birkhäuser Verlag/Switzerland

Theory and practice of parallel direct optimization

Daniel A. Janies and Ward C. Wheeler

Division of Invertebrate Zoology, American Museum of Natural History, New York, NY 10024, USA

Summary. Our ability to collect and distribute genomic and other biological data is growing at a staggering rate (Pagel, 1999). However, the synthesis of these data into knowledge of evolution is incomplete. Phylogenetic systematics provides a unifying intellectual approach to understanding evolution but presents formidable computational challenges. A fundamental goal of systematics, the generation of evolutionary trees, is typically approached as two distinct NP-complete problems: multiple sequence alignment and phylogenetic tree search. The number of cells in a multiple alignment matrix are exponentially related to sequence length. In addition, the number of evolutionary trees expands combinatorially with respect to the number of organisms or sequences to be examined. Biologically interesting datasets are currently comprised of hundreds of taxa and thousands of nucleotides and morphological characters. This standard will continue to grow with the advent of highly automated sequencing and development of character databases. Three areas of innovation are changing how evolutionary computation can be addressed: (1) novel concepts for determination of sequence homology, (2) heuristics and shortcuts in tree-search algorithms, and (3) parallel computing. In this paper and the online software documentation we describe the basic usage of parallel direct optimization as implemented in the software POY (ftp://ftp.amnh.org/pub/molecular/poy).

Introduction

The first step in phylogenetic analysis is to establish putative homology statements for characters observed among study species. When considering morphology, putative homology statements result from comparative analysis by a trained specialist. However, the establishment of homologies across many sequence positions and species are not easily or optimally conducted by eye (see Giribet et al., this volume). In the analysis of molecular sequence data, multiple alignment algorithms can assign provisional homologies among residues (e.g., nucleotides, amino acids). Putative statements of morphological and molecular homology are tested by phylogenetic analysis. Cladograms are constructed from those putative homologues that are shown to be shared derived features.

The problem

The number of cells in a multiple alignment matrix are exponentially related to the number of taxa and sequence length. An alignment of m sequences of length N nucleotide bases will require N^m elements of storage (Needleman and Wunsch, 1970). One commonly used heuristic approach is to provide an initial topology of relationships among the taxa (guide tree) for accreting sequences into a matrix of provisional homologies (Sankoff et al., 1973). In theory, an alignment procedure could be repeated for each possible set of relationships among taxa. However, the number of topologies is combinatorially dependent on the number of taxa. Multiple alignment of more than a few short sequences requires heuristics.

Furthermore, the results of heuristic multiple alignment are dependent on the order in which the sequences are accreted and the functions chosen for the relative costs of insertion-deletion and substitution events in sequences. For evolutionary studies, the objective of performing a multiple alignment is often to proceed to a phylogenetic analysis with a set of putative homologies unbiased by initial assumptions. Clearly, computationally efficient and assumption-minimizing alternatives to the existing paradigm of multiple alignment are essential for evolutionary and molecular biology.

A solution

Direct optimization is a novel method of comparing putatively homologous sequence residues during cladogram diagnosis, thus obviating multiple alignment (Wheeler, 1996). Alignment algorithms create correspondences between sequence strings of various lengths by inserting gaps. In multiple alignment the relative costs of insertion-deletion and substitution events determine the number and position of gap characters inserted in sequences. Direct optimization works by creating parsimonious hypothetical ancestral sequences at internal cladogram nodes. The key difference between direct optimization and multiple alignment is that evolutionary differences in sequence length are accommodated not by the use of gap characters but rather by allowing insertion-deletion events between ancestral and descendant sequences. Evolutionary base substitution and insertion–deletion events between ancestor and descendant sequences are treated with the same cost functions (e.g., Sankoff matrices) as in multiple alignment.

Theory

Determination of DNA sequence homology

The phylogenetic analysis of DNA sequences, like that of all other comparative data, is based on schemes of putative homology that are then tested via congruence to determine synapomorphy and cladistic relationships. Unlike some other data types, however, putative molecular homologies or characters are not directly observable. DNA sequences from various organisms are often unequal in length. Hence, the correspondences among sequence positions are not evident and some sort of procedure is required to determine which regions are homologous. This procedure is typically multiple sequence alignment. Alignment inserts gaps to make the corresponding (putatively homologous) nucleotides line up into columns. These columns (characters) comprise the data used to reconstruct cladograms. Many investigators try to hand-align raw data or hand-edit algorithmic alignments to reduce "errors" and ambiguity, but this is certainly a subjective and unrepeatable process. Whether alignment is accomplished manually or algorithmically, the resultant characters are then submitted to phylogenetic analysis as column vectors in the same manner as other forms of data, such as morphological characters scored by an investigator. Whatever the analytical pathway, alignment is an artificial manipulation of DNA observations via the insertion of gap characters that are not data but rather just place-holders. The primary reason in phylogenetics to create an alignment has only an operational basis—to make it possible to submit these data to standard phylogeny programs that were designed to handle column vectors of morphological characters. This is not a reason to believe that construction of an alignment followed by a separate tree search procedure is the only or the best way to do phylogenetics.

Limitations of multiple alignment

Alignment-based homology schemes rest on a notion of base-to-base correspondence in which individual nucleotide bases transform among five states (A, C, G, T or U, and gap) within a single character. The use of a base-to-base framework to view DNA homology is in large part responsible for the the phenomenon of long-branch attraction because of the paucity of character states (A, C, G, T/U, or –) in a column. A method available in POY, fixed-state optimization, can be used to avoid this pitfall because the method views the whole sequence, not individual bases, as characters (Wheeler, 1998, 1999b). In a fixed-states approach the number of possible character states are related to the length (n) of the sequences (up to 4^n) thus reducing the chance of random non-historical similarity to a negligible probability.

Static versus dynamic homology

In standard phylogenetic analysis, once an alignment is created it is not revised during or as a result of subsequent phylogenetic analysis. In this sense the putative homologies defined in the alignment are static. Reexamination is often done by hand but users will likely fall prey to biases and rearrange bases in favor of preferred groups. However, as implemented in MALIGN (Wheeler and Gladstein, 2000), randomization of an alignment's guide tree can achieve reexamination of putative homology and the alignments can be judged by an optimality criterion applied to the trees produced from the alignment via phylogenetic analysis.

As pointed out by Phillips et al. (2000), Mindell (1991) advocated using "known" phylogenies to guide alignments but the required phylogenetic information is often unavailable. In most evolutionary studies, the object of performing a multiple alignment is to allow phylogenetic analysis with a set of putative homologies unbiased by initial assumptions of relationship. Topology-based alignment comes at the cost of results that are dependent on the addition order of sequences as determined by the guide tree (Fitch and Smith, 1983). Thus, preconceived notions of relationships will bias the analysis. Randomization of the alignment topology is the most objective course of action.

The most significant advantage of direct optimization is that homology assessment is dynamic. In direct optimization, nucleotide homologies are fluid in the sense that they change not only when different guide trees are used, but also when various data are combined. Statements of putative homology depend not only on the addition order of sequences during the initial build of a cladogram and base transformation costs (as with standard alignment) *but also* on congruence among characters. In direct optimization, many optimization schemes, each implying a distinct set of putative homologies, can be examined via variable sequence alignments that occur concurrently with initial cladogram building. The diagnosis of each cladogram involves finding the lowest-cost hypothetical ancestral sequences possible. Direct optimization is accomplished by examining all possible homologies between the nucleotide bases of two descendant vertices. Dynamic programming is used (in a step akin to pairwise sequence alignment) to optimize each hypothetical ancestral sequence for the minimum weighted number of insertion-deletion events and base substitutions. At each vertex in a cladogram, all possible hypothetical ancestral sequences are implicitly constructed and their costs determined. The minimum cost ancestral sequence is retained and used to optimize the next vertex down the cladogram. Wheeler formally describes the algorithm's downpass in this volume.

Dynamic homology and combined analysis

A logical assumption is that there is one phylogeny of a natural group of organisms because there is one evolutionary history. Comparative data of various sorts reflect the phylogeny of groups under study with different levels of support. No one type of data has been demonstrated to have a high fidelity record of evolutionary history across groups of very different ages. The basic strength of the combined analysis approach lies in the ability of synapomorphies from different types of data to provide additive support for related groups. Dynamic homology takes combined analysis one step further by allowing co-optimization of molecules and morphology. Putative sequence homologies are tested and revised via optimization of their congruence with morphological synapomorphies. This contrasts sharply with standard combined analyses in which prealigned sequences are attached to morphological characters. Standard analysis is restricted by static alignment to seeking for a common signal at the level of the tree search. It has been demonstrated that, in terms of character congruence and topological congruence, combining pre-aligned datasets produces cladograms which are suboptimal to those produced when the same raw data are analyzed with direct optimization (Wheeler, 1998).

Computational complexity of phylogenetics and heuristic solutions

Alignment

As introduced earlier, the number of cells in a multiple alignment matrix are exponentially related to the number of taxa and sequence length. Furthermore, the number of multiple alignments becomes very large with a small number of short sequences (Slowinski, 1998). As a consequence, exact solutions are intractable and heuristics are required to produce multiple alignments. Heuristic alignment algorithms get the job done at the cost of alignment ambiguity. As discussed above, one common heuristic is the use of a guide tree to direct the addition order of sequences in multiple alignment (Sankoff et al., 1973). In theory, an alignment procedure could be repeated for each possible set of relationships among the taxa. However this is intractable because of the large number of evolutionary trees with just a few taxa (discussed below). Alignment heuristics are reviewed in detail in Phillips et al. (2000). In common practice, one topology is used (e.g., as implemented in CLUSTAL [Thompson et al., 1994] and in TREEALIGN [Hein, 1990]). As discussed above, topology-based alignment comes at the cost that results are dependent on the addition order of sequences as determined by the guide tree (Fitch and Smith, 1983). This bias can be addressed by increasing the number of random additions performed which increases runtime (e.g., as implemented in MALIGN, [Wheeler and Gladstein, 2000]). Furthermore, various parameter sets for base transformation costs in alignment may lead a limited set of groups or few groups in common. In many cases when results of many parameter sets are compared, phylogenies share few groups (e.g., W.C. Wheeler, 1995; O'Leary, 1999; Giribet, 1999; Giribet et al., 2000; Giribet and Ribera, 2000; Janies, 2001). However, some analyses have shown consistent results despite parameter variation (Edgecombe et al., 1999). The implementation of topology-based alignment can be improved by concurrent examination of many guide trees and can be explored in reasonable time with an inexpensive cluster of PCs using POY or MALIGN (discussed below). Parallellization of software implemented on inexpensive computing clusters and evermore popular multi-processing PCs provide a very efficient (in terms of maximizing analytical rigor within available time and money) means to rationally address large phylogenetic datasets at the level of sequence alignment.

Topologies

The number of networks facing a topology-based alignment or a phylogenetic tree search is combinatorily dependent on the number of taxa. Thus the number of possible topologies becomes astronomical as taxa are added to the analysis.For example, the number of possible rooted topologies increases as a power series (let y = the number of rooted topologies, let i = the starting point of 3 taxa, let t = the total number of taxa) (Cavalli-Sforza and Edwards, 1967).

$$y = \prod_{i=3}^{t} (2i - 3)$$

The number of possible rooted topologies reaches 34,459,425 with only 10 taxa, 8.2×10^{21} with 20 taxa, and 2.75×10^{76} with 50 taxa.

Practice

The challenges presented by alignment of DNA and phylogenetic tree search have prompted research in heuristics and parallellization. There are several operational reasons to do parallel direct optimization as implemented in POY. Commonly used alignment algorithms produce one (or sometimes many) alignments based on a single parameter set and distance-based addition sequence. Then the investigator has to run a phylogenetic tree search algorithm. POY produces trees, is reasonably fast under a variety of platforms and runs very fast in parallel on inexpensive clusters of PCs (Gee, 2000; Sterling et al., 1999; Janies and Wheeler, 2001). This new paradigm offered by POY permits the investigator to examine many alignment topologies (up to millions of trees per second) and ratchet and swap replicates. Furthermore, the speedup that parallelism affords permits searching a wide parameter space in reasonable time. New, fast phylogenetic search algorithms will produce short trees from a single alignment at unprecedented speed (e.g., Goloboff, 1999; Nixon, 1999). However the speed and quality of the phylogeny produced by these algorithms is dependent on the speed and quality of the alignment(s). Multiple alignment can take weeks of processing time on desktop computers. POY challenges the existing paradigm of alignment followed by a separate tree search, by unifying these steps into a single algorithm that is efficiently scalable to the large datasets necessary to make sense of the large amounts of data being produced by high-throughput DNA sequencing and character coding.

Efficiency of parallel direct optimization

Four major algorithms of POY were tested for parallel efficiency: two types of initial cladogram building and two types of branch swapping. Random replicates of initial cladogram builds were distributed to several processors via a one-processor-per-replicate strategy (via the commands -parallel -multibuild n). Alternatively, single cladogram builds were partitioned across many processors (via the command -parallel). Branch swapping jobs were partitioned across many processors (via the commands -parallel -tbr -spr). These algorithms were tested on several datasets comprised of DNA and morphology ranging from 40–500 taxa (Janies and Wheeler, 2001).

The results of these studies are straightforward and very informative on the scaling properties of POY on large and small clusters. The results on the large cluster (256 processor cluster comprised of Intel Pentium 500 MHz PIIIs networked via 100 Mbps switched Ethernet) contrast significantly (for some but not all algorithms) with those derived from similar studies on a small cluster (11 processor cluster comprised of Intel Pentium 200 Mhz PIIs networked via a 10 Mbps Ethernet hub) previously in service at the AMNH. Various algorithms in POY show fundamentally different properties within and between clusters.

The multibuild command exhibits excellent parallel efficiency in the large cluster (Fig. 1). Speedup (trees examined per second) is very close to linear with the addition of processors regardless of dataset or cluster size. In contrast, parallel building shows poor parallel efficiency in the large cluster with only slight speedup up to 128 slave processors (Fig. 2). This result is similar in large and small clusters. Branch swapping commands show excellent speedup for 10

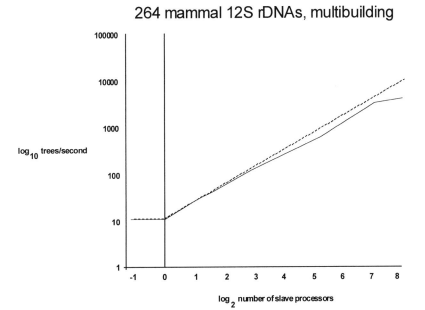

Figure 1. Parallel efficiency of the one-processor-per-replicate strategy using the POY commands (-parallel -multibuild n). The dotted line represents perfect parallel speedup. The solid line represents actual speedup. The multibuild command exhibits excellent parallel efficiency for 264 mammal 12S rDNA and results are similar for other large datasets.

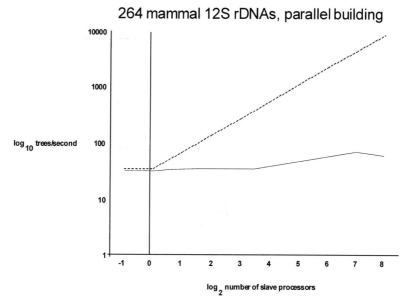

Figure 2. Parallel efficiency of a strategy in which work of each single cladogram build is partitioned across many processors using the POY command -parallel. The dotted line represents perfect parallel speedup. The solid line represents actual speedup. The parallel command exhibits poor parallel efficiency for 264 mammal 12S rDNA and results are similar for other large datasets.

slave processors on the small cluster and excellent speedup for 32 slave processors on the large cluster (Fig. 3.). However, there is no appreciable speedup with the addition of slave processors and this result is independent of dataset size.

These results are fundamental to improving the algorithms for hierarchical parallelism and multi-user load balancing to achieve maximum performance per unit investment. Furthermore, the excellent parallel efficiency of the multi-build command is very encouraging. This result demonstrates the viability of building clusters comprised of several hundred of processors without investing in expensive, non-standard, network hardware. Also, it will be important to invest resources in obtaining higher clockspeed processors to shorten per–node runtimes when using multibuild.

Progress in phylogenetic analysis of DNA sequence data is limited by computational capacity. Advances in DNA sequencing technology have permitted the accumulation of phylogenetic data sets with hundreds to thousands of taxa, each with thousands of nucleotides. Parallelism offers a tractable means to create the computational power required for aggressive heuristic searches. The ongoing development of parallel algorithms combined with the low cost and simplicity of off-the-shelf hardware make cluster computing a revolutionary technology for evolutionary biology.

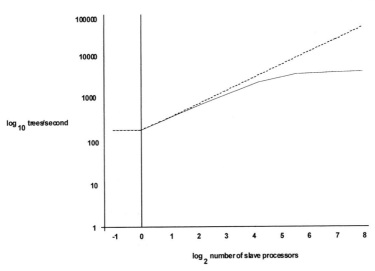

Figure 3. Parallel efficiency of branch swapping using the POY commands -parallel -tbr -spr. The dotted line represents perfect parallel speedup. The solid line represents actual speedup. Branch swapping on trees based on 264 mammal 12S rDNAs in parallel shows excellent speedup for 32 slave nodes but additional processors provide no appreciable speedup. This result is independent of dataset size.

Acknowledgements
The National Aeronautics and Space Administration, the American Museum of Natural History and the New York City Department of Cultural Affairs provided research funding. Lisa Gugenheim, Tim Mohrmann, Pete Makovicky, Diego Pol, Estelle Perrera, Al Phillips, Julian Faivovich and Rebecca Klasfeld of the AMNH were instrumental in the procurement and construction of the parallel cluster. Valuable comments on the manuscript were provided by Gonzalo Giribet, Susan Perkins and Marc Allard. Test dataset for mammals was provided by Ginny Emerson.

Some statistical aspects of the maximum parsimony method

Mike Steel

Biomathematics Research Centre, University of Canterbury, Christchurch, New Zealand

Summary. The last three decades have seen considerable debate concerning the relative merits and problems associated with two competing approaches to phylogeny—approaches based on the parsimony principle *versus* maximum likelihood methodology. Although the two approaches may seem quite opposed, there are in fact some close relationships between them. For example, we describe a recent result that shows how maximum parsimony can be regarded as a type of maximum likelihood estimator when there is no common mechanism between sites (such as might occur with morphological data and certain forms of molecular data). Distinguishing between this and other implementations of maximum likelihood helps clarify some of the dispute that has surrounded the two methodologies. We also provide a brief overview of some mathematical and statistical properties of the maximum parsimony criterion.

Introduction

Many techniques now exist for reconstructing phylogenetic trees from genetic sequence data. Two of the most popular approaches are usually referred to as 'maximum parsimony and 'maximum likelihood'. We will abbreviate these approaches here as MP and ML respectively. Although MP continues to be widely used, it is often criticised as being statistically unsound and as failing to make explicit an underlying 'model' of evolution. Indeed there is little agreement on how, or even whether, MP should be justified. According to Edwards (1996), who prefers to call MP the 'method of minimum evolution', the method was introduced in his joint 1963 paper with Cavalli-Sforza (in the context of continuous characters) merely as a computational approximation for ML, and by not as a method of choice in its own right. The discussion is further complicated by claims that MP variously is, or is not, a form of ML, and by the discussion of 'zones' within which either method performs worse than the other in recovering the true tree.

Several authors (for example, Farris, Kluge and Eckardt, 1970; Sober, 1988) claim MP is the preferred method of tree reconstruction, citing Willi Hennig's writings on phylogenetic inference, or alternatively the *Principle of Parsimony*. The latter is a minimalist principle, also referred to as 'Ockham's razor'. It states that one should prefer simpler explanations, requiring fewer assumptions, over more complex, *ad hoc* ones. In phylogeny reconstruction, this principle has been applied in two ways: (i) to assume as little as possible about any underlying model or mechanism for evolution (however, this can also be used as an argument in favour of the more usual forms of ML), or (ii) to emphasise the feature that MP favours the tree requiring the fewest evolutionary events (such as mutations) to explain the observed data, and so is, in some sense, the 'simplest' or an 'optimal' description of the data.

Several authors (e.g., Farris, 1973; Sober, 1985, 1988) have also presented explicit statistical arguments in favour of MP, based on underlying evolutionary models. Still others have undertaken the more modest task of providing a statistical framework for using MP (Cavender, 1978; Kishino and Hasegawa, 1989; Maddison and Slatkin, 1991; Archie and Felsenstein, 1993; Steel, Hendy and Penny, 1992; Steel, Lockhart and Penny, 1993b, 1995).

The simplicity of a method like MP and variations that allow weightings on characters and transition types, together with its apparent lack of assumption involving underlying models, has made it popular in phylogeny, particularly in the 1970s and 80 s. However, model-based approaches have come to rival, and even dominate, phylogenetic methodology, particularly over the last decade. While ML is the leading alternative, other approaches include distance-based methods that use transformed or inferred distances, for example logdet/paralinear distances (see Swofford et al., 1996 for a review of distance methods which are outside the scope of this overview of parsimony and likelihood). One justification for model-based approaches was the classic and much-cited statistical inconsistency of MP due to Felsenstein (1978). This paper demonstrated that if sequence sites evolved under certain models and combinations of rates, then MP would favour an incorrect tree. Furthermore, the probability of selecting an incorrect tree would tend to 1 as the sequence length grew (this phenomenon of statistical inconsistency will be discussed further in Section 4). The particular combination of short and long branches that Felsenstein used has become known as the 'Felsenstein Zone'.

Both Felsenstein (1973) and Yang (1994) informally claimed the nonexistence of any such zone within which ML would be statistically inconsistent (though this was questioned by Sober (1988, Ch. 5)). Indeed, the statistical consistency for ML (when the underlying model had no rate distribution across sites, and this same model was then also used in the ML method to reconstruct the tree) was rigorously established recently by Chang (1996b). Note that the use of the 'correct' model (the same as the model used to generate the data) is essential to the proof that maximum likelihood is consistent, and ML can be inconsistent if the model used to analyse the data differs from that which generated it (see Chang, 1996a). Although one may seldom know the correct model of evolution, the more one knows about the evolutionary process, the more likely one is to avoid a zone of inconsistency by analysing the data correctly.

Nevertheless, objections to ML have arisen on a number of fronts, which we now describe. First, there is concern about the validity and exact form of any underlying stochastic model (for example, there is concern as to the choice of underlying parameters/distributions), and that by selecting the appropriate model one could perhaps reconstruct any favoured tree. There is also concern that ML estimation of a tree (and statistical tests between different trees) that involves optimizing 'nuisance (supplementary) parameters' is statistically problematic. There are also suggestions that the Felsenstein zone rarely if ever arises for real data and claims for the existence of a 'Farris zone' where MP outperforms ML. Another factor is the increasing analysis of aspects of genome data that extend beyond site substitution—for example, gene order, SINEs (short interspersed nuclear elements) for which MP may be more appropriate. Finally there is some concern about the computational complexity of ML. Even on a *given* tree, optimising the likelihood can be problematic (unlike MP, where Fitch's algorithm (Fitch, 1971a) provides a linear time algorithm for computing the parsimony score).

In this chapter we will explore some of these objections and survey some recent theoretical results that shed light on the interplay between the two methodologies and on the limits of what one can hope to achieve in phylogeny reconstruction. We also describe some statistical properties of the parsimony score function.

It is useful to make a three-way division of the model of evolution. This consists of a tree T (or more generally a graph when median networks or splits graphs are considered), a stochastic mechanism of evolution (such as whether or not it is neutral, Kimura 3ST, exhibits rate heterogeneity, etc.) and the initial conditions (for example, inter-speciation times or rates on each edge (branch) of the tree).

Often researchers will seek to recover different aspects of the model. Most frequently perhaps it is just the unweighted tree, regardless of the amount of mutation on each edge of the tree. In addition, the tree will usually be unrooted unless an outgroup or an assumption about a molecular clock is used. Frequently, however, the rates of mutation will be required in order to estimate times of divergence. Others will also wish to estimate the character states at the internal nodes. It is thus too simple just to compare 'parsimony' and 'likelihood'. Indeed likelihood itself comes in many flavors and these will be discussed next. The usual form of ML is 'maximum average likelihood', an example of 'maximum relative likelihood'.

Varieties of forms of ML in phylogenetics

According to Edwards (1972), the *likelihood* of the hypothesis H, given data D and a specific model, is proportional to $P(D|H)$, the conditional probability of observing D given that H is correct. A ML method of inference selects the hypothesis H that maximises the likelihood function for the data D (given the specified mechanism). In the context of phylogeny reconstruction from sequences, the data D typically counts the number of 'site patterns' that occur in a collection of aligned sequences. The order in which these patterns occur and the phylogenetic information that this might convey is usually discarded; however, some authors have recently incorporated this also (for example,

Thorne, Goldman and Jones, 1996; Giribet and Wheeler, 1999b). The hypothesis H is usually the discrete phylogeny (unweighted tree) T, and the model is some stochastic process for site substitution (or, more generally, genome transformation if insertions and deletions are allowed).

What complicates matters is that $P(D|T)$, and hence the likelihood of T, requires more information to specify it than just the data D and the parameter T. More precisely, the probability of evolving D depends on further parameters, sometimes referred to as 'nuisance parameters'. In order to talk about $P(D|T)$ we either need to specify these parameters, or place some prior distribution on them. The word 'nuisance' is a little misleading. It does not imply that these parameters are of no interest, but rather that they need to be considered even if all one wants to know about is the tree T. Examples of such parameters in molecular phylogenetics are the edge lengths (inter-speciation times and rates of mutation on the edges), parameters associated with the substitution matrix (for example, transition/transversion bias) and parameters that describe how rates vary across sites.

Nuisance parameters arise widely in many statistical settings and have been discussed in the phylogeny setting by several authors, for example Goldman (1990). Nuisance parameters may further be classified into 'structural' and 'incidental' parameters. The former are parameters that influence all (or nearly all) of the sites; incidental parameters influence only one or a few. Structural parameters typically correspond to the edge (branch) lengths and parameters that constrain the substitution process (for example, the transition/transversion bias). Typically, such parameters are either selected to maximize the likelihood or estimated directly from the data. Incidental nuisance parameters arise either if (i) we wish to hypothesise a particular choice of sequences to appear at internal vertices of the tree, in which case we need to specify states for each site, or if (ii) the process varies from site to site. We will discuss both these situations below. In any case, for a model of sequence evolution we will represent nuisance parameters collectively by the Greek letter θ.

Two frequent assumptions concerning substitution models are that aligned sites evolve *independently* and according to an *identical* process—the so-called 'i.i.d.' assumption. Note that the i.i.d. assumption still allows sites to evolve at different rates by regarding the rate of a site as being randomly and independently selected from an appropriate distribution (such as a gamma distribution). Of course in real sequences one has clustering of 'conserved' and 'hypervariable' sites (so the real process is definitely not i.i.d. across sites) but when one passes to the frequencies of site patterns (i.e. the data D) the process can be modelled by an i.i.d. process. Similarly, certain covarion-style mechanisms (where sites can alternate between invariable and variable during evolution) can be modeled using an i.i.d. process (Tuffley and Steel, 1997a), even though the original covarion model (e.g., Fitch, 1971b) implied explicit dependency between sites.

The i.i.d. assumption allows one to readily compute $P(D|T, \theta)$ by identifying this with the product of the probabilities of evolving each particular site.

Occasionally, more intricate models have been proposed and analysed. These include models that allow a limited degree of non-independence between sites (for example pairwise interactions in stem regions, Schöniger and von Haeseler, 1994), and models that work with non-aligned sequences and explicitly model the insertion-deletion process as well as the site-substitution process (Thorne, Kishino and Felsenstein, 1992).

Maximum integrated likelihood versus *maximum relative likelihood (MIL versus MRL)*

If the nuisance parameters θ and the phylogeny T are generated according to some known prior distribution (for example, a Yule pure-birth process) one can formally integrate out these nuisance parameters, and thereby take $P(D|T)$ to be this average value. That is, if $\Phi(\theta|T)$ denotes the distribution function of the nuisance parameters, conditional on the underlying tree T, then

$$P(D|T) = \int P(D|T,\theta)d\Phi(\theta|T).$$

This approach is sometimes referred to as 'integrated likelihood', and a tree T that maximizes $P(D|T)$ we will refer to as a *maximum (integrated) likelihood tree*. Maximum integrated likelihood (MIL), and, more generally, the assignment of posterior probabilities to trees based on sequence data (using Markov chain Monte Carlo technique to approximate the integral in the above equation) has been independently developed by several authors recently, in particular Yang and Rannala (1997) and Mau, Newton and Larget (1999).

Assume for the moment that one possesses such a prior distribution. A natural question arises, namely, in what sense is MIL an optimal method for selecting a tree? In particular, is it the method that is most likely (on average) to return us the true tree? In order to formalize this question, suppose we have a tree reconstruction method, and we apply it to sequences that have been generated by a model with underlying parameters T and θ. The *reconstruction probability* denoted $\rho(M,T,\theta)$ is the probability that the sequences so generated return the correct tree T when method M is applied. Since we have a distribution on trees and the nuisance parameters, let $\rho(M)$ denote the *expected reconstruction probability* of the method M, obtained by integrating $\rho(M,T,\theta)$ over the joint parameter space. That is,

$$\rho(M) = E[\rho(M,T,\theta)] = \sum_{T} p(T)\int \rho(M,T,\theta)d\Phi(\theta|T)$$

where *p(T)* is the probability of the tree T under the prior distribution (we will assume that only binary trees have positive probability). The following proposition describes precisely the method that maximizes the expected reconstruction probability:

Under the conditions described, the method M that maximizes the expected reconstruction probability $\rho(M)$ is precisely that method that selects, for any data D, the tree(s) T that maximizes $p(T)P(D|T)$.

For a proof of this last assertion, see Székely and Steel (1999). The tree(s) that maximizes $p(T)P(D|T)$ is sometimes referred to as the *maximum a posteriori* (MAP) estimate. This is precisely the maximum (integrated) likelihood tree(s) whenever the prior distribution on binary trees is uniform (i.e., when all binary trees are equally likely). Consequently, assuming that the prior distribution assigns equal probability to all binary trees, MIL maximises one's average chance of recovering the correct tree. However, if the distribution on binary trees is not uniform—for example, if the trees are described by a Yule process —then the optimal selection criteria are slightly different. In any case, it is clearly a difficult problem to find (let alone agree upon!) a compelling and biologically reasonable distribution on trees and parameters.

The alternative approach, which is more widely adopted, is sometimes called *maximum relative likelihood* (MRL). One simply assumes that the nuisance parameters take values that, simultaneously with an optimal tree T, maximize $P(D|T,\theta)$. Usually one then discards θ and outputs just the tree(s) T. Such an approach can be problematic in general statistical settings where data D depend on both continuous (nuisance) parameters and a discrete parameter x of interest. In this situation, there may be one 'unlikely' value of θ that for $x = x_1$ gives a higher $P(D|x,\theta)$ value than $\max_\theta P(D|x_2,\theta)$, yet for most 'likely' values of θ the probability $P(D|x_1,\theta)$ is less than $P(D|x_2,\theta)$. This property means that MRL may make different selections from MIL and it seems to have been a fundamental issue in the exchange between Felsenstein and Sober (Felsenstein and Sober, 1986) on the relative merits of MP and ML. Moreover, in the phylogenetic setting, MRL may select different trees from the MIL method described above even when all binary trees are equally likely (at least for certain distributions on the edge parameters of the tree). An example of this is described later.

For the remainder of this chapter we will generally assume there is no prior distribution given for trees and edge parameters, and so all forms of ML involve MRL. With this in mind we review some further distinctions.

Three forms of maximum relative likelihood

In fitting sequence data to a tree, the sequences at the leaves (tips) of the tree are given, but those at the internal vertices (speciation or branching points) of the tree are not. In the usual implementation of maximum (relative) likelihood in molecular phylogenetics, one effectively averages over all possible assignments of sequences to these internal vertices. Following Barry and Hartigan (1987) we call this *maximum average likelihood*, and we denote it as $M_{av}L$.

However, one could also assign sequences to the internal vertices (along with the other parameters) so as to maximize the likelihood. Such an approach was suggested explicitly by Barry and Hartigan (1987) who called it *most parsimonious likelihood*, to distinguish it from $M_{av}L$. They remarked that most parsimonious likelihood 'is therefore similar to the maximum parsimony fitting technique'. However, it differs slightly from MP in that the other parameters (e.g., edge-lengths) must be fixed across all the characters. Likelihood calculations that place sequences at the internal vertices of a fixed tree have also been explored by other authors (Koshi and Goldstein, 1996; Pagel, 1999) where the interest has been primarily in reconstructing, say, ancestral sequences of proteins (or other characters), rather than in selecting an optimal tree. Goldman (1990) described a link between MP and most parsimonious likelihood. He showed that, under a symmetric 2-state mutation model, and with the artificial constraint that all mutation probabilities on each edge of any binary tree are equal to some value p, then the MP tree(s) are exactly the most parsimonious likelihood trees.

Given the most parsimonious likelihood approach, it might seem natural to carry the approach of assigning ancestral sequences further. That is, one could select sequences for each time interval right through the tree (jointly with the other parameters) to maximise the probability of observing the given sequences at the leaves. Thus, one would associate along each edge of the tree a series of sequences, corresponding to their evolution at frequently sampled time intervals.

Such an approach was suggested by Farris (1973), and it was subsequently referred to as an *evolutionary pathway* approach—since it is a complete specification of the sequences through time. Farris showed that the tree(s) that maximizes the likelihood in this sense are *exactly* the maximum parsimony trees. Indeed, the argument is straightforward and requires few assumptions regarding the underlying model—in particular, it does not require any assumption about mutations occurring at a slow rate (only that they occur at a continuous rate) or edge lengths that are constrained in any way. Also, the equivalence with MP holds with the edge lengths either specified or allowed to be optimised. Of course there will generally be a huge (potentially infinite) choice of possible evolutionary pathways of maximal probability; however, this is not a problem if the value of this maximal probability is all that is being used to select trees.

As noted by Felsenstein (1978) (see also Sober, 1988, p. 160) the distinction between $M_{av}L$ and Farris's evolutionary pathway likelihood is crucial for reconciling the apparent paradox between Felsenstein's claim that ML (but not MP) is statistically consistent and with Farris's claim that MP is a ML method. Both claims are correct; they are simply referring to different forms of ML. Figure 1 illustrates the three forms of ML we have just discussed.

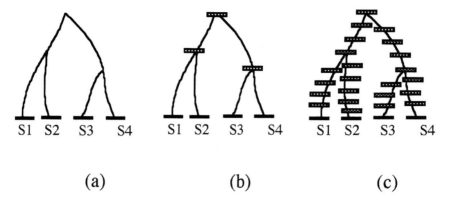

Figure 1. Three forms of ML. (a) *Maximum average likelihood* ($M_{av}L$): all possible sequences at the internal vertices contribute to the likelihood; (b) *Most parsimonious likelihood*: sequences to maximize the likelihood are placed at the internal vertices; (c) *Evolutionary pathway likelihood*: sequences to maximize the likelihood are placed at each position throughout the tree.

A model for which maximum parsimony is a maximum (average) likelihood estimator

Most parsimonious likelihood and evolutionary pathway likelihood both involve the specification of a choice of sequences to points inside the tree. Although a particular selection of sequences may be the most probable, the attraction of $M_{av}L$ is that it effectively allows all possible assignments of sequences to the interior of the tree. These are weighted according to their probability, and then summed up to give the marginal probability of evolving the sequences observed at the leaves. The question arises then as to whether MP can be regarded as a $M_{av}L$ method under some model.

Suppose we take the simplest type of substitution model—the Jukes-Cantor type model—in which each of the possible substitutions at a site occurs with equal probability. Now suppose the rates of evolution on each branch of the tree can vary freely from site to site. In this case we have some constraints on the underlying type of substitution model (i.e., Jukes-Cantor type), but no constraints on the edge parameters from site to site. We might refer to this as *no common mechanism*. This is even more general than the type of approach considered by Olsen (see Swofford et al., 1996, p. 443) in which the rate at which a site evolves can vary freely from site to site; however, the ratios of the edge lengths are equal across the sites. In the Jukes-Cantor style model with no common mechanism (not even the same rates for different characters) the following theorem applies.

Under the model described (with no common mechanism) the $M_{av}L$ tree(s) are precisely the maximum parsimony tree(s).

A proof of this result is given by Tuffley and Steel (1997b) who generalised an earlier special case by Penny et al. (1994). The significance of the result should not be taken as any special justification of MP over usual implementations of ML; neither does it imply that MP trees are the same as those that ML would produce under the 'usual' models (e.g., Jukes-Cantor with fixed edge lengths). Rather, the significance is of a more philosophical nature, as it describes a model in which MP can be regarded as a ML method in the usual 'average' ML setting (that is, where one does not select particular sequences for the internal vertices as part of the optimisation step).

The argument used to establish the above theorem also shows that, under the Jukes-Cantor type model, if we are given just a tree and a single character (and no information as to the edge lengths) the ML estimate of the state at any internal vertex of the tree (given the states at the leaves of the tree) is precisely the MP estimate. For a further link between ML and MP suppose we take any sequence data and add a sufficiently large number of unvaried sites. Then, under a Jukes-Cantor style model, the ML tree of this extended data set is always an MP tree. For details and justification of these last two results see Tuffley and Steel (1997b).

Of course this type of underlying model (in the above theorem) is almost certainly too flexible, since it allows many new parameters for each edge. It might be regarded as the model one might start with if one knew virtually nothing about any common underlying mechanism linking the evolution of different characters on a tree (for example, as with some morphological characters).

For processes like nucleotide substitution, as one learns more about the common mechanisms involved, it would seem desirable to use this information. This would lead towards the more usual implementations of maximum (average) likelihood where the model parameters (such as edge lengths) are constant across sites. Indeed, advocates of Ockham's razor (the Principle of Parsimony) might well invoke the principle at this point, as illustrated by the following example. Consider sequences of a pseudogene, each sequence being many thousands of nucleotides long. As a first approximation there is no selection at any of the sites and therefore it is more 'parsimonious' to assume one common mechanism for all sites, rather than several thousand different mechanisms, one for each site. In such a case, the Principle of Parsimony would support the usual maximum (average) likelihood over using data uncorrected for multiple changes.

This conclusion should, however, be taken with care. Such a model may not apply to other sequence data and would not often apply to morphological data (for example, where the evolution of numbers of legs may differ from that of wing colour). It is clear that we still need to learn more about the processes leading to different types of insertion and deletion events in sequence data to postulate a common mechanism.

Regions where MP may outperform ML

It is easy to construct examples where $M_{av}L$ will be inconsistent if the model used in the ML analysis differs from the model that generated the sequences. What is perhaps more surprising is that MP can perform better than $M_{av}L$, even when the underlying model matches the generating model. These regions of parameter space have been called the 'Farris Zone' (Siddall, 1998) and the 'anti-Felsenstein Zone' (Waddell, 1996); this phenomenon has been noted by others (for example Huelsenbeck, 1998; Yang, 1996).

Here the 'performance' of a tree reconstruction method M (on sequence data generated under a tree-indexed Markov model) is again taken to mean the reconstruction probability $\rho(M,T,\theta)$ described in Section 2 (the probability that the method will correctly return the true tree T). This depends not just on M but also on T and the parameters on the edges of the tree. Now there exist trees T and parameters where MP will have a higher probability of returning the 'true tree' T than $M_{av}L$. In more detail, consider a fully resolved tree T on four species a, b, c, d, with species a, b on one side of the central edge, and species c and d on the other. Consider the simple symmetric 2-state model with mutation probability $p(e) = \varepsilon$ on the two edges incident with leaves a,b; while $p(e) > 0.5 - \varepsilon$ on the other three edges, where ε is small but positive. Thus three edges have long interspeciation times (or, alternatively, high mutation rates) and so are near site saturation, while two sister taxa are recently separated (or, alternatively, have low mutation rates on their incident edges). Note that such a situation is entirely possible under a molecular clock, though we need not insist on this.

Suppose we evolve k sites independently on this tree. Let $P_1(k)$ be the probability that MP recovers the true tree T and let $P_2(k)$ be the probability that $M_{av}L$ recovers T from the k sites. Then, as ε converges to 0 (with k fixed) we have:

$$P_1(k) \cong 1 - (\frac{3}{4})^k; P_2(k) \le \frac{2}{3}$$

A proof of this result is presented in Steel and Penny (2000) (a similar result was stated without proof in Székely and Steel (1999)). Notice that for ε very small (but positive), MP will recover T with 99% probability with just 16 sites, yet $M_{av}L$ could take potentially millions of sites to achieve the same probability of correctly reconstructing T. In that case, for realistic length sequences, other effects, for example deviations from the model, might have more effect on the reconstructed tree than the sequence data.

It is tempting to dismiss this example as a triviality by noting that one could also outperform $M_{av}L$ in this example by simply disregarding the data and always outputting the tree that groups species a and b together, and c and d together. However, there is a fundamental difference here, since MP will outperform $M_{av}L$ for *any* of the three possible underlying trees on four species, when the parameters are in the right range. Clearly a trivial method, like the one described, cannot achieve this.

While the example described above is somewhat extreme, it still shows there are cases where we would expect $M_{av}L$ to require much longer sequences to recover the true tree than MP needs. In fact we actually only require $p(e) > 0.5 - \varepsilon$ on two of the three edges, but we have opted to allow three edges to be near site saturation, since then the example can arise under a molecular clock. In contrast, the Felsenstein Zone cannot arise under a molecular clock with four species; yet to be fair, if we want to impose a molecular clock, we should implement ML with a molecular clock, and then ML no longer behaves as described above.

The significance of this example should not be overstated—it does not mean that one 'should' be using MP—it may well be that 'on average' (under some prior distribution on trees and their parameters) $M_{av}L$ outperforms MP, but it does not globally outperform (in the sense described above) MP. This example also does not demonstrate statistical inconsistency of $M_{av}L$, since if the edge mutation probabilities are fixed (and strictly between 0 and 0.5), then $M_{av}L$ will eventually recover the true tree with probability converging to certainty as k tends to infinity. This example can also be modified to demonstrate that $M_{av}L$ can differ from MIL, even when all trees have equal prior probabilities (provided the prior distribution on the edge lengths is sufficiently contrived). Specifically, suppose that each of the three binary trees on sequences a,b,c,d has equal probability, and that the prior distribution on the edge lengths allows all possible values for the mutation probabilities, but with probability $1 - \delta$, we have $p(e) \leq \varepsilon$ on two edges incident with two sister leaves and $p(e) > 0.5 - \varepsilon$, on the other three edges. Then it can be shown that for ε, δ sufficiently small (but positive), MIL can select a different tree than $M_{av}L$ on certain data.

The statistics of parsimony under a null model

In order to carry out hypothesis tests using the parsimony score of a tree, one needs to know the distribution of this score on a given tree under a suitable null model for generating characters. This approach has been adopted by a number of authors, for example Archie and Felsenstein (1993), Maddison and Slatkin (1991), Goloboff (1991), Steel et al. (1992, 1993b, 1995), Kishino and Hasegawa (1989).

In this section we describe some exact formulae for this problem under certain simple null models. Although these results have been in the literature for several years now, they are not well known, yet they are surprisingly simple and explicit. We consider first the very simplest such null model. In this case there are just two character states and each leaf in the binary tree T has equal probability of being assigned either of the two states. The resulting parsimony score of the character on T is then a random variable, which we denote here as

$L(T)$. Let $P[L(T) = k]$ denote the probability that this parsimony score takes the value k. For example, for any binary tree with 4 leaves, we have $P[L(T) = 2] = {}^4/_{16}$ since there are $2^4 = 16$ binary characters and exactly four of them require two mutations on T. One would like to determine this probability distribution, as well as its mean $\mu(T)$ and variance $\sigma^2(T)$. Several authors (Maddison and Slatkin, 1991; Goloboff, 1991; Archie and Felsenstein, 1993) have constructed recursive formulae for $\mu(T)$. However, it is possible to give exact and explicit formulae, not just for the mean (and variance) but for the entire probability distribution, as we describe shortly. All of these formulae depend only on the number of leaves of the binary tree T, and not on its shape (this surprising, and pleasing property does not extend to characters with more than 2 states). The explicit formulae for the probability distribution and its mean (from Steel, 1993) are:

$$P[L(T) = k] = \frac{(2n - 3k)(n - k - 1)! \, 2^{k-n}}{k!(n - 2k)!}$$

$$\mu(T) = \frac{(3n - 2 - (-0.5)^{n-1})}{9}$$

where n is the number of leaves of the binary tree T.

Notice that for at least modest-sized binary trees (that is, when $n > 6$) we have the close approximation $\mu(T) \sim n/3$ (here and below 'close' means that the difference between the true value and its approximation goes to zero exponentially fast with n). It is instructive to contrast this with the expected value of $L(T)$ when T is star-shaped (fully unresolved). In that case it can be shown that $\mu(T) \sim n/2$ (Steel, 1993). Thus, the additional edges present in a binary tree allow one to reduce the expected number of mutations required to fit random data from approximately $n/2$ per character (for an unresolved tree) to $n/3$ (for a binary tree), a difference of $n/6$ mutations per character. There is also a slightly more complicated but exact formula for the variance $\sigma^2(T)$ (see Steel, 1993), from which one obtains the close approximation:

$$\sigma^2(T) \sim 2n/27.$$

One can extend this very simple null model in three ways—(i) by allowing more than two character states (ii) by allowing the probability distribution of the states to be non-uniform and (iii) by allowing the probability distribution of the states at the leaves to vary between leaves. Extension (ii) recognises that some states may be more frequent than others, while extension (iii) allows for phenomena such as GC-variation between different genetic sequences.

Even if we allow all of these three extensions ((i)–(iii)) simultaneously, one can still efficiently compute the probability distribution of $L(T)$. An algorithm to do this is described in Steel et al. (1996) and this paper also shows that the limiting distribution of $L(T)$ converges to a normal distribution as the number of leaves in the binary tree T becomes large. This again applies under the extended null model (allowing (i)–(iii)) subject to a mild technical condition[1]). Of course if we take the cumulative sum of a large number of characters generated independently under this (extended) null model, then the parsimony score of these data on T will also be normally distributed (by the central limit theorem) regardless of whether T has few or many leaves (though if the tree is large the approximation should be much better for a small number of characters).

One can also consider the statistics of the "dual" setting where a character is given, and we wish to find the probability that a binary tree chosen uniformly at random has a given parsimony score for that character. Determining these probabilities provides, for example, a simple formula for the average parsimony score of a collection of characters over all binary trees (Hamel and Steel, 1997). Once again, in the case of binary characters there is, surprisingly, an exact formula for these probabilities, which we now describe.

First, it is easily seen that the number of binary trees having a given parsimony score on a given character depends only on the numbers of species assigned the two states. Consequently, if the number of species assigned the two states is a and b we can denote the probability that a randomly selected binary tree has parsimony length k by $p_k(a,b)$. For example, $p_2(2,2) = 2/3$, since two of the three binary trees on 4 leaves require exactly two mutations to fit a character of type 0011. The *bichromatic binary tree theorem* gives an exact formula for $p_k(a,b)$ as follows.

$$p_k(a,b) = \frac{(k-1)!(2n-3k)N(a,k)N(b,k)}{B(n-k+2)}.$$

In this formula, $n = a + b$ is the total number of species, and $N(m,k)$ is the number of forests consisting of k rooted trees on a total of m leaves, and this is given exactly by the formula:

$$N(m,k) = \frac{(2m-k-1)!}{(m-k)!(k-1)!2^{m-k}}$$

(for $1 \le k \le m$), while

$$B(m) = \frac{(2m-4)!}{(m-2)!2^{m-2}}$$

is the number of binary trees on m labelled leaves (and so $B(m) = N(m-1,1)$).

This remarkable formula for $p_k(a,b)$ was first established by Carter et al. (1990) using complicated generating function techniques. However, the combinatorial nature of the quantities in the above formula suggested there should

---

[1] For some fixed $\varepsilon > 0$, and each state α, each leaf has probability at least ε of being assigned state α.

be a constructive proof of this theorem based on matching up forests of trees. Such a constructive proof was given by Steel (1993) and subsequently simplified by Erdös and Székely (1993). These proofs, and some others involving parsimony lean heavily on a fundamental property of 2-state parsimony, that follows from Menger's theorem in graph theory. For completeness we state this property (established formally as Lemma 1 in Tuffley and Steel, 1997b) as follows:

The parsimony score of a 2-state character on a tree T equals the maximum number of paths that can be placed in T so that (i) each path joins leaves that are assigned different states by the character, and (ii) no two paths share any edge of T.

An example of such a path packing is illustrated in Figure 2. This result is an example of a "min-max" theorem since it relates a quantity we seek to minimize (the number of mutations) to a quantity we maximize (the number of allowed paths in a packing). A clever extension of this theorem to r-state characters has been obtained by Erdös and Székely (1992).

Furthermore, the distribution of trees according to their parsimony score on a fixed character becomes normally distributed as n (the number of species) becomes large, at least for 2-state characters (it is likely also to hold for r-state characters, though this has not yet been rigorously established). More precisely Moon and Steel (1993) show that as n grows, $p_k(a,b)$ becomes normally distributed (as k varies) with mean μn and variance $s^2 n$ where

$$\mu = \frac{2}{3}\left\{1 - \sqrt{1 - 3\frac{ab}{n^2}}\right\}, s = \frac{\mu\sqrt{1-\mu}}{2-3\mu} \ .$$

There has been only limited success in generalising the bichromatic binary tree theorem to non-binary characters. However, one noteworthy and pleasing result is an exact formula for the probability that a randomly selected binary

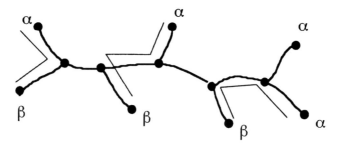

Figure 2. An illustration of the min-max theorem for parsimony score, provided by a tree and a 2-state character with parsimony score 3. Indicated is a maximal system of three edge-disjoint paths, each of which joins a pair of leaves assigned different states by the character.

tree displays no homoplasy for a given r-state character (that is, the character has parsimony score $r - 1$ on the tree). For details see Carter et al. (1990) or Steel (1993).

Conclusion

This chapter highlights two contrasting points: First, the parsimonious approach suggested by Ockham's razor can, given information on a common mechanism, support the usual forms of ML over MP for sequence data. Second, when we generalise traditional substitution models (like Jukes-Cantor) sufficiently far—namely to allow different edge parameters at different sites—the usual ML approach arrives back at MP. Indeed, as models become increasingly sophisticated and parameter-rich, one risks losing the ability to discriminate between different underlying trees. Essentially, this is because the data may be able to be described perfectly by any underlying tree, by adjusting the other parameters appropriately. This is a real possibility for site-substitution models that allow a distribution of rates across sites. Indeed there are situations where all trees could perfectly describe the same data, provided one can select, for each tree, a corresponding distribution of rates across sites (Steel, Székely and Hendy, 1994). The model we described earlier (no common mechanism), where MP can be regarded as a ML method, clearly would also have this non-identifiability problem. An interesting problem for future investigation would be to determine the extent to which a stochastic model needs to be constrained in order that the underlying tree can be recovered from sufficient data.

Acknowledgments
I thank the New Zealand Marsden Fund for supporting this research and the Isaac Newton Institute (Cambridge, UK) for its hospitality during the 1998 BFG program. I also thank David Penny, Peter Lockhart and Charles Semple for helpful comments on an earlier version of this manuscript.

Molecular Systematics and Evolution: Theory and Practice
ed. by R. DeSalle, G. Giribet and W. Wheeler
© 2002 Birkhäuser Verlag/Switzerland

'Pluralism' and the aims of phylogenetic research

Gonzalo Giribet[1], Rob DeSalle[2] and Ward C. Wheeler[2]

[1] Department of Organismic and Evolutionary Biology, Museum of Comparative Zoology, Harvard University, Cambridge, MA 02138, USA
[2] Division of Invertebrate Zoology, American Museum of Natural History, New York, NY 10024, USA

Summary. In science, and particularly in the field of phylogenetic systematics, investigators may choose among different methods to analyze their data. These methods include neighbor-joining (or other genetic distance approaches), maximum-likelihood, and cladistic parsimony, among others. These distinct methods of analysis differ considerably in how they process information from the observed data. However, many published molecular analyses utilize trees generated under more than one of these methods, which we will call a 'pluralistic' approach. Here, we explore the statistical, philosophical and operational aspects of the pluralistic approach. We suggest that the pluralistic approach is misguided from all three perspectives and we propose an alternative, logically consistent, strategy as an aim of phylogenetic research.

Pluralism and 'statistical support'

Some authors have advocated pluralism as a severe test for phylogenetic hypotheses, in the sense that hypotheses may be more robust if they are supported simultaneously by different methods of data analysis. This is based on an analogy with statistical analysis of multiple samples. Any result found significant in multiple samples is thought to be more convincing than a single test. This is true. It is, however, also misleading. When multiple samples are used in statistics, they are assumed to follow the same distributional model, hence the sample error is decreased with each additional datum. The data sets are multiple, not the analytical procedures. When multiple analytical techniques are used, the "significance" attached to results can be highly variable. Results that are "significant" via a standard Fisher-type approach may not be so convincing in Likelihood or Bayesian analyses.

The outcome of a pluralistic approach

There are two possible outcomes when hypotheses are generated using different methods—either they agree (they are congruent) or they disagree (they are incongruent). When three methods are used (such as neighbor joining, maximum-likelihood and parsimony) there are five different competing outcomes: they can all agree, all can be different, or there are three ways that any two can agree. In addition each of the three major kinds of methods could also give any number of hypotheses dependent on the parameters/models used. When the results of different analyses are incongruent two routes are usually taken: (1) some method is used to settle on a consensus result, very much resembling a taxonomic congruence approach, or (2) criteria are established that result in the choice of one hypothesis over another.

Suppose that 'a favorite method' yields a different hypothesis from that of other methods because it is able to account for something that the alternative methods do not. For example, gaps are usually treated as missing data in phylogenetic analyses. A situation could arise where the best hypothesis of a method that uses gap information contradicts all the other methods, precisely because they differ in the way of treating this certain set of characters. In addition, the alternative methods could converge on a hypothesis that differs from the one based on the unique information of gap characters. In such a case, pluralism is agnostic with respect to defending a chosen method *versus* competing ones, because the different assumptions of the various methods are indeed what make one 'superior' to the others. We ask why try everything if we are going to choose the hypothesis generated by 'our favorite method' anyway? This is why, as investigators, we should be compelled to choose a single method based on philosophical criteria. In addition, an increasing number of empirical and philosophical papers have been published that discuss the different methods of phylogenetic analysis in terms of hypothesis testing (for parsimony [Kluge, 1997] and for maximum-likelihood [Huelsenbeck and Bull, 1996; Huelsenbeck and Crandall, 1997; Huelsenbeck and Rannala, 1997; Cunningham et al., 1998]), accuracy (Hillis et al., 1994; Hillis, 1995; Swofford et al., 1996) and consistency (Felsenstein, 1978; Hillis, 1996; Kim, 1996; Swofford et al., 1996; Huelsenbeck, 1997, 1998; Siddall, 1998). None of these discussions allow for a pluralistic approach.

An example: the phylogeny of the arachnid order opiliones

The arachnid order Opiliones (daddy-long-legs or harvestmen) has been classically divided into three suborders: Cyphophthalmi, Palpatores and Laniatores. The internal phylogeny of the Opiliones was studied by Giribet et al. (1999) using sequence data from the 18S rDNA and 28S rDNA loci and morphology. In this analysis, two alternative hypotheses were supported by the molecular data, the 'suborder Palpatores' (= Eupnoi and Dyspnoi) was monophyletic (Cyphophythalmi ((Eupnoi + Dyspnoi) Laniatores)), or was paraphyletic (Cyphophthalmi (Eupnoi (Dyspnoi + Laniatores))) (Fig. 1). The first hypothesis (Palpatores monophyletic) will be referred to as topology '*A*', while the hypothesis supporting Palpatores paraphyly will be referred to as topology '*B*'. The analyses of the molecular data sets were consistent with either topologies *A* or *B*, depending on the cost matrix used, while the morphology and the combined analyses (molecular + morphology) supported topology *B*.

topology A   topology B

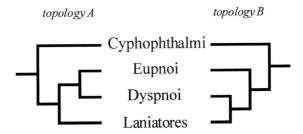

Figure 1. Two alternative hypotheses for the phylogeny of the arachnid order Opiliones, involving the arrangement among the three suborders Cyphophthalmi, 'Palpatores' and Laniatores. Palpatores (Eupnoi and Dyspnoi) result in either a monophyletic (topology A) or paraphyletic (topology B) phylogeny, depending on the method used or depending on the gap information (Giribet et al., 1999; Giribet and Wheeler, 1999b).

The 18S data set was reanalyzed (manual alignments were replaced with computer-generated alignments, using MALIGN [Wheeler and Gladstein, 1995]), and a congruence analysis was conducted for alignments generated under different insertion/deletion regimes (Giribet and Wheeler, 1999b). In this reanalysis, topology A was obtained when gaps were coded as 'missing data', as well as for the analyses at gap costs = 1 and 2 (with gaps treated as a character state). Topology B was obtained for gap costs = 4 and 8 (when gaps were coded as a character state). Using congruence with the morphological data set as a measure to decide the best-supported hypothesis (using the ILD metrics; Mickevich and Farris, 1981; W.C. Wheeler, 1995), all the analyses (at four gap regimes) were more congruent (lower ILD) when gaps were coded as a character state than when they were coded as missing data, regardless of the topology they supported. Maximum-likelihood analyses (using several models of DNA substitution, from the simplest model to models including among-site rate variation with gamma distribution) of the four alignments were consistent with topology A.

If an external criterion had not been available to judge these trees, one would have chosen topology A over topology B because it is the topology obtained by both methods, parsimony (for low gap costs) and maximum-likelihood. If data exploration for higher gap cost regimes had not been undertaken, topology A would again have been favored. But, if one appeals to congruence among data sets as the external criterion for hypothesis testing (i.e., W.C. Wheeler, 1995; Wheeler and Hayashi, 1998), topology B is favored.

At this point, our decision for choosing among topologies A and B is purely philosophical (whether to use parsimony or maximum-likelihood, whether to code gaps as missing data or as character states, whether to combine data sets or not, etc.), and does not have anything to do with the number of methods that yield any of the solutions.

Congruence among methods, or incongruence within a single method?

Congruence among different methods (as shown in the example presented above) may mean nothing if we apply similar 'models' to evaluate the data under each method. Suppose that we impose a simple Kimura 2-parameter (Kimura, 1980) model with transversions weighted twice as much as transitions for the maximum-likelihood and distance calculations, and we use an identical step-matrix for the parsimony calculations (gaps as missing data; tv/ts ratio = 2). In the absence of autapomorphic changes, we should obtain nearly identical solutions under the three criteria. However, if sequences are evolving in a very different manner than the designated model, the three methods could recover identical, but probably non-optimal solutions. In this case, pluralism tells us nothing about the best-supported hypothesis, but rather how similar the conditions explored in each method are. Or imagine another situation where we use parsimony and alphabetical order as criteria to formulate a phylogenetic hypothesis, and that for a particular data set, inferences using both criteria agree. Does agreement in this case really mean anything? Does congruence between results mean that alphabetical order would therefore be a defensible approach for systematic analysis? Certainly not!

A sensitivity analysis framework

The term 'sensitivity analysis' derives from the idea that trees generated from molecular data are sensitive to the parameters (indels, tv/ts ratio, etc.) used to generate the alignments (Fitch and Smith, 1983; W.C. Wheeler, 1995), the primary homology hypothesis. These authors recommended studying alignments generated under different models. In principle, we could obtain as many hypotheses as starting alignments. If we accept that, those nodes recovered under a wide range of parameters are more supported than those nodes obtained for only particular parameter sets (W.C. Wheeler, 1995). Then, alignments generated under different regimes of parameters should be studied, or our results are in peril of being based on very narrowly chosen parameter regimes.

Cunningham et al. (1998) also acknowledged that choosing among multiple models of DNA sequence evolution (in a maximum-likelihood or distance analysis framework) remains a major problem in phylogenetic reconstruction. The choice of appropriate models is especially important when there is large variation among branch lengths. This approach extends the idea of 'sensitivity analysis' to a model-testing framework. The next step in deciding on one model over another (for maximum likelihood) or a particular parameter set (for parsimony) should be based on a non-arbitrary decision, an external criterion.

Specifically, in the field of cladistic parsimony, character-based congruence can be used as an external criterion to choose a tree generated under a particular parameter set, if more than one data partition exists (see W.C. Wheeler,

1995; Whiting et al., 1997; Wheeler and Hayashi, 1998; Giribet and Wheeler, 1999b). This idea is based on the notion that the parameter set that minimizes incongruence among the different partitions would be preferred, in the same way that the shortest tree (the one that minimizes homoplasy) is to be preferred for a particular data set. Thus, the ILD metric (Mickevich and Farris, 1981) can be used for this purpose. In a maximum-likelihood framework, data exploration can be done by imposing different models, while the external optimality criterion can be defined as the likelihood itself by testing different models via the likelihood-ratio test (Huelsenbeck and Bull, 1996; Huelsenbeck and Crandall, 1997; Huelsenbeck and Rannala, 1997; Cunningham et al., 1998). This approach is accomplished by adding new sets of parameters in a hierarchical manner, and the likelihood-ratio test is performed to determine whether the more complex model can be favored over simpler models (Goldman, 1993; see Cunningham et al., 1998). The idea here is that if the log likelihood increases with parameter addition, there is a better fit to the model which is preferred by the optimality criterion. Both of these strategies, character congruence and the likelihood-ratio test, constitute external criteria by which we can choose among competing hypotheses, a property that should be present in all logically consistent methods of data analysis.

Understanding the behavior of data within a phylogenetic method as an alternative to pluralism

Incongruence of hypotheses generated under different parameters/models for a given method is well known (W.C. Wheeler, 1995; Cunningham et al., 1998). This is why data exploration (i.e., 'sensitivity analysis' [Fitch and Smith, 1983; W.C. Wheeler, 1995]) as a phylogenetic tool to study the behavior of the data is a reasonable test for robustness of phylogenetic hypotheses. We also propose here that data exploration is a preferable alternative to pluralism, specifically because sensitivity analysis within a method using different parameters/models often results in contradictory hypotheses that need to be examined. In addition, criteria can be defined that allow hypothesis testing (in a sensitivity analysis framework). In a parsimony framework, character congruence (Mickevich and Farris, 1981) is used as an optimality criterion in the case of multiple partitions (W.C. Wheeler, 1995; Whiting et al., 1997; Wheeler and Hayashi, 1998). In the case of maximum-likelihood, the likelihood-ratio test is used as an optimality criterion (Huelsenbeck and Bull, 1996; Huelsenbeck and Crandall, 1997; Huelsenbeck and Rannala, 1997; Cunningham et al., 1998). The criteria in these two different methods allow for data exploration and hypothesis testing, two extremely important (but often ignored) issues in phylogenetic systematics. However, they apply to two philosophically irreconcilable methods, and cannot be combined in a pluralistic framework.

The bottom line is that when inferences from different methods are congruent (for comparable parameters/models), but data exploration within a single method (using different parameters/models) yields incongruence, the pluralistic approach hides inherent conflict within the data set. Even more important, when the results of different methods are incongruent, there is no criterion to choose one hypothesis over another, other than the justification of an analytical method on first philosophical principles.

Conclusions

Our intention here is not to preach a particular method of data analysis, but to advocate data exploration (through different parameters, models, and weighting schemes) and more fundamentally for philosophical consistency. Certainly, data sets may be extremely sensitive to parameters or models, which can be stable to different methods under certain assumptions. Since there are no criteria for choosing parameters/models *a priori*, the hypothesis testing issue (i.e., congruence in parsimony, likelihood-ratio test in maximum-likelihood) is extremely important. But more importantly, we stress that there is no justification for pluralism in phylogenetic systematics. Investigators should choose a method based on a philosophical rationale and carry on with their choice to the end. Non-pluralism, however, does not necessarily require distrust of other justifiable positions.

Acknowledgements
We thank Miguel Angel Arnedo, Dan Janies, Pete Makovicky and Lorenzo Prendini for discussion and suggestions.

Molecular Systematics and Evolution: Theory and Practice
ed. by R. DeSalle, G. Giribet and W. Wheeler
© 2002 Birkhäuser Verlag/Switzerland

Molecular systematics and the origin of species: new syntheses or methodological introgressions?

Paul Z. Goldstein[1] and Andrew V.Z. Brower[2]

[1] *Division of Insects, Field Museum of Natural History, Chicago, IL 60605, USA*
[2] *Department of Entomology, Oregon State University, Corvallis, OR 97331, USA*

> Biologists and philosophers have long recognized the importance of species. Yet species concepts serve two masters, evolutionary theory on the one hand and taxonomy on the other.
> Cracraft (1987b: 329)

Summary. The advent of molecular phylogenetics stimulated the need to reprise for many discussions surrounding species concepts. The interpretation of cladograms as accurate representations of phylogeny, when the characters upon which they are based exhibit a reticulate pattern, is inconsistent with the epistemological axiom of hierarchy we assign to the cladistic method (Brower, 2000c). Discrepancies in the interpretation of cladograms would appear to account for differences in the kinds of questions to which they are applied. The philosphical and empirical issues surrounding this subject are examined in this chapter.

Phylogenetic species concepts: an introduction

One of the more vexing problems in biology has been the search for a species concept adequate to the task of describing the most basic units in the hierarchy of life. The "Species Problem" presents to biologists a logical and philosophical challenge involving several choices. These include whether we should view species as strictly operational constructs, or require instead that they correspond to "real" entities in nature, i.e., whether the designation of species should make ontological claims about the entities being named and, if so, what kinds of ontological claims are acceptable (Frost and Hillis, 1990; Frost and Kluge, 1994). If species are viewed as real things that exist regardless of our ability to discern them, we must choose between a pluralistic approach, in which a near-endless array of entities may be recognized as species (Mishler and Donoghue, 1982; Holsinger, 1984; Mishler and Brandon, 1987), and a universal approach that requires some form of biological equivalence among all species (Cracraft, 1987b, 1997). We must also ask whether our criteria for species recognition should be process-oriented (mechanistic), presupposing modes of speciation and mechanisms of species integrity, or as independent of such suppositions ("theory neutral") as possible. Caused in part by the accessibility of DNA sequence data and the advent of explicit tree-building methods, these issues have resurfaced, sometimes more explicitly than not, and the debate over species concepts has shifted into the phylogenetic-historical arena. Numerous phylogenetic species concepts or criteria have been proposed since the (1966) translation of Hennig's (1950) book, all of them claiming foundations set forth in that work (cf. Baum, 1992; de Queiroz, 1998). Table 1 summarizes the primary features and implications of several major phylogenetic species concepts along with the popular alternative, the Biological Species Concept (BSC) for comparison (Tab. 1).

Given the vigorous literature surrounding the interpretation of cladograms during recent decades, it is perhaps not surprising that many of these concepts overlap imperfectly with respect to the salient issues. For example, both the "evolutionary species concept" (Simpson, 1961; Wiley, 1978) and the internodal species concept of Hennig (1966), Brundin (1972) and Ridley (1989) require time-extended visions of species, as opposed to atemporal "snapshots." But while Simpson's and Wiley's concepts are prospective, relying on predictions about historical fates, Hennig's species concept is retrospective, relying on the identification of ancestral species. Likewise, Baum and Shaw (1995), Baum and Donoghue (1995), de Queiroz and Donoghue (1988, 1990), Nixon and Wheeler (1990), Luckow (1995) and Cracraft (1983) all share the view that species should serve phylogenetic inference as efficiently as possible and be consistent with the evolutionary research program, yet many of these authors disagree on the role of cladistic evidence in the illustration of species. These issues have been discussed at length elsewhere (e.g., Frost and Hillis, 1990; Frost and Kluge, 1994; Rieppel, 1994b; Luckow, 1995; Goldstein et al., 2000; Brower, this volume; Goldstein and DeSalle, 2000).

Although these concepts vary in several parameters (e.g., their generalizability, their interpretation of individuality, the kinds of evolutionary change they incorporate, etc.), the focus of our discussion will be the dichotomy between those concepts based on character state distributions alone, i.e., on aggregative operations prior to phylogenetic analysis, and those based on phylogenetic analyses that presuppose a correspondence between cladograms and relationships among terminals both within and across species. Because DNA sequences frequently exhibit variation within populations that can be employed in cladistic analyses, this dichotomy bears critically on lower-level studies that address relationships among closely related species or populations.

Primary among the differences between character- and tree-based approaches is the epistemological distinction between species as observations, and species as entities supposed to have had unique, but in some sense comparable evolutionary histories. Delimiting phylogenetic species with character state distributions requires that species are the minimal units of *phylogenetic analysis*. (The term *cladistic analysis* refers to a more general operation that can be applied to any class of objects, biological or otherwise [Davis and Nixon, 1992], without invoking common ancestry as a causal explanation [Brower,

Table 1. Primary features of various species concepts and criteria rooted in phylogenetic theory, with the Biological Species Concept (BSC) for comparison

Species concept	Author/proponent	Criterion	Primary features					
			Biological emphasis	Ontological emphasis	Criteria general vs context-dependent	Mechanistic vs theory-neutral	Pluralistic vs monistic	Discrete vs continuous
BSC	Mayr, Dobzhansky, Avise	Potential interbreeding	Maintenance	Essentialist	Context-dependent	Mechanistic	Monistic	Continuous
EvSC	Simpson, Wiley	Lineage cohesion	Maintenance	Essentialist	General	Theory-neutral	Monistic	Continuous
ISC	Hennig, Brundin, Ridley	Lineage cohesion	Maintenance	Essentialist	General	Theory-neutral	Monistic	Discrete
PSC	Cracraft, Nixon, Davis, Wheeler, Luckow	Diagnostic characters	Origin	Nominalist	General	Theory-neutral	Monistic	Discrete
PSC*	Brower	Haplotype networks	Origin	Nominalist	General	Theory-neutral	Monistic	Discrete
ASC	de Queiroz, Donoghue, Olmstead, Hill and Crane	Monophyly	Origin	Nominalist	General	Mechanistic	Pluralistic	Continuous
GSC	Baum, Donoghue, Shaw	Monophyly	Origin	Nominalist	General	Mechanistic	Pluralistic	Continuous
GCSC	Mallet	Phenetic similarity	Origin	Nominalist	General	Mechanistic	Monistic	Continuous

1999]). Although Mallet (1995) lumped character- and tree-based criteria together in his discussion of phylogenetic species, he correctly emphasized the backwards nature of using phylogenetic relationships to define species, namely the idea that we should examine how things are related in order to determine what the related things themselves are. Among the several versions of "the" phylogenetic species concept (PSC) (see de Queiroz, 1998), it is the original, character-based aggregative version of Eldredge and Cracraft (1980), Nelson and Platnick (1981), and Cracraft (1983) that presupposes the least about species origins. Subsequent versions purporting to be more phylogenetic in spirit (e.g., Donoghue, 1985; Baum and Donoghue, 1995) have precipitated important discussions, but appear to have led to some confusion, particularly outside the systematics community (e.g. Meffe and Carroll, 1997). In our terminology, the PSC refers to the original, character-based descriptor: a species is an irreducible (basal) cluster of organisms that is diagnosably distinct from other such clusters, and within which there is a parental pattern of ancestry and descent (Cracraft, 1989).

Phylogeneticists who would weave a historical perspective into the identification of species are confronted with the question of whether or not to invoke cladograms as arbiters of competing species concepts and definitions. This problem requires clarification of the limits of cladistic analysis in interpreting patterns of historical relatedness. In our view, species are concepts that owe their recognition to particular distributions of empirical evidence. Character diagnosis (Nixon and Wheeler, 1990) is the preferred criterion for species recognition, as well as the most efficient means of enhancing resolution and providing an independent framework for testing evolutionary scenarios. Species viewed from within this framework stand in sharp contrast to concepts demanding that species exist irrespective of our knowledge of them (e.g., Baum and Donoghue, 1995; Ghiselin, 1997). As has been pointed out elsewhere (de Queiroz, 1998; Goldstein and DeSalle, 2000), the PSC as such is more of an operational criterion for organizing data than an ontological conception of nature. Some authors (e.g., Frost and Kluge, 1994) have objected to such operationalism on grounds that it constrains the study of nature to fall within pre-ordained criteria devoid of biological correspondence. Proponents of tree-based criteria have argued that the general application of phylogenetic recovery methods regardless of level of inference has the unique ability to incorporate history into the delimitation of species (Baum, 1992; Baum and Donoghue, 1995). Character-based diagnostic operations, so the argument goes, are "ahistorical." By contrast, it is our contention that species "defined" by non-empirical criteria represent untestable metaphysical claims that lie outside the realm of science (Brady, 1985).

This philosophical debate sits at the forefront of current discussion concerning the phylogenetic mechanics of species. Aspects of the problem distill to straightforward questions of graphic representation, touching on at least two broad ontological aspects of the longstanding debate on species concepts: the relationship between the origin of species and their maintenance, and the boundary between hierarchical (phylogenetic) networks and reticulating (genealogical) networks. This chapter discusses the relationship between cladistic networks and their interpretation as phylogenetic trees, and assesses technical and analytical approaches to the delimitation of species that are suitable for phylogenetic inference.

Pattern and process

Molecular phylogenetics, as it is commonly practiced today, represents an amalgamation of ideas drawn from the disparate intellectual traditions of systematics and population genetics. Both systematics and population genetics meet the requisite criteria of "normal sciences" (Brower et al., 1996), but their methods and assumptions are quite divergent. Systematics, which may be thought of as the elucidation of macroevolutionary patterns, has at its core the goals of documenting the branching pattern of species' relationships and, to the extent possible, of enhancing the information conveyed by resultant classifications. The two primary products of systematics, cladograms and classifications, serve both as means of efficient information retrieval and as the fundamental frameworks for macroevolutionary scenario-testing, that is, for exploring evolutionary processes at and above the species level. Systematics deals with nested patterns of holomorphological variation implying degrees of evolutionary relationship, and relies on direct observation of characters to corroborate or refute competing historical phylogenetic hypotheses. Thus, systematics deals with the historical mechanics of specific, historically unique macroevolutionary events. In contrast, population genetics exploits more general theoretical paradigms for studying microevolutionary changes that occur prior to speciation. Population genetics is primarily concerned with four phenomena: mutation, selection, drift, and gene flow, evolutionary processes that are inherently intractable without the explicit use of models and simulations.

For many (e.g., Avise et al., 1987), molecular phylogenetics represents a bridge between population genetics and systematics, a bridge based not only on the shared focus on DNA, but also on the assertion of conceptual and methodological continuity. Species represent the ontological nexus for these fields. Systematists, whose job it is to partition variation and, if so desired, to use those partitions to explore macroevolution, require a boundary that is discrete with respect to characters in taxonomic and phylogenetic inference. Population geneticists, whose focus is inherently mechanistic, require a process-based boundary among groups of populations. That boundary is most commonly equated with reproductive isolation (Coyne, 1994), and for that reason the BSC retains significant popularity in the population genetics community. The BSC of Mayr (1969: 26) defines species as "groups of interbreeding natural populations that are reproductively isolated from other such groups." Although it was coined by an advocate of systematics and endorsed by the primary architects of the modern synthesis, the BSC enjoys dominance primarily in the population genetics literature.[1]

Thus there appears a philosophical schism in how biologists view species. Population genetic studies, whether theoretical or empirical, tend to be time-extended or rely at some level on equilibrium conditions for most variables (selection, population size, etc.) over periods of time. As such they require that notions of origin and maintenance of species be addressed. Systematic studies, which explore the hierarchical pattern of groups nested in groups, are by *implication* time-extended, but are empirically rooted in relatively straightforward analyses of shared character state distributions. While most if not all systematists wish to explain cladograms in evolutionary terms, that interpretation is applied to the phylogenetic pattern which relies fundamentally not on underlying mechanistic theories, but on analysis of empirical data embodied by the character matrix.

Although advances in phylogenetic theory coupled with those in molecular techniques have occasioned a renaissance in the appreciation of phylogenetic studies (e.g., Harvey and Pagel, 1991; Harvey et al., 1996; Martins, 1996; Page and Holmes, 1998), much of that appreciation appears to derive from the crossing over of methods from one field to the other: the application of the paradigms of population genetics to systematic questions, and the application of the historical paradigm of phylogenetic systematics to intra-populational questions (Goldstein et al., 2000). Unfortunately, the cavalier application of phylogenetic methods may lead to interpretations that are inconsistent with phylogenetic premises.

Systematists have long argued that the independence of systematics from considerations of evolutionary process is critical to the use of cladograms in exploring and testing such processes (Rosen, 1978; Brady, 1985) while others have advocated the incorporation of information we "know" about general evolutionary processes in specific cases of phylogenetic inference (e.g., Swofford et al., 1996). Much the same sort of discussion has surrounded the demarcation of species, and until recently discussions of species have been dominated by notions of reproductive continuity rather than phylogenetic continuity or cohesion. Such suggestions date to the Modern Synthesis, and some of the more pointed treatments argue that any attempts to partition the fields of systematics and population genetics amount to nothing less than anti-evolutionary views and empirical heresy. To those who would fuse systematics and population genetics, each field is seen as a separate forum for erecting phylogenetic hypotheses, although an area of broad agreement seems to be the utility of cladograms in testing evolutionary scenarios. Cladograms are not generally viewed as tests of evolution *per se*, but as schemes of presenting natural

---

[1] In one of his more recent defenses of the BSC, Mayr (1992) dismissed phylogenetic species concepts on the grounds that their primary proponents were not "naturalists." Notwithstanding the authoritarianism, it is ironic to note that the BSC was never widely taken up by the vast majority of botanical and entomological systematists, those charged with the responsibility of dealing with approximately three-quarters of life on earth.

order that can, if so desired, be used to test *specific* hypotheses of character evolution. Evolution, more specifically descent with modification and clado-genesis, is an explanatory theory that provides a material cause for observed hierarchical patterns of relationship among species and more inclusive taxa.

To the extent that these debates have been scientific rather than semantic or political, many have been at cross-purposes, and we will not discuss those here. More central to the distinction—and the strengthening of each field—are the implications of viewing nested hierarchies as phylogenies and the role of species as terminals in those hierarchies. This is a philosophical problem, to be sure, but to a critical scientist one which bears on how we interpret biolog-ical data. Two broad questions present themselves: Is there a correspondence between reproductive continuity and phylogenetic continuity; and is there cor-respondence relevant to the delineation of species? In other words, does the use of cladograms represent a panacea that approximates or describes both reproductive continuity and phylogenetic continuity?

Hierarchies, continuity and monophyly

Central to the debate over phylogenetic species concepts has thus been the dis-tinction between hierarchical and reticulating networks. The term "network" is used variously (and correctly) in the systematics literature to refer to rooted and unrooted trees (Page, 1987). We begin this discussion by pointing out that a resolved cladogram is a type of network in which no terminus is connected to more than one node. Fully resolved cladograms are networks that exhibit completely dichotomous branching; incompletely resolved cladograms are networks in which more than two termini are associated with some nodes. When cladograms are interpreted as hypotheses of phylogeny, they become claims of relative recency of common ancestry. The implication of the clado-gram graphic is that *relative* statements of relationship can be made, and in order for such statements to convey the most information, a given terminal must be placed unambiguously closer to another terminal or group of termi-nals than to any other. But hierarchical (cladistic) hypotheses of ancestry do not accurately reflect evolutionary relationships below the level at which rela-tionships among sexually reproducing organisms are reticulate, due to inde-pendent assortment of alleles and recombination. In such systems, individual organisms are not legitimately represented as termini with unambiguous hier-archical relations to one another. A child is by definition no more or less relat-ed to his mother than his father, so to portray the relationships among three such individuals as a nested hierarchy is a categorical error.

In order to appreciate the conflict at the intersection of modern population genetics and phylogenetic systematics, it is not sufficient merely to explore the alternative interpretations of cladograms as organismal phylogenies *versus* gene trees (see Brower et al., 1996). It is necessary to address the broader issue of the applicability of cladograms themselves to different kinds of hierarchical systems. The phylogenetic interpretation of cladograms is a complicated and, at times, contentious matter involving a seemingly bottomless morass of epistemological issues. Without getting too enmired in a discussion of systematic philosophy, one primary issue deserves reiteration: the distinction between cladistic analysis, a hierarchical grouping operation that produces cladograms, and phylogenetic analysis, the interpretation of those cladograms, which translates into hypotheses of relative recency of common ancestry. Although the distinction may seem trivial to some, it is of considerable importance when dealing with groups of organisms whose relationships are reticulate.

Prominent among the theoretical and technical advances in molecular evolution that have fueled the popularity of phylogenetics is the understanding of unique modes of inheritance for organellar DNA, especially the maternally inherited animal mitochondrial DNA, and the discovery of other highly variable systems (e.g., microsatellites) that enable resolution of ostensibly phylogenetic hypotheses at much finer levels than have hitherto been possible. An implication of such uniparental inheritance is that mitochondria never exhibit patterns of reticulation, regardless of the pattern of relationships implied by other characters in the organisms bearing them. Beginning with the neologism of Avise et al. (1987), the paradigm of infraspecific phylogeography[2]—which relies on relationships inferred through mitochondrial DNA to explore historical biogeographic processes—has gained wide endorsement (e.g., Moritz and Bermington, 1998; Avise, 2000).

But the pursuit of phylogenetic studies directed towards population-level questions that reside below traditional, biological species boundaries has not always been associated with critical treatment of potential pitfalls. Many authors have argued that it is useful to explore hierarchies below traditional species boundaries, applying cladistic methods both to hierarchical systems and to demonstrably reticulate networks. Relying on organellar genes that are uniparentally inherited, phylogeography is at least partially immune to the criticism of applying hierarchy-generating methods to reticulate systems; mitochondrial elements are necessarily related in a hierarchical fashion. In fact, Avise has argued at length that mitochondrial trees allow phylogenetic methods to bridge the gap between systematics and population genetics by allowing strictly hierarchical methods to be applied with accuracy to reticulate systems (Avise et al., 1987; Avise, 2000). So long as the distinction between cladistic analysis and phylogenetic interpretation emphasized above is borne in mind, and so long as it is recognized that the ontological claims of phylogeographic studies relate to hierarchical relationships among genes and not species, this paradigm is quite valuable in studying evolutionary history.

Resolved cladograms generated by analyzing data, molecular or otherwise, imply unambiguous relationships among individual termini. Unless coupled with other forms of evidence to arbitrate whether the underlying system of

---

[2] The term "intraspecific phylogeography" explicitly relies on the BSC, and it should be recognized that the term "intraspecific" is intended by its authors to refer to species identified under that rubric.

organismal relationships is reticulate or hierarchical, the uncritical application of the phylogeographic paradigm may fail by lending itself to over-interpretation. Unfortunately, resolution may be spurious in cases of interbreeding groups where uniparentally inherited genetic elements are used to infer relationships among sexually reproducing organisms. The first problem is a simple issue of shoehorning non-hierarchically related individuals into hierarchical hypotheses of relationship. The second and related issue is one of equating a legitimate, organellar hierarchy with the organismal phylogeny to which its correspondence may be imperfect. The "phylogeny" of mitochondrial haplotypes, even if inferred correctly, may not correspond to the phylogeny of the organisms bearing those genetic elements (cf. Doyle, 1992, 1995). This problem is most likely to be encountered at precisely the shallow levels of divergence where phylogeography is purported to recover historical pattern. That is not to say that mitochondrial or any other organellar data are not valuable tools for phylogenetic reconstruction. The message is simply one of caution when exploring relationships among organisms with shallow levels of divergence.

As Avise and others (e.g., Eldredge, 1985; Rieppel, 1988; Maddison, 1995, 1997; Harrison, 1998) have pointed out, hierarchies are present throughout nature. Hierarchical statements may conflict without either being "wrong", if they represent different kinds of ontological statements. But here it becomes important to specify what question is being asked, and how broadly the answer is supposed to inform us about evolutionary history. Frost and Kluge (1994) emphasized the distinction between scalar hierarchies, in which the levels are non-transitive (e.g., human organ systems), and specification hierarchies, in which the levels are transitive, or nested (e.g., taxonomic hierarchies). It is necessary to understand the nature of the hierarchy in question before presupposing transitive levels. Statements of relationship among mitochondrial elements are by no means mutually exclusive of statements of organismal relationships. But one must distinguish between reticulate systems and hierarchical ones when asking historical questions that are in any sense absolute. That distinction distills to a deceptively simple question of when a reticulate system becomes hierarchical.

Both systematics and population genetics offer tests designed to assess the nature of the underlying relationships in question, and several methods are available to researchers who wish to explore these issues critically for their own data. Some approach the problem in a "top down" way, using aggregative methods to identify phylogenetic species (Davis and Nixon, 1992). Others have suggested "bottom up" approaches that draw on a mixture of haplotype networks and aggregative statistics (e.g., Crandall et al., 1994; Templeton, 1998). There would appear to be some broad agreement between the conclusions of coalescence theory drawn by population geneticists and the conclusions surrounding the importance of character fixation to speciation drawn by systematists. Recently, Brower (1999, this volume) has described a cladistic approach for determining on which side of the phylogenetic species boundary

a particular group lies. Brower's Cladistic Haplotype Analysis uses cladograms to test *a priori* hypotheses of grouping among putative phylogenetic
species, but does not depend on reciprocal monophyly or any other evolutionary assumptions.

Cladograms and species boundaries

How well, then, do cladograms lend themselves to the task of arbitrating species boundaries? To address this question we must explore the vocabulary of
cladograms, specifically the implications of monophyly. The related concepts
of monophyly and character polarity may be seen as the two most fundamental advances that enabled the practice of phylogenetic systematics. A monophyletic group—one including a given ancestor and all its descendants—is
hypothesized on the basis of synapomorphies, the identification of which relies
in turn on the most parsimonious hypotheses of character state origins. The
inferred cladogram (or branching diagram produced by some other phylogenetic inferential method) lends itself, perhaps too easily, to interpretation as a
simulacrum of the true branching pattern of phylogeny, and for this reason it
makes sense to explore the meaning of monophyly itself, as applied by various
authors in discussions of species boundaries.

Paralleling the application of phylogenetic methods to infraspecific questions, terms such as monophyly and apomorphy have been extended beyond
their original meanings as biologically rooted character-based concepts for
discussing relationships among taxa, to refer to descriptors of any cladogram
or hierarchical network. Drawing on established principles of population
genetics, Avise has observed that, over time, genes achieve monophyly by first
passing through stages of polyphyly and paraphyly. Another way of looking at
this phenomenon is through the timing of fixation events: as characters responsible for polyphyletic and paraphyletc groupings disappear, monophyletic res-
olutions obtain. Over time, two or more genes are expected to be found to be
"reciprocally monophyletic." But as Goldstein and DeSalle (2000) discuss, the
achievement of reciprocal monophyly of two groups of organisms need not be
coincident with the achievement of symmetrical or alternative character fixation. Two groups can appear to be reciprocally monophyletic, particularly
when analyzed by distance or probabilistic methods, without either being distinguished by fixed, unique traits.

Thus, while a systematist would characterize a monophyletic group as one
including an ancestor and all its descendants, a phylogeographer might choose
to abandon the ontological implications of monophyly and synonymize it with
its graphic depiction. This reduces the term to encompass any group of termini traceable to a single node to which no other termini can be traced, and
indeed monophyly is now commonly used in the molecular phylogenetic literature to refer to a cladistic result rather than a statement of common ancestry.
Of course, this uncritically expanded interpretation of monophyly has ramifi-

cations for the concept of common ancestry. Among other things, it requires that there be graphic *and* theoretical congruence between reticulate, inter-breeding networks of organisms and non-reticulate cladistic statements of relationship. At its most fundamental, this issue resolves to a simple question of continuity. The premise of applying hierarchical graphic representations to the relationships among organisms requires that there be a fundamental discontinuity between the reticulating network of relationships that unites organisms within a species, and the (presumably) non-reticulate, nested hierarchy of relationships among species.

We would argue that the distinction between reticulate and hierarchical networks can only be meaningfully addressed by the analysis of character state distributions, because it is character states that are distributed hierarchically, not individual organisms. The "boundary" between the reticulate relationships among individuals and the hierarchical distributions of character states corresponds to character state fixation. Stated another way, character state distributions exhibit hierarchy only when shared polymorphisms are extinguished. While the origin of discontinuity is usually accompanied (eventually) by reproductive discontinuity, knowledge of the latter is not necessary to infer or elucidate discontinuities among character state distributions, and is not necessarily responsible for driving them.

The meaning and interpretation of monophyly is thus central to discussions of species, because species occupy a unique position at the nexus of hierarchical and reticulate systems. If we accept that species reside at this "boundary", then the logical asymmetry between species and "higher" taxa, as well as the discontinuity between species and populations, are inescapable corollaries to the practice of phylogenetic systematics. The PSC attempts to make use of this boundary in species formulations, but the PSC does not purport to pinpoint that boundary, a potentially serious shortcoming. Brower (1999) has argued that such absolute pinpointing may be unfeasible. If the boundary corresponds to a character fixation event, and if we accept that such events are instantaneous, corresponding to the death of the last individual responsible for a given polymorphism in a population, then the boundary is almost infinitely narrow. That is to say, it is a point, not a line, and the endeavor of finding it is not only unfeasible but without useful purpose, a fool's errand.

Avise envisioned a generalized ontological scenario whereby the relationships among populations endure a phyletic progression of sorts, achieving monophyly only after passing through polyphyletic and paraphyletic stages. Meanwhile, the point of performing phylogenetic analyses is to develop statements of relationships among taxa, which are necessarily ambiguous in reticulate systems. This seeming paradox, or a version of it, was first identified by J.H. Woodger (1937), and has been referred to (by Medawar and Medawar, 1983; Rieppel, 1994b) as Woodger's paradox. Woodger presented it as follows: assuming that an evolutionary hierarchy is legitimate, and given that no individual organism can belong to more than one species, then the transition from one species to another must, by logical inference, take place in a single

generation. This observation was seen by many as a major obstacle to speciation theory, since it is counter-intuitive to suppose that a parent and its offspring could belong to separate species.

But that is precisely what is required of both systematics and population
genetics if we are to document and explain cogently the evolutionary history
of life on earth. The species to which we assign an individual organism (other
than a type specimen) must logically be able to change depending on the available information on character state distributions. The biological discontinuity
noted by Woodger as an obstacle to species theory is in fact necessary to any
hierarchical description that is intended for evolutionary interpretation. This is
so for a simple reason: if the discontinuity between reticulate and hierarchical
networks did not exist, then unambiguous statements of ancestry would never
be justified. Without such a discontinuity, all organisms might in effect be considered one species (Nixon and Wheeler, 1990). More importantly, if "the"
hierarchy of life is to be interpreted as an unbroken stream of descent with
modification and phylogenesis, then any boundaries we draw are necessarily
arbitrary. If the boundaries are arbitrary, then there are no natural units of
study, biological or otherwise. The classes delineated by such boundaries are
also arbitrary, and cannot therefore serve as objects of scientific exploration. It
may be safely said that depicting life as an unbroken stream of reproduction
does nothing for the empirical exploration of nature.

Cladograms and evolution

The premises of interpreting a monophyletic group as having a unique common ancestor are twofold: cladogenesis and descent with modification (Davis
and Nixon, 1992). But this is not to say that cladograms should be interpreted
as representations of direct ancestor-descendant relationships. Mayr's (1942)
model of allopatric speciation provides an interesting touchstone for discussing the implications of phylogenetic species. Consider an ancestral population A from which a series of smaller populations break off and, ultimately,
speciate allopatrically (Fig. 1a). So long as none of the smaller populations is
characterized by a diagnostic feature—one that does not occur in the ancestral
population or any of the other smaller populations—none of them may be
considered phylogenetic species, and it would not be legitimate to treat them
as separate terminals in a phylogenetic analysis (Davis and Nixon, 1992). The
achievement of "phylogenetic speciation", for lack of a better phrase, coincides with the fixation of alternative character states in ancestral and daughter
populations. If fixation events accumulate rapidly enough, then successive
daughter populations are expected to share enough features with the ancestral
population for the relationships among all the phylogenetic species—the
sequence in which the daughter populations split from the ancestral one—to
be resolved. In practice, we can never know *a priori* that a population is ancestral, and as long as it is extant we should include it in the cladistic analysis

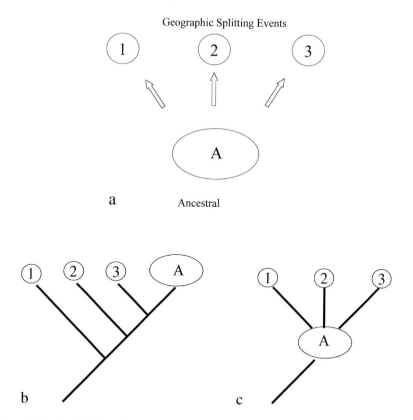

Figure 1. Graphic depiction of three daughter populations (1–3) splitting allopatrically, in sequence, from an ancestral population A (1a), and becoming phylogenetic species. Assuming that sufficient character evidence exists to resolve these aggregates cladistically in a way that reflects the actual sequence of splitting events, the "ancestral" population is united most closely with phylogenetic species 3 (1b). Attempting to synonymize cladograms with direct statements of ancestor-descendant relationships leads to a representation of this scenario in which the ancestral population occupies the node from which all three species diverge (1c). This kind of statement cannot obtain from empirical cladistic methods, and does not provide any information on relative recency of common ancestry, despite the visual appeal of making the cladogram superficially consistent with the geographic scenario.

regardless. Assuming the character information reflects history accurately, the ancestral population would appear on our cladogram as a terminal united most closely with the daughter population with which it shared most recent affinities (Fig. 1b).

This stands in contrast to the depiction of the ancestral population as a node, rather than a terminus (Fig. 1c). Commonly, not enough character information has accumulated to resolve relationships among closely related species, and the resultant cladogram will be unresolved (Fig. 2a). Such polytomies can be interpreted in one of two ways: as situations in which the available character data do not afford a resolution ("soft polytomies"; Coddington and Scharff, 1996), or literally, as accurate portrayals of simultaneous species radiations in

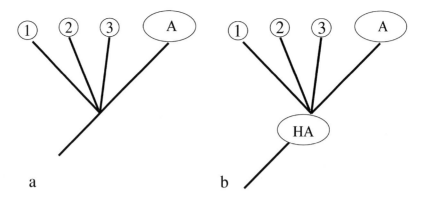

Figure 2. When character evidence does not allow for resolution of terminals, there are no empirical grounds for preferring any cladistic hypothesis, and this ambiguity is represented by a polytomy (2a). To interpret the polytomy as a "true polytomy" or "hard polytomy", that is, a burst of simultaneous speciation events from a hypothetical common ancestor, an unjustified ontological leap is made whereby the cladogram is transformed from an admission of empirical ambiguity to an assertion concerning the timing of speciation events that cannot be supported empirically, and that requires an additional *ad hoc* hypothesis of a single ancestral population (2b).

nature ("hard polytomies" or "true polytomies" or "star phylogenies"; e.g., Hoelzer and Melnick, 1994). The latter interpretation amounts to an ontological claim based on the lack of informative data. But more importantly, it requires that nodes be thought of as hypothetical ancestral populations (Fig. 2b). An implication of phylogenetic species is that they form the foci for statements of relative degree of relationship, not direct ancestry. Hence nodes on cladograms do not correspond to taxa in the cladistic framework.

This example illustrates two further implications of the PSC. First, because the formation of phylogenetic species coincides with the extinction of polymorphism, "speciation" is an instantaneous event. Any ontological claim of a resolved cladogram comprises the sequence of those events. But regardless of whether we choose to append ontological status to results of cladistic analyses in the abstract, if the available data do not afford a resolution, then we simply have no empirical means of choosing among alternative phylogenetic hypotheses. Second, because discontinuity among character state distributions does not necessarily coincide with the discontinuity of reproductive compatibility, groups of organisms identified as phylogenetic species may re-unite (if geographically adjacent, for example), in effect erasing any evidence of their separate histories. This has been used as an argument against the phylogenetic species (Hoelzer and Melnick, 1994). However, if our species criterion is to serve the purpose of recovering statements of relationship, rather than the nomenclatural or reproductive stability over time, this concern is not well founded.

Conclusion

Discontinuity between reticulate networks and cladistic depictions of relationship sits at the heart of phylogenetic species concepts. Clearly, hierarchical discontinuity can have no ontological status within the framework of reproductive or mechanistic species concepts; parents and their offspring are by definition members of the same "gene pool", "reproductive community", and "field for recombination." In phylogenetic systematics, discontinuity is only understood with reference to the extinction of shared polymorphism, the fixation of characters used to diagnose species and delimit higher taxa, and whose distributions are, by definition, tied to hierarchical representations of common ancestry. Indeed, evolutionary biologists have known that extinction is an important part of macroevolution for a long time.

Perhaps one of the central shortcomings of the Synthesis was the failure to identify the species as the level across which the fields of systematics and population genetics best complement one another. Instead, the architects of the synthesis felt that each field would be served by integrating their respective components into a composite paradigm. Thus was the genesis of the BSC: rooted in reproductive isolation and intended to provide the framework for systematics. But it was the practice of systematics that became corrupted by this marriage. The rise of cladistic methods re-directed discussions of species such that species definitions began to incorporate historical components. To be sure, most cladists were more interested in higher-level phylogenetics than in lower-level questions, and to the extent this is the case the species debate has been largely orthogonal to progress in systematics. It might even be argued that the New Synthesis's emphasis on species retarded the progress of systematics (Carpenter, 1992; Q.D. Wheeler, 1995). Nonetheless, the species question still occupies an important space in systematic and phylogenetic studies, particularly those with evolutionary components to them.

Phylogenetic systematics may be thought of as the historical partitioning of variation. Molecular techniques allow us to draw partitions along ever more fine lines, but the fundamental issue relating character fixation to hierarchies remains the same. Characters, molecular and otherwise, are what allow us to describe and delineate species. Analysis of characters towards phylogenetic reconstruction relies on certain assumptions concerning the nature of the entities involved, not the least of which is that cladogram resolution not be an artifact of analysis. Shared similarities are observations; shared derived similarities are analytical inferences. Perhaps an ironic implication of the PSC—that is, the original character-based version of it—is a relationship between character fixation and reproductive cohesion, the fulcrum between systematics and population genetics. The inference of cohesion or coalescence is not necessary to the business of describing species and elucidating their relationships. But it is a side-effect of the logical and graphical requirements of hierarchical character state distributions that the PSC may well serve the interests of both fields.

Molecular Systematics and Evolution: Theory and Practice
ed. by R. DeSalle, G. Giribet and W. Wheeler
© 2002 Birkhäuser Verlag/Switzerland

Is morphology still relevant?

Richard H. Baker[1] and John Gatesy[2]

[1] *The Galton Laboratory, University College London, London NW1 2HE, UK*
[2] *Department of Molecular Biology, University of Wyoming, Laramie, WY 82071, USA*

Summary. The utility of morphological data in modern systematics has recently been challenged because strong selection pressures are thought to create widespread patterns of convergent evolution at this level. This concern has led to suggestions that morphological data should be excluded either from all analyses or at least from analyses where there is conflict with molecular data. These concerns, however, are generally unwarranted and excluding data is not a defensible strategy for dealing with problems that do exist. We emphasize the importance of empirical responses, such as collecting additional and diverse data and exploring taxa and data set interactions, rather than the implementation of *a priori* assumptions, to overcoming many of the concerns associated with combining morphological and molecular data. Numerous factors may create biases in both molecules and morphology. While these biases are prevalent enough to cause widespread incongruence, they highlight the importance of combining, rather than separating, data. Morphological data also offer distinct advantages over molecular data, such as the inclusion of fossil taxa, cost-effectiveness and the presence of biases different from those in molecular data.

Introduction

As the ease of collecting molecular sequence data increases, the continued relevance of morphology in systematics is being questioned. Current debates among systematists are driven primarily by issues engendered by molecular data, and the 'data partition', in particular, has become one of the major focal points in modern systematics. The wealth of character data currently available for study is increasingly being divided into different categories (e.g., larval, nuclear, ribosomal, third position, hydrophobic, to list a few), with significant attention being focused on the phylogenetic behavior of these various categories. One of the goals of this type of analysis is to find general patterns that identify more reliable types of data and to use this information to obtain a more accurate picture of evolutionary history. Following this line of thinking, several authors have recently challenged the utility of morphology in systematics (Lamboy, 1994; Hedges and Maxson, 1996; Givnish and Sytsma, 1997, 1998). Therefore, while several treatments of general differences between molecules and morphology already exist (Hillis, 1987; Patterson, 1988; Donoghue and Sanderson, 1992; Patterson et al., 1993; Larson, 1994; Parker, 1997), it is important to reexamine how morphological data fit into the conceptual systematic framework currently being shaped by molecular issues. We organize this review around the question of whether morphological data should be combined with molecular data and argue against both discarding morphological evidence and, alternatively, keeping morphological and molecular data separate in phylogenetic analyses.

Reasons to exclude morphology

Homoplasy and convergence

Recently, several authors (Lamboy, 1994; Hedges and Maxson, 1996; Givnish and Sytsma, 1997, 1998) have argued that morphological data should not be combined with molecular data because they exhibit greater convergence and homoplasy than do molecular data. Morphology is regarded as being more susceptible to directional selection pressures that may cause a functional complex of characters to evolve several times within a lineage. To empirically support this view, Givnish and Sytsma (1998) surveyed over 40 studies of each type of data and showed that, for a given number of taxa, molecular data had a significantly higher consistency index than morphology. Based on these results, they recommended that morphological data be relegated to "recognizing species and suggesting broad patterns of relationships," with molecular data providing the actual guide to phylogenetic relationships. There are several reasons, however, why their analysis does not provide convincing criteria for not combining morphology with molecules.

First, the actual difference in homoplasy between DNA sequence data and morphology demonstrated by Givnish and Sytsma (1998) has not been supported by other studies. Their analysis compared consistency indices for studies based exclusively on one type of data, but an examination of 31 studies that combined molecules and morphology for the same sets of taxa (Baker et al., 1998) found that, in the majority of cases (26), morphology had higher consistency than molecules. In addition, Sanderson and Donoghue (1989), in an analysis of 42 morphological studies and 18 molecular studies, found no difference in the consistency index between these types of data. Givnish and Sytsma (1998) stressed that their analysis was more relevant than Sanderson and Donoghue (1989) because it examined more recent molecular studies, but they used only one morphological study published after 1987. Given the computational advances made in tree searching programs, a more comprehensive comparison using a similar selection of morphological and molecular studies is required to firmly establish differences in homoplasy levels.

Second, even if morphological data do generally exhibit greater homoplasy than molecular data, this is not sufficient grounds for disregarding their signal. The logic of Givnish and Sytsma's (1998) argument requires that only characters with the highest consistency be included in an analysis. Therefore, many molecular characters would have to be removed from an analysis along with all of the morphological evidence. Following this logic to its extreme results in clique analysis in which only the largest group of characters with no homoplasy is given weight in an analysis. This methodology has been heavily criticized (Farris, 1983) and is no longer used by systematists. In addition, numerous studies (Philippe et al., 1996; Baker and DeSalle, 1997; Olmstead et al., 1998; Bjorkland, 1999; Källersjö et al., 1999; Baker et al., 2001) have now demonstrated that homoplasious characters may still contain substantial phylogenetic information. For example, inclusion of rapidly evolving third position nucleotide sites has been shown, in some cases, to both increase the number of strongly supported groups (Källersjö et al., 1999) and reduce the incongruence among data partitions (Baker et al., 2001).

Finally, there is so much variation in levels of homoplasy across studies of morphological and molecular data (see Fig. 2.2 in Givnish and Sytsma, 1998), that any general pattern is hardly relevant to a specific study. Many morphological data sets have high consistency, and even in relatively homoplasious data sets there will be both consistent and inconsistent characters that are highly informative. Systematists interested in giving greater weight to more consistent characters will surely be better off using a method that discriminates among characters individually (e.g., successive approximation [Farris, 1969; Carpenter, 1988] or Goloboff weighting [Goloboff, 1993]), than to always exclude all morphological evidence.

Homoplasy that is directed rather than random is a more significant problem for reconstructing evolutionary history. Adaptive syndromes of morphological characters with similar homoplasious phylogenetic distributions have been highlighted in several recent studies (Dubuisson, J-Y, 1997; Georges et al., 1998; Harry et al., 1998; Álvarez et al., 1999; Givnish et al., 1999; McCracken et al., 1999; Quicke and Belshaw, 1999) and used to caution against the reliability of morphological data. For instance, in a recent study of duck phylogeny, McCracken et al. (1999) proposed that functional relationships among pelvic characters associated with multiple origins of diving had caused two taxa not closely related to each other to be mistakenly grouped together.

This problem, however, is not restricted to morphological data. There are several phenomena such as compensatory changes, long branches, and base composition bias that can create convergent patterns of sequence evolution in molecular data. While examples of convergent evolution at the molecular level have proven elusive, cases are accumulating, and this search is too preliminary to make any definitive statement concerning differences between molecules and morphology in this regard. Convergent sequence evolution associated with similar selection pressures has been identified in a few studies (Messier and Stewart, 1997; Roux et al., 1998; Crandall et al., 1999), and a recent examination of laboratory evolution in bacteriophage found convergent evolution along replicate lineages beyond that expected by random substitutions (Bull et al., 1997). This mutation pattern affected phylogenetic inference such that the likelihood for the true topology was significantly worse than that for the estimated tree. Similarly, Cao et al. (1998), in a phylogenetic study of relationships among eutherian mammalian orders using the mitochondrial genome, found evidence that contradictory relationships supported by the ND1 gene were caused by convergent evolution. Overall, statements that a given type of data is more susceptible to non-random convergence cannot be supported simply by comparison of consistency indices (Givnish and Sytsma, 1998) because this measure by itself is not an adequate criterion for detecting this phenomenon. Currently, no test exists that can adequately identify differences in the homoplasy distributions of molecular and morphological data. Comparisons of consistency indices and examples of morphological characters coevolving on a tree topology are not sufficient evidence for such broad statements.

Furthermore, the presence of systematic bias in character data indicates the need to be more, not less, inclusive in terms of analysis because this is the only method for resolving the conflict that bias creates. Biased characters cannot generally be identified as such without the presence of a true (as in the case of the bacteriophage study) or strongly supported hypothesis. Strongly supported hypotheses, however, can withstand the inclusion of these convergent characters so there is little cause for concern. Alternatively, weakly supported hypotheses provide no empirical grounds for directly identifying problematic data (Brower et al., 1996). Therefore, in such cases, excluding data requires a reliance on information only indirectly relevant to the reliability of characters involved in these weak statements. For instance, the apparent rapid evolution of certain lineages, misleading homoplasy elsewhere on the tree, or misleading homoplasy in other studies must be used as justification for downweighting or excluding entire classes of character data. This evidence may have little bearing on the specific phylogenetic statement at hand as characters may exhibit different levels of homoplasy on different parts of the tree (Kluge, 1997; Siddall and Kluge, 1997). Inductive extrapolations from limited computer simulations may not be justified, and weighting schemes based on previous studies may not be general (e.g., Allard and Carpenter, 1996; DeSalle and Brower, 1997). More importantly, separating data leaves no opportunity for arbitration, as the conflict will always exist and can only be resolved by the *post hoc* arguments of the investigator. On the other hand, if misleading results are suspected, collecting more data and including them in a combined analysis provide a viable means for testing the proposition that certain characters are convergent.

Incongruence with molecular data

While *a priori* exclusion of morphological data is still not widespread, a more common position is to separate data in specific instances where significant conflict between partitions is evident. A considerable debate has emerged concerning the use of incongruence as a criterion for separating data (Kluge, 1989; Barrett et al., 1991; Bull et al., 1993; Kluge and Wolf, 1993; de Queiroz et al., 1995; Nixon and Carpenter, 1996; DeSalle and Brower, 1997; Siddall, 1997) and several tests have recently been developed to identify data heterogeneity (Rodrigo et al., 1993; Larson, 1994; Farris et al., 1994, 1995; Huelsenbeck and Bull, 1996). These tests have allowed for a more rigorous assessment of the disagreement between molecular and morphological data than was previously allowed by comparing topologies. So far they have revealed that significant incongruence between molecules and morphology is not uncommon. A recent survey of 25 studies (Baker et al., 1998) found 12 with significant conflict, and more examples of this level of disagreement continue to accumulate in the literature (Dubuisson et al., 1998; Graham et al., 1998: Harris and Disotell, 1998; O'Grady et al., 1998; Wetterer et al., 1998; Gatesy et al., 1999b; Harrison and Crespi, 1999; Littlewood et al., 1999; Liu and Miyamoto, 1999; McCracken et al., 1999; O'Leary, 1999; Porterfield et al., 1999; Quicke and Belshaw, 1999). This result may cause concern among investigators worried about the perils of combining heterogeneous data and contribute to notions that morphological data are unreliable. There are several factors that must be considered when evaluating both this general pattern and any specific instance of morphological and molecular disagreement.

First, comparisons of molecular data suggest there may be similar levels of conflict among the many types of partitions present in large molecular data sets. We examined incongruence in 7 studies containing at least four different gene fragments and for which all pairwise gene comparisons were examined for incongruence (Baker and DeSalle, 1997; Gatesy, 1998; O'Grady et al., 1998; Remsen and DeSalle, 1998; O'Grady, 1999; Baker et al., submitted, Durando et al., 2000). Of 134 tests, 60 showed significant conflict, a proportion that is comparable to the molecular *versus* morphological comparisons (see Baker et al., 1998). Given the numerous causes discussed earlier for biased character information at both the molecular and morphological level, this considerable disagreement is perhaps not surprising. As yet, contrasting molecular and morphological signals do not represent a more common form of conflict than is found among different molecular partitions.

Second, significant incongruence does not necessarily indicate that conflict exists throughout the tree. Differences in the placement of a single taxon are sufficient to create statistically significant incongruence, and several examples of this type of local incongruence now exist (Mason-Gamer and Kellogg, 1996; Poe, 1996; Baker and DeSalle, 1997; Brower and Egan, 1997; Cao et al., 1998; McCracken et al., 1999). For example, disagreement between morphological and molecular data in an analysis of crocodilian relationships (Poe, 1996) was isolated to two taxa (*Gavialis* and *Tomistoma*) that are united by the molecular data but separated by morphology. Excluding morphology in this case would have prevented this data set from having an influence over all the remaining relationships on the tree for which there is no conflict with the molecular data.

Third, there are often strong empirical reasons for combining significantly incongruent morphological and molecular character partitions. The assertion that combining incongruent data will cause misleading results is founded on limited theoretical and empirical examples. The simulations of Bull et al. (1993) that initiated the concern for data heterogeneity examined an unrealistic case (Brower, 1996b; Sullivan, 1996; Olmstead et al., 1998) in which there are extreme rate differences between partitions, homogeneity within partitions, and very few taxa. Alternatively, empirical investigations have demonstrated that combining heterogeneous data can produce resolved and strongly supported hypotheses (Baker and DeSalle, 1997). Congruence is an important concept, but it is certainly not the only important concept in systematic biology. A simple example illustrates this point.

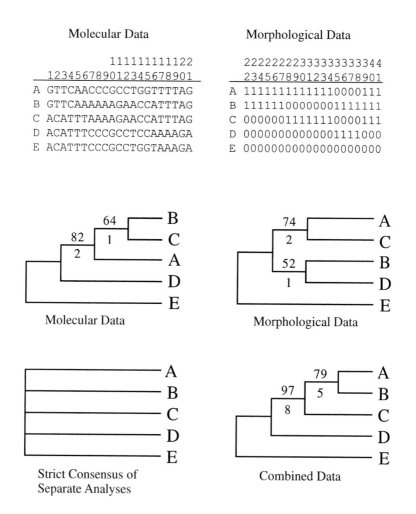

Molecular Data

```
                 111111111122
      123456789012345678901
A  GTTCAACCCGCCTGGTTTTAG
B  GTTCAAAAAAGAACCATTTAG
C  ACATTTAAAAGAACCATTTAG
D  ACATTTCCCGCCTCCAAAAGA
E  ACATTTCCCGCCTGGTAAAGA
```

Morphological Data

```
      2222222233333333344
      23456789012345678901
A  11111111111110000111
B  11111100000001111111
C  00000011111110000111
D  00000000000001111000
E  00000000000000000000
```

Molecular Data

Morphological Data

Strict Consensus of
Separate Analyses

Combined Data

Figure 1. Two hypothetical data sets, one molecular (characters 1–21) and the other morphological (characters 22–41), for five taxa (A–E). Separate cladistic analyses of the two data sets produce unstable, conflicting topologies that share no groups. The data sets are significantly incongruent according to the ILD test, but combination of the data sets in simultaneous analysis yields a single, fully resolved, well-supported topology. Bootstrap scores are above internodes, and Bremer support scores are below internodes. All analyses were done with PAUP 4.0b3a (Swofford, 1999). Searches were branch and bound. Bootstraps and ILD tests included 10,000 randomizations. See text for further explanations.

A hypothetical molecular data set, characters 1–21, for five taxa, A–E, and a hypothetical morphological data set, characters 22–41, for the same five taxa are shown in Figure 1. The two data sets are significantly incongruent ($p = 0.038$) according to the incongruence length difference (ILD) test of Farris et al. (1994). Separate analyses of the two data sets show that no common groups are supported; the strict consensus tree derived from the two separate analyses is totally unresolved. Furthermore, each of the separate analyses produces minimum length trees that are unstable; Bremer support (BS; Bremer, 1994) ranges from one to two for each supported node (Fig. 1). The significant ILD might suggest to a researcher that the two data sets should not be combined in simultaneous analysis. However, when the two data sets are combined, the resultant tree has greater support, stability and resolution. The combined analysis supports a single minimum length topology; there is full resolution. Bootstrap scores (Felsenstein, 1985a) and BS are higher in the combined analysis than in either of the separate analyses (Fig. 1). Furthermore, partitioned BS scores (Baker and DeSalle, 1997) are positive for both molecular and morphological data sets at both nodes favored by the combined data set. This pattern suggests that each data set contributes positively to support at these nodes, even though the two data sets are significantly incongruent.

The clade A + B is not supported by either separate analysis, and the clade A + B + C is supported more solidly in combined relative to separate analysis. Clearly, there is much hidden support (Barrett et al., 1991; Chippindale and Wiens, 1994) for each of these nodes. There are numerous synapomorphies for these relationships that are hidden in each of the separate data sets. The hidden support also derives from the different distributions of homoplasy in the morphological and molecular data sets. The combined character set supports ((((A + B) C) D) E). Most of the incongruence in the molecular partition supports a grouping of B + C, while most of the incongruence in the morphological partition supports A + C and B + D. These different distributions of homoplasy in the individual data sets imply significant incongruence according to the ILD, but the common support for ((((A + B) C) D) E) in both data sets renders this incongruence "insignificant." The combined tree is stable, well resolved, and supported by both data sets. None of this would be apparent unless the significantly incongruent data sets were actually combined in simultaneous analysis. Such hidden support, hidden resolution, and hidden corroboration among data sets are not readily apparent in separate analyses of data partitions. Despite the extreme ILD, we contend that it would be ill advised to keep the data sets separate. In fact, whenever morphological and molecular data sets are examined for the same taxa, combined analyses of the data sets should be executed. Otherwise hidden support and resolution may go unnoticed.

Some real data sets show analogous but less extreme patterns (e.g., Sullivan, 1996). For example, a molecular data set and a morphological data set for 18 bovid artiodactyls (antelopes) support radically different topologies (Fig. 2). Each of the separate analyses is relatively unstable; bootstrap scores are below

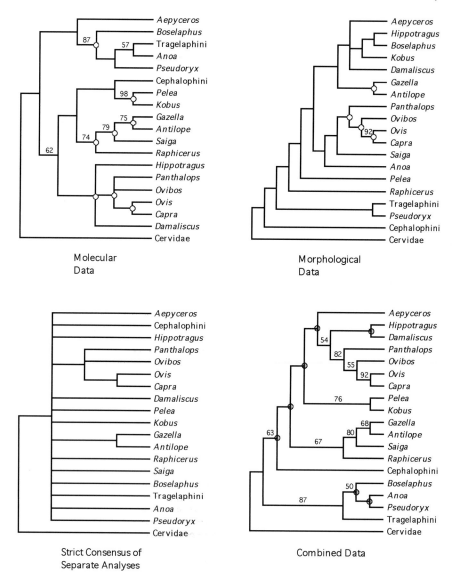

Figure 2. Two data sets for bovid artiodactyls support radically different topologies. The molecular data set, 12S + 16S ribosomal DNA and β-casein exon 7 (Gatesy and Arctander, 2000), and the morphological data set (Gentry, 1992; Thomas, 1994) are significantly incongruent according to the ILD test (p = 0.001), and the strict consensus of topologies supported by separate analysis is almost totally unresolved. Nevertheless, combination of the data sets in simultaneous analyses results in a single fully-resolved hypothesis that is fairly well supported. Open circles at nodes mark clades that are supported in separate analysis and in combined analysis. Gray circles at nodes mark clades that are supported in combined analysis but not in either separate analysis. All analyses were done with PAUP 4.0b3a (Swofford, 1999). Searches were heuristic with TBR branch swapping and random taxon addition. Bootstraps and ILD tests included 1,000 randomizations. Data sets are available at the web site for the AMNH molecular lab. See text for further explanations.

50% for most nodes. The strict consensus of topologies supported by these separate analyses is very poorly resolved, and the data sets are significantly incongruent according to the ILD ($p = 0.001$). Nevertheless, combined analysis of the two highly incongruent data sets supports a single topology that is more resolved and stable than results for either of the separate analyses (Fig. 2). Several inconsistencies in the combined analysis, however, still remain as indicated by conflicting PBS scores at various nodes. To resolve these inconsistencies, a staunch molecular systematist might argue against the morphological data on the grounds that they are more homoplasious than the molecular data. Similarly, a staunch morphologist might counter that the smaller morphological data set contains certain characters that are of critical significance and that these characters are being swamped in the combined analysis by the larger molecular data set. As an alternative to these speculative arguments, a more productive approach would be to collect another data set (e.g., transposons, behavior, chromosomal rearrangements, other molecular or morphological data), and let these new data arbitrate the conflict between data sets. Appeals to character data are much more powerful than appeals to authority, small data sets, specific models of evolution, unstable data sets, or computer simulations (Allard and Carpenter, 1996; Brower et al., 1996; Kluge, 1997; Siddall, 1997). Congruence is an important concept, but the dismissal of clique methods nearly 20 years ago (see Farris, 1983; Farris and Kluge, 1985) demonstrates that an obsession with congruence is unwarranted. The removal of incongruent data is not the primary goal of systematics, and combined analyses should at the least be explored whenever morphological and molecular data are available.

Advantages of morphological evidence

While molecular approaches to systematics have become very popular, morphological data still have several advantages over molecular data. First, the collection of morphological characters is generally less expensive than the collection of molecular characters. Costly equipment, such as centrifuges, PCR machines, incubators, automated sequencers, and reagents are not required to observe gross anatomy. In many cases, morphological characters can be scored from museum specimens without the aid of anything more than a dissecting microscope. In terms of cost to benefit, a strong case can be made for the superiority of morphological data over molecular data (Nixon and Carpenter, 1996).

Second, extinct taxa usually are scored only for morphological characters. The great majority of species are extinct, and many long-lived and ecologically important clades, such as Trilobita, Ammonoidea, Ornithopoda, Multituberculata, and Pterosauria, to name just a few, are wholly extinct. If morphological evidence is ignored, the phylogeny of over 99% of life is ignored (Novacek and Wheeler, 1992). DNA molecules are only rarely preserved in ancient fossils. So, in order to have comprehensive, or perhaps even just adequate, taxonomic coverage in a particular systematic study, morphological information must be considered. Computer simulations and empirical work have shown that improved taxonomic sampling can have a profound influence on systematic results (Gauthier et al., 1988; Huelsenbeck, 1991; Eernisse and Kluge, 1993; Halanych, 1998). Given that many molecular systematists fear the potentially negative effects of "long branches" in phylogenetic analysis (e.g., Felsenstein, 1978; Hwang et al., 1998), extinct taxa that can subdivide such long branches should be embraced by proponents of molecular data. Nixon and Carpenter (1996) have outlined common-sense procedures for merging fossils, that may be particularly incomplete, with extant taxa in combined phylogenetic analyses (also see Gauthier et al., 1998; Eernisse and Kluge, 1993; Brochu, 1997; O'Leary, 1999 for some empirical cases).

Because fossil taxa are readily incorporated into morphological systematic studies, phylogenies based on morphology can be calibrated relative to the geological time scale. Fossils are only rarely integrated directly into molecular phylogenies, so systematists who wish to place minimum times of divergence for various taxa in their DNA-based trees are at some level dependent on the fossil record of morphological characters. Even extensive molecular clock studies that use the number of DNA mutations to estimate the timing of evolutionary events are reliant on at least a few fossil calibration points (e.g., Kumar and Hedges, 1998). Therefore, if time is seen as a critical component of evolutionary studies, the incorporation of morphological data into phylogenetic analysis is a necessity.

Finally, morphological character information may be used to test molecular data; morphological characters are unlikely to have all of the same biases that may negatively affect phylogenetic analyses of DNA sequence data. Gross anatomical characters will never be characterized by nucleotide base composition bias, transition/transversion bias, compensatory nucleotide substitution, or other mutational inequalities at the DNA sequence level (e.g., Collins et al., 1994; Naylor and Brown, 1998). The addition of morphological evidence to molecular data in combined analyses may offset these unique biases of molecular data. That patterns of homoplasy can cancel each other out in a combined analysis has been shown in several studies (Olmstead and Sweere, 1994; Gatesy et al., 1999b, also see Fig. 1).

Fear of the biases of molecular evolution has promoted the use of evolutionary models for improving phylogenetic reconstruction (Felsenstein, 1978; Huelsenbeck, 1997, 1998). These methods use class statements to give higher weight to some types of character changes relative to others in phylogenetic analysis. This scheme, however, has been viewed as discarding evidence and several arguments supporting equal weighting have been put forth (Farris, 1983; Nixon and Carpenter, 1996; DeSalle and Brower, 1997; Kluge, 1997; Siddall and Kluge, 1997; Allard et al., 1999). Many complicated evolutionary models employing differential weighting, particularly maximum likelihood, preclude the incorporation of morphological data. This is a drawback of the likelihood approach. If morphological data cannot be combined with molecular data in a simultaneous maximum likelihood framework, proponents of the likelihood approach only have two choices: 1) entirely ignore morphological evidence or 2) analyze morphological data separately from molecular data and then compare results for molecules and morphology by consensus techniques. For the many reasons outlined previously in this chapter, neither of these two options is ideal.

The recent exchange in the literature regarding the phylogenetic position of the holometabolous insect order, Strepsiptera, illustrates these difficulties (Whiting and Wheeler, 1994; Carmean and Crespi, 1995; Whiting et al., 1997; Huelsenbeck, 1997, 1998; Hwang et al., 1998; Siddall and Whiting, 1999). There has been controversy over whether Strepsiptera is most closely allied to Diptera, Coleoptera, or some other insect group. Whiting et al. (1997) collected 18S and 28S ribosomal (r) DNA sequences for over 50 species of insects, and compiled morphological evidence for 26 taxa. Cladistic analyses of these combined data consistently and robustly favored a grouping of Strepsiptera with Diptera to the exclusion of other holometabolous insects. Whiting et al. (1997) noted that the molecular branch lengths that lead to Diptera and Strepsiptera are long, but argued that the combined analysis of all available data represented their best-supported estimate of insect phylogeny.

In contrast, Huelsenbeck (1997) suggested that these long molecular branches may be problematic. Computer simulations previously had shown that such branches sometimes could attract distantly related taxa in phylogenetic analyses of four taxa (Huelsenbeck and Hillis, 1993). Huelsenbeck (1997) analyzed a small subset of the 18S rDNA data using likelihood methods that correct for multiple substitutions, and concluded that the data are ambiguous as to the placement of Strepsiptera. Similarly, Huelsenbeck (1998) and Hwang et al. (1998), again only using subsets of the total molecular data base, argued that long branch attraction may have misplaced Strepsiptera in parsimony analyses.

Because they utilized maximum likelihood methods to analyze their data, Huelsenbeck (1997; 1998) and Hwang et al. (1998) did not choose to include the extensive morphological data that Whiting et al. (1997) had compiled. Indeed, it would be difficult to include these morphological data in a combined maximum likelihood analysis. What stochastic model of evolution could be used to describe both morphological and molecular characters? Instead, these authors excluded the only other available data partition that could empirically test the worth of the heterogeneously evolving and potentially biased rDNA data! A comparison of topologies supported by likelihood analyses of rDNA sequences (Huelsenbeck, 1997, 1998; Hwang et al., 1998) and parsimony analyses of morphological characters (Whiting et al., 1997) shows conflict as to the placement of Strepsiptera. So, separate analyses of morphological data and molecular data in this case leads to lack of resolution and support for the placement of this taxon. In contrast, a combined cladistic analysis of rDNA and morphology, characters that have very different evolutionary constraints, leads to a well-supported and well-resolved tree for holometabolous insects in which Diptera and Strepsiptera cluster (Whiting et al., 1997; Siddall and Whiting, 1999). Of course, this phylogenetic hypothesis could be overturned with the collection of additional relevant data (e.g., Rokas et al., 1999).

If data sets are separated as in the likelihood approaches of Huelsenbeck (1997, 1998) and Hwang et al. (1998), common support that is hidden in different data sets cannot emerge (Barrett et al., 1991), and data sets with very different evolutionary biases cannot directly test each other. The inability to incorporate valuable morphological data represents a significant limitation of maximum likelihood approaches. However, this does not preclude the future development of combined likelihood approaches that incorporate different stochastic models of evolution for very different character partitions such as morphology and molecules. Unfortunately, we doubt that modeling morphological evidence in phylogenetic studies will ever be tractable. Any generalities in morphological evolution are even less obvious than in molecules. There does not appear to be any common pattern such as a transition/transversion bias in almost all known DNA sequences. If there are no generalities in the evolution of morphological characters, each character (or character transformation) would need its own set of model parameters (Farris, 1999). As combined data sets get larger and larger and include more diverse data (i.e., transposons, introns, ribosomal DNAs, exons, chromosomal rearrangements, behavior, gross anatomy, etc.), the number of model parameters and branch length estimations may be beyond computational feasibility.

Conclusion

Only one generalization can be made about the phylogenetic utility of entire classes of systematic characters. That is, no generalities about the phylogenetic utility of an entire class of characters, including morphology, are universally true. So, morphological evidence should not be dismissed or isolated from other systematic data. Most of the concerns associated with combining molecular and morphological data are unfounded. Furthermore, problems that do exist are best addressed through empirical responses rather than the implementation of *a priori* assumptions. In many cases these concerns will be dispelled by a thorough analysis of all the data equally weighted. If concern for misleading results persists, then collecting additional data provides the most legitimate means for resolving these issues. Characteristics such as the existence of fossils and character data not containing molecular biases ensure morphological information will continue to be a valuable tool in systematics.

Acknowledgements
We thank Andy Brower and Alan de Queiroz for helpful comments on the manuscript. Richard Baker was supported by a NERC grant and John Gatesy was supported by NSF grant DEB-9985847.

Development, homology and systematics

Ranhy Bang[1], Ted R. Schultz[2] and Rob DeSalle[3]

[1] *Graduate Training Program in Arthropod Systematics, Department of Entomology, Cornell University, Ithaca, NY 14853, USA*
[2] *Department of Entomology, MRC 188, National Museum of Natural History, Smithsonian Institution, Washington, DC 20560, USA*
[3] *Division of Invertebrate Zoology, American Museum of Natural History, New York, NY 10024, USA*

Summary. The basic issue of recognizing and delimiting characters derived from different levels of biological organization, including molecular, cellular, morphological and behavioral levels, has been addressed previously in the cladistic literature. But when considering new sources of information (such as the proliferating Evo-Devo data), it is critical to review traditional theoretical and methodological approaches to their interpretation. This is especially important because the conclusions of a phylogenetic analysis are dependent upon the initial recognition and definition of the characters, the basic units of comparison in phylogeny reconstruction. This chapter explores the role of recent Evo-Devo studies in systematics and attempts to place the Evo-Devo literature into a systematics context.

Introduction

In the history of science, novel sources of data are frequently regarded as, *a priori*, more reliable than traditional sources. Unchecked enthusiasm for utilizing such new sources, however, can divert attention from the theoretical bases for their proper interpretation. Beginning in the 19th century, embryological data provided hope for distinguishing "true" homologies of morphological structures in related taxa from those structures that are similar due to parallelism and convergence (see Gould, 1977, for one overview of ontogeny and phylogeny). During the past few years, a similar hope has driven several attempts to incorporate molecular developmental data into evolutionary studies. The rapid proliferation of such studies has even prompted the creation of a new name—"Evo-Devo" (Akam, 1998)—for the field. This chapter attempts to understand these new Evo-Devo data from the perspective of phylogenetic systematics.

Evo-Devo researchers have concentrated their greatest efforts on using the tools of molecular biology to detect the expression of developmentally important gene products in the developing embryo. The results of these studies are used to support statements about the homology of morphological structures in related organisms. In fact, the concept of homology has been a major focal point of the Evo-Devo literature, and Janies and DeSalle (Tab. 1; 1999) list several Evo-Devo studies in which hypotheses of homology are proposed, based on the expression patterns of developmental genes. Unfortunately, precisely what is meant by "homology" varies among authors, and this is a problem that is by no means unique to Evo-Devo research. At the beginning of the Evo-Devo onslaught, for example, Panchen (1992) pointed out that at least nine homology concepts were already in use in the evolutionary and systematic literature. By our count at least five distinct homology concepts have been used in the Evo-Devo literature (Box 1), leading one author to suggest that the concept of homology is "ripe for burning" (Tautz, 1999).

In what follows, we will examine the bases for the homology concepts used in Evo-Devo studies and in phylogenetic systematics, outlined in Box 1. We

---

Box 1. Homology concepts in the Evo-Devo literature

At least five distinct homology concepts are identifiable in the Evo-Devo literature. For all of these, two features in two different species are considered homologous because the two species also share an additional feature, X, where X is some causal factor. We summarize the five different causal factors cited in these five different homology concepts as follows:

1) Phylogenetic origin (Darwin 1859) of a feature, or corresponding feature, derived from a corresponding feature in a common ancestor (Mayr, 1982) ("historical homology")
2) Structural organization (Geoffroy Saint-Hilaire, 1824)
3) Developmental origin (Roth, 1984)
4) Developmental constraints ("biological homology") (Wagner, 1989a)
5) Continuity of information (van Valen, 1982).

---

will distinguish between homologous and non-homologous attributes, or *characters,* of organisms and we will discuss how their proper recognition impinges on competing hypotheses of evolutionary change. We will take the point of view that homology is an implicitly phylogenetic concept, and that an initial hypothesis of homology about a character state shared by two species must precisely delimit that character, especially in terms of its biological level (genetic, developmental, morphological, etc.). We will further suggest that such initial hypotheses of homology must subsequently be accepted or rejected based on the results of a phylogenetic analysis that maximizes the patterns of congruence among multiple putative homologies.

Raw similarity, homology and symplesiomorphy

The concept of homology predates both evolutionary and developmental biology. In its most general (and oldest) biological usage, a hypothesis of homology identifies the "same" attribute, or character, in different species (Aristotle, 1991). Abouheif (1997) argues that the "sameness" of a character that is being compared across species can have many aspects, including structure, developmental origin, developmental constraints and genetic information. He also argues that in some cases these aspects may conflict in the information that they provide about homology. We would suggest that, in actual practice, the phylogenetic systematist is rarely if ever confronted with idealistic, broadly defined characters that span multiple biological levels and that can thus have "conflicting aspects of 'sameness'" (Abouheif, 1997: 405). Rather, the phylogenetic systematist encounters suites of narrowly defined characters drawn from the various biological levels of organization, each of which must be separately sorted out according to a practically applied, consistent criterion of sameness (see below), so that some characters (of developmental origin, genetic information, anatomy, etc.) may be found to be the same while others may not. This criterion of sameness is not always based solely on absolute, raw similarity but is instead frequently applied to strikingly different forms of a character, e.g., the pectoral fins of a salmon and the forelimbs of a lemur. Even in its pre-evolutionary form, sameness due to homology was understood to be fundamentally different from superficial likeness due to similarity in function, such as occurs in features of organisms with otherwise hierarchically disjunct suites of characters, e.g., the gills of a stonefly and the gills of a mudpuppy. Today this latter kind of similarity would be attributed to convergence, which, together with parallelism and reversal, constitute the causes of non-homologous similarity. In cladistic terminology, these non-homologous similarities are ultimately regarded as errors in the scoring of characters in the course of phylogenetic analysis, and are referred to as *homoplasy*. Both types of similarity—homology and homoplasy—fulfill the criterion of raw similarity, but only the former is additionally supported by the weight of correlated, hierarchically distributed suites of characters that are elucidated during the course of an explicit phylogenetic analysis.

If "Homology as a phenomenon is a manifestation of replication and of continuity of biological information" (Roth, 1991), then, obviously, the term to describe that phenomenon should be distinct from a term that simply describes raw similarity. Indeed, "similarity is not a test of homology because nonhomology also implies similarity; instead, similarity is the factor that compels us to postulate homology" (Cracraft,1981). In short, similarity provides the initial impetus for proposing a hypothesis of homology, whereas phylogenetic analysis employing multiple characters provides the test of that hypothesis (Box 2). Only those shared character states that withstand such testing should be considered homologous.

In evolutionary terms homology, then, should be defined as sameness due to an unbroken chain of descent from a common ancestor. This chain of descent can only be discovered, and thus the validity of a hypothesized homology can only be judged, through phylogenetic analysis and the resulting recognition of synapomorphies. Unless we are able to examine directly every organism in a chain of descent, our way of confirming homology is indirect and subject to the same error as all indirect inference. Such inference is only possible if, for any

Box 2. Synapomorphy and homology

This box shows a comparison of discontinuous *versus* continuous distributions of two character states for a phylogenetic relationship of 5 taxa, A–F, and an outgroup, O (a group relatively closely related to the group of organisms under consideration in a study, where all members of the ingroup are more closely related to each other than any member is to the outgroup) as shown in Figure a below. In Figure a, the character transformations depicted are independent gains, i.e., this character is the product of two parallel evolutionary origins, one for taxon A + B and one for taxon E + F. The character is homoplasious and shows a discontinuous distribution on the cladogram in Figure a below. In Figure b, the character has arisen only once, and therefore shows a continuous distribution among taxa A–F. It is thus a homology, i.e., a synapomorphy, for this group. The outgroup taxon, O, defines the phylogenetic level at which the hypothesis of homology is posed for a character state shared by at least some of the members of the ingroup taxa, A.–F. A summary of the character states is shown in the matrices above their respective trees.

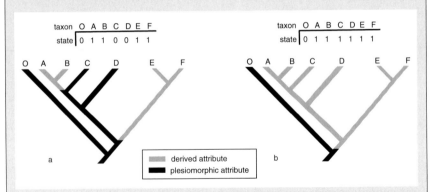

Similarity due to homology can be distinguished from similarity due to convergence only with a reference to a well-supported phylogeny and corroborating, congruent characters, including gene expression, among others. That is, by analyzing the distribution of states in the character of interest on a reconstructed phylogeny, it can be determined whether a character state arose only once and is therefore a homology for the taxa in which it occurs, or whether a character state arose multiple times and is therefore a homoplasy (see figure in this box). In turn, a phylogeny is deemed well supported when there is high congruence among multiple characters. Congruence of multiple, homologous characters, such as can be found on a well supported phylogeny, "...represents a regularity of phenomena which calls for a causal explanation such as common descent" (Rieppel, 1992).

given character, an unambiguous concept of what this sameness constitutes is used, or else it will not be possible to properly assess the indirect evidence for tentative acceptance or conclusive rejection of a hypothesis of homology.

The biological homology concept and the transformational approach

Derived from what was originally intended as a purely developmental framework for epigenesis (Holland, 1996; Waddington, 1940), the biological homology concept ("bhc") relies on the similarity of developmental processes as the arbiter of homology. In its various interpretations and applications, the bhc recognizes homologies as features of organisms possessing "similar patterns of individuality and developmental constraints" (Shubin, 1994). Wagner (1989a) describes homologous characters as sharing a "set of developmental constraints, caused by a locally acting self-regulatory mechanism of organ differentiation, and are thus developmentally individualized parts of the phenotype." Roth (1991: 169) modifies Wagner's definition to also include "relatively simple characters (or features or aspects of development) that may be retained in a phylogenetic lineage, or iteratively expressed in various forms within a single individual, but which have not accumulated about them the self-regulatory mechanisms of developmental constraints, and more complex structures consisting of more than one developmentally individuated part, which collectively may not behave as an individual." Because developmental genetics can be used to address the processes that generate morphology and can suggest hypotheses for how changes in those processes might underlie morphological evolution, the bhc has been accepted as a framework for interpreting developmental data in an evolutionary and (especially) macroevolutionary context.

In those Evo-Devo studies that do not strictly adopt the bhc, other, related methods are used that share with the bhc a recognition criterion that ultimately depends on similarity. These methods posit transformations of morphological characters derived from a prior or underlying level of biological organization, such as the primordia from which the character develops, the commonality of the genes involved in forming the character, gene expression, etc. (see below for discussion on levels of biological organization). In the bhc, developmental constraints are ultimately recognized according to conserved morphology, the conservation of which is inferred from similarity. Likewise, continuity of biological organization across species is inferred from similar morphological structures or developmental processes.

Completing statements of comparative relation, or avoiding two taxon statements to infer homology

Ghiselin (1976) states, "…homology is not a property such as mass, which organisms may be said to possess. It is simply a relationship; the word is used to express the idea that particular structures stand to each other in a particular way within the context of a specific theory." Homology describes a relationship of characters and their states that either exists, or does not, within a specific hypothesis of evolutionary relationship. It is not a statement of degree; two taxa cannot share a most recent common ancestor by 20% or 80%, nor can two explicitly defined character states in those taxa be descended from a state in the ancestral taxon in some fractional way.

Many developmentally based definitions of homology fail to specify clearly the requirement of a third reference point in making homology statements (e.g., Collazo and Fraser, 1996; Holland, 1996; Abouheif, 1997) and thus inadvertently encourage the use of two-taxon statements. The Mayrian and Darwinian homology definitions require shared phylogenetic origins or ancestry, but omit explicit mention of a third reference point. The requirement of ancestry alone is insufficient, however, because all life forms share origins and ancestry, and therefore continuity, at some phylogenetic level or another. Some authors attempt to use some kind of "conditional" to make a homology comparison (e.g., Ghiselin, 1976; Roth, 1988), but also fail to stress the need for an explicit third term. The resulting statements of homology are often statements of juxtaposition or redundancies. As Nelson (1994) points out, Bock's (1974) "the wing of birds and the wing of bats are homologous as the forelimb of tetrapods" (c.f. Bolker and Raff, 1996) and Ghiselin's (1966) "the wings of eagles are homologous as wings to those of hawks, and as derivative of forelimbs to those of bats," do not specify any level of comparison but "merely confirm that wings are not arms (Bock) or, oddly, that wings are not arms but are derivatives of them (Ghiselin)."

Homology does not describe an intrinsic property of similarity possessed by a character at any level of biological organization, or a resemblance shared by two characters (c.f. Abouheif, 1997); rather, it describes a relationship between two character states that is unique with respect to those states excluded from the relationship. That unique relationship is one of most-recent common ancestry and relies upon the phylogenetic hierarchy of taxa, which, of course, can be established through recognition of synapomorphies. Indeed, "homology does not exist independent of synapomorphy, any more than a set exists independent of its members (subsets)" (Nelson and Platnick, 1981).

The various developmental homology concepts that attempt to incorporate phylogeny (Abouheif, 1997; Abouheif et al., 1997; Bolker and Raff, 1996; Box 1) rightly argue for distinctly defined biological levels of comparison (e.g., character states should be homologized across the same life stages), but require substantiation by previously "known" morphological homologies. Unfortunately, when confronted with novel data, one often does not have the luxury of "known" homologies on which to rely, and an alternative method is required for constructing hypotheses of homology. We suggest that in such cases the proper approach is not to presume to define morphological homology on the basis of biological processes, but rather to treat each datum as discrete. After all, homologized structures or functions may be derived without change from a most recent common ancestor but be produced by different ontogenetic pathways (Abouheif, 1997; Abouheif et al., 1997), and a given morphological character may thus be one of a number of levels of information about phylogeny. Unless the units of comparison are drawn from the same level, the comparison is invalid (Hennig, 1966).

Separately assessing homologies at different biological levels runs counter to the bhc-based concepts (Collazo and Fraser, 1996; Gilbert et al., 1996; Roth, 1988; Wagner, 1989b), in which two given features are recognized as homologous if both derive from a similar developmental precursor (e.g., cell type, tissue type, gene expression pattern, or developmental constraint). The bhc criteria do not directly homologize morphological characters; rather, they push the question of homology back to something developmentally antecedent to those characters. Thus, eyes in two taxa become homologous because genes involved in their formation are homologous, a wing and a forelimb are homologous because their evolutionary constraints are homologous or because they have a common evolutionary origin, etc. This problem derives from a conflation of separate characters that are both logically and (by experiment) demonstrably independent. Realizing this, one can say that the genes underlying animal segmentation are (probably) homologous and body segmentation is not, contrary to the conclusions obtained by Gilbert et al. (1996) in their discussion of *Drosophila* and vertebrate segmentation (Box 4). Approaching this problem with the understanding that homology is dependent upon a hierarchical relationship of two taxa in relation to a third, we are unlikely to consider body segmentation in insects and vertebrates to be homologous because of the discontinuous distribution of this character state across the minimal clade that includes both insects and vertebrates.

If homology describes the relationship of two character states of two taxa as more closely related to each other than either is to the state in a third taxon, this relationship necessitates the comparison of more than two taxa and eliminates the ambiguity inherent in "homology" statements based on pairwise similarity. The non-hierarchical relationship of a two-taxon statement, in contrast, neglects propinquity of relationship. If homology is supposed to be about biological relatedness, then it is better considered in terms of organismal hierarchy rather than in terms of pairwise comparison.

Levels of homology and homoplasy when using developmental characters

For developmental characters to serve as the sole arbiters of phylogeny, it must be shown that they are immune to homoplasy (parallelism, convergence and reversal). If, however, they are not so immune, then, as two lineages diverge from their most recent common ancestor, homologous genes may produce non-homologous structures and non-homologous genes may produce homologous structures because gene function may have been co-opted for new pur-

Box 3. Segmentation in animals: homology or not?

Gilbert et al. (1996) argue for the homology of segmentation in *Drosophila* and vertebrates based on the finding that the genetic control of segmentation in both sets of organisms shares at least one homologous gene (i.e., an ortholog). In contrast, the homology of segmentation in these two taxa is contradicted by various lines of evidence that suggest that the character of segmentation was not present in an unbroken chain of ancestors linking the two groups, including:

1) Other classes of metazoans, including those that are not segmentally arranged, have arisen following the initial divergence of the lineages that eventually yielded Insecta and Vertebrata.
2) A wide range of additional characters (including other developmental ones) suggests that vertebrates and *Drosophila* share an unsegmented, planarian-like most recent common ancestor.
3) Taxa possessing a radial body symmetry and lacking segmentation (e.g., Echinodermata) are more closely related to vertebrates than to the insects.

Thus, based on the sum of the data available at this time and despite the presence of a shared developmental gene, segmentation in *Drosophila* and vertebrates is most likely due to convergence. The segmentation control gene itself, however, may be considered homologous if it can be shown that this gene existed (modified or unmodified) in all "ancestors" linking *Drosophila* and Vertebrata, regardless of its function. Arguing the homology of *Drosophila* and vertebrate segmentation based on the homology of a particular gene misses the point of "homology," which requires that *Drosophila* and Vertebrata acquired their segmentation from their most recent common ancestor. Put another way, defining genetic architecture as decisive of morphological homology ignores the possibility of convergence in gene function.

poses in one or both (divergent) lineages (Abouheif, 1997; Lowe and Wray, 1997). Thus, homologous morphological structures may remain unchanged (i.e., identical structures may be present in all descendant species), whereas the genetic system that produces those structures may be partly or entirely replaced by a different set of genes. Examples of such developmental divergence include the induction of Meckel's cartilage and the non-homologous origins of regenerating eye lenses in amphibians (DeBeer, 1971; Sander, 1983). Regarding the use of developmental criteria for homology, Bolker and Raff (1996) argue, "this scheme accepts 'homologies' that previous generations would have used as counterexamples." We agree with Bolker and Raff (1996) that a developmental redefinition of morphological homology is unwarranted,

Box 4. Levels of biological organization and homology

This figure, redrawn from Abouheif (1997), attempts to demonstrate how a combination of historical and biological homologies works toward homology recognition. An outgroup has been added. The characters, derived from several levels of biological organization, have been renamed with the addition of numerals in parentheses to indicate alternate forms of the characters: morphological feature, M; gene structure, G; gene expression pattern, GE; embryonic origin, EO. EO(0) and GE(0) are homologies (synapomorphies) for clade A + B. EO(1) and GE(1) are homologies for clade C + D. Likewise, EO(2) is a homology for E + F, EO(3) is a homology for G + H. G(0) is a homology for A + B + C + D; G(1) and GE(2) are homologies for E + F + G + H, and M is a homology for A + B + C + D + E + F + G + H.

This example demonstrates the hierarchical nature of homology. At the taxonomic level of A–H, morphology is shared, but not derived (symplesiomorphic). If A–H were compared with an outgroup (0), which is not more closely related to any one of A–H (the ingroup, or group of interest) than any of these are to each other, M would be homologous under a synapomorphy-based homology concept. Taxic relationships, taking on different significances based on taxonomic level, mediate homology recognition.

A homology concept that incorporates both hierarchical relatedness and biological organization is possible only if homology is synapomorphy. "Homology implies generality (that there is a set that includes...), and synapomorphy implies relative, or restricted, generality (that there is a subset included in...)" (Nelson and Platnick, 1981:158).

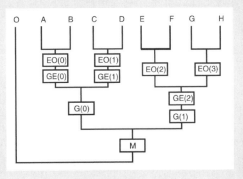

and we further suggest that, before we reject the hypotheses of previous generations of biologists, explicit and empirically-based criteria should be employed to assess the accuracy of hypothesized homologies.

Phylogenetic systematics aims to reconstruct genealogies of species and higher taxa utilizing conjunctions of multiple hypotheses of homology drawn from all the available data. The resulting topologies (unrooted networks and rooted cladograms) serve to test those hypotheses by congruence. Those hypotheses that survive such testing show "topological relations of similarity" and are best explained by common ancestry, i.e., their similarity is consistent with "descent, between different organisms, or parts of different organisms" (Nelson, 1994).

Those characters for which hypotheses of homology fail have discontinuous topological distributions on cladograms and require parallel evolutionary origins. They are non-homologous and their perceived similarity requires an explanation other than common descent. In this two-step approach, the term "homology" is applied only to those characters that withstand the test of phylogenetic analysis and that are objectively linked by descent from a most recent common ancestor (based on all the available character information) (dePinna, 1991; Eldredge and Cracraft, 1980; Patterson, 1982; Patterson, 1988; Rieppel, 1992; Brower and Schawaroch, 1996). "Homology" thus distinguishes those characters that have been tested by phylogenetic analysis from those characters that are similar but that do not, according to the weight of the evidence, share a common origin.

Certainly, characters drawn from developmental biology have much to offer in the way of new evidence on phylogeny, and Gilbert et al. (1996) are fully justified in emphasizing their neglect by the Modern Syntheticists. However, given that developmental genes and the morphological structures they participate in producing are capable of evolving independently, developmental and morphological characters initially should be regarded as potentially providing separate evidence on phylogeny and analyzed in the context of a data set that, ideally, includes other characters as well. This approach is reflected in Bolker and Raff's (1996), Abouheif's (1997) and Abouheif et al.'s (1997) presentations of homology, which center on distinguishing among the various levels of biological organization (Box 4).

Conclusion

Wheeler's (Q.D., 1990) definition of a character emphasizes the difference between Mayr's (1942) definition, "...any attribute of an organism (or better, of any group of organisms) by which it may differ from other organisms," and that of Platnick (1979), which defines a character as an original attribute and its subsequent modifications. By recognizing the originality of a character, character comparisons gain a greater biological significance. The originality, or uniqueness of origin, of characters can only be inferred with knowledge of the hierarchical relations of the taxa that bear them.

Process-based, transformational concepts attempt to incorporate "biological content" into the definition of the relationship of observed features of organisms, ostensibly to increase the empirical content of the posited relationship. Unfortunately, this approach depends upon a subjective assessment of the similarity of two developmental processes and/or on ambiguous appeals to common ancestry or phylogenetic origin. Furthermore, because developmental processes may be similar due to convergence, their similarity may not reflect the continuity of information among taxa in phylogenetic lineages due to shared ancestry. One example of this kind of phenomenon is the vertebrate lens crystallins and enzymes that have undergone gene sharing, where taxon-specific crystallins and enzymes have been co-opted from other functions outside of the lens multiple times (Piatigorsky and Wistow, 1989). Transformational homology concepts are the avatar of Aristotelean "essential similarity"; at best, they serve to identify "structural correspondences" in molecular, developmental and other biological processes (Bang et al., 2000). The rapidly accumulating data of modern developmental biology undoubtedly contain powerful new evidence on the evolutionary origins of diverse comparative biological features, so it is unfortunate that some modern homology concepts still define homology in terms of similarity. Not only do these concepts fail to distinguish homology from homoplasy (cf. Hunter, 1964; Abouheif, 1997), but they fail to distinguish among distinct levels of evolutionary processes at discrete taxonomic levels. The following are our suggestions for how Evo-Devo research can most efficiently contribute to evolutionary studies in general and to phylogenetic systematics in particular:

1. Homology should be viewed as an inherently phylogenetic concept. Prior to the theory of evolution, biologists did not have an explanation for morphological correspondences across species. Evolution provided the explanation that these correspondences were the result of common ancestry. Pre-Modern-Synthesis biologists did not have an objective methodology for corroborating hypotheses of homology, and largely relied on similarity-based assertions of homology. Modern phylogenetic analysis, which produces phylogenies constructed from all available character data, supplies a means for testing hypotheses of homology and thus provides an alternative to transformational approaches.

2. Developmental data must be coded as character data in phylogenetic analyses. The establishment of phylogenetic homology is an empirical endeavor. The challenge for synthesizing developmental biology with evolutionary biology, then, is to view developmental data within the context of a phylogeny in order to reconstruct evolutionary origins and thus to infer the degrees of generality of developmental features.

3. Hypotheses of homology, including hypotheses arising from molecular-developmental studies, must be tested empirically. Like all hypotheses of homology, homologies of developmental processes can only be falsified in light of a phylogeny, contrary to the proposition that "...there may be no real distinction between the biological and historical homology concepts" (Abouheif, 1997). Proponents of the bhc acknowledge that phylogeny is necessary to homology: "The biological basis of homology can only be examined if we know the phylogenetic pattern on which evolutionary

processes can be traced," (Roth, 1991:173); "...true homologs must be derived from a common ancestor and are unique by definition," (Abouheif, 1997; Abouheif et al., 1997; Raff and Kaufman, 1983; Wagner, 1989b: 1162). The bhc and like concepts, however, provide an effectively essentialistic homology criterion that recognizes homologues by putatively similar developmental processes.

4. Hypotheses of homology must be precisely defined, particularly with regard to biological level, because developmental genes and the structures they participate in producing are capable of evolving independently. Although similarity of development can provide additional evidence on homology (as well as on phylogeny), it cannot *define* homology. Two homologous morphological characters may be shown to have different underlying developmental processes (Abouheif, 1997; Bolker and Raff, 1996; Collazo and Fraser, 1996; DeBeer, 1971; Wagner, 1989a), but may be homologous nonetheless.

5. The possibility of homoplasy in developmental characters must be recognized. In addition to the point made in Item 4, two non-homologous morphological characters may share strikingly similar developmental-genetic processes that are, nonetheless, similar due to convergence.

No doubt some authors regard the bhc as a means for describing a universal process that underlies all character-state transformations or as a means for formulating hypotheses of homology prior to phylogenetic analysis (Shubin, 1994). In these interpretations, the concept embraces all types of similarity without discriminating between those that are due to common descent and those that are due to convergence. For the sake of clarity, and in order to integrate developmental biology and phylogenetics, such similarity should not be framed as homology. Similar developmental processes certainly suggest the possibility of shared ancestry, but homology can only be separated from homoplasy in light of a phylogenetic analysis.

Acknowledgements
TRS was supported by National Science Foundation Grants DEB-9707209 and by Smithsonian Scholarly Studies Grant 140202-3340-410. RB was supported by an American Museum of Natural History Graduate Traineeship. The kindness and tolerance of William L. Brown, Jr. is fondly remembered. A.V. Z. Brower, C. Chaboo-Michalski, D. Kosmun, M. Luckow, M. Wolfner and Q.D. Wheeler read earlier incarnations of the manuscript and offered many helpful suggestions.

Part 3
New approaches to
molecular evolution

Introduction to part 3

New approaches to molecular evolution

The final section looks at five new areas of interest in molecular evolution spawned by the proliferation of molecular data and by the development of new analytical techniques. Thornton begins the section by examining the methods and approaches that have been developed to examine gene family evolution. This topic has become more and more important recently due to the completion of various genome projects and the need to annotate the products of these projects. Thornton points out that phylogenetic methods can be used to detect specific evolutionary processes such as exon shuffling and tracing of amino acid changes and how these correlate to functional change. The second chapter in the section, takes a specific look at the evolution of a gene family—the spider silk proteins. Hayashi examines the evolutionary history of repeated motifs in spider silks that confer the mechanical attributes of the silks. The next two chapters describe the utility of the comparative method in examining evolutionary questions. Vogler and Purvis summarize the methods used by the comparative method and Nishiguchi describes in detail approaches to the use of physiological data to corroborate co-speciation events. We conclude this section of the book with a chapter by Planet that examines the phenomenon of horizontal transfer in microbial evolution and systematics.

R. DeSalle, G. Giribet and W. Wheeler

Molecular Systematics and Evolution: Theory and Practice
ed. by R. DeSalle, G. Giribet and W. Wheeler
© 2002 Birkhäuser Verlag/Switzerland

Gene family phylogenetics: tracing protein evolution on trees

Joe Thornton

Columbia University Earth Institute, Columbia University #2430, New York, NY 10027, USA

Summary. How have proteins taken on the remarkable diversity of biochemical and physiological functions necessary to create and maintain complex organisms? The majority of proteins are organized hierarchically into families and superfamilies, reflecting an ancient and continuing process of gene duplication and divergence. The techniques of molecular phylogenetics, developed to recover the nested hierarchy of taxa from character information in their gene sequences, can also reconstruct the evolutionary relationships among genes and provide a conceptual foundation for comparative evolutionary analysis of proteins and their functions. In this review, I outline the application of phylogenetic approaches to issues in gene family studies, beginning with the inference of phylogeny and the assessment of the two types of homology by which genes in a family can be related: orthology (common descent from a cladogenetic event) and paralogy (common descent from a gene duplication event). I show how the phylogenetic approach makes possible novel kinds of comparative analysis, including detection of exon shuffling, reconstruction of the evolutionary diversification of gene families, tracing of evolutionary change in protein function at the amino acid level, and prediction of structure-function relationships. A marriage of the principles of phylogenetic systematics with the copious sequence data being generated by molecular biology and genomics promises unprecedented insights into the nature of biological organization and the historical processes that created it.

A phylogenetic approach to gene families

With the advent of rapid nucleic acid sequencing and whole-genome analysis, it has become clear that the coding portions of the genome are organized hierarchically into families and superfamilies[1]. More than 50 percent of the genes in the bacterium *Escherichia coli* are members of identified gene families (Koonin et al., 1998), and the proportion of gene family members in eukaryotes may be in the same range or even higher (Semple and Wolfe, 1999; Chervitz et al., 1998). The hierarchy of genes, like the nested organization of living organisms, has been produced primarily by Darwinian processes of duplication and divergence (Ohno, 1970; Graur and Li, 1999), so the concepts and analytical tools used in phylogenetic systematics can and should be used to reconstruct the evolutionary relationships among genes in genomes. Just as these techniques allow the overwhelming diversity of taxa in nature to be systematized into a

---

[1] Traditionally, a gene family has been defined as a group of genes all of whose members have more than 50 percent pairwise amino acid similarity, and a superfamily as an alignable groups of genes with similarity below this threshold (Graur and Li, 1999); in this review, I use the term gene family to encompass both types of groups.

concise and historically meaningful conceptual framework, they provide a powerful way to make sense of the ever-increasing body of gene sequence data.

A phylogenetic approach to gene families makes it possible to predict the functions of newly sequenced genes and draw detailed inferences about the processes and patterns of protein evolution. Because comparative biological analysis can be carried out only in the context of a phylogeny (Harvey and Pagel, 1991), a sound classification of gene family relationships is a prerequisite for virtually all types of inference about the evolution of genes and the proteins for which they code. With a reliable gene phylogeny in hand, we can predict the structure and function of uncharacterized proteins, infer the evolutionary processes by which new genes appeared and took on novel functions, reconstruct the biochemical pathways and gene complements of ancestral organisms, analyze coevolutionary relationships and dynamics among proteins, and understand links between genomic change and morphological innovation (Koonin et al., 1996, 1998).

Despite the power of phylogenetics for comparative analysis, a clear understanding of its principles has been lacking from most gene family studies. Since the 1980s, biological systematists have widely accepted the superiority of phylogenetic to phenetic methods, for both theoretical and practical reasons (Hull, 1988; Farris, 1982, 1983; Swofford et al., 1996); today, virtually no systematist would classify taxa on the basis of quantitative measures of pairwise similarity. The same cannot be said of molecular biologists studying the relationships among genes. With some exceptions (e.g., Goodman, 1979; Beintema and Neuteboom, 1983; Sanderson and Doyle, 1992; Page, 1993b, 1994; Song and Fambrough, 1994; Agosti et al., 1996; Chiu et al., 1999; Thornton and DeSalle, 2000), most papers in gene family evolution have been based on trees inferred from pairwise distances among sequences, using neighbor-joining (NJ) or unweighted pair-group (UPGMA) methods (examples include Amero, 1992; Laudet et al., 1992; Brendel et al., 1997; Bonci et al., 1997; Coulier et al., 1997; Chervitz et al., 1998).

Reliance on phenetic methods has become particularly acute with the rise of high-throughput DNA sequencing. As the sequences of whole genomes have become available, computationally sophisticated informatics techniques, all based on phenetic criteria, have been implemented with the stated goal of recovering evolutionary relationships among genes in genomes (Koonin et al., 1996; Tatusov et al., 1997; Chervitz et al., 1998; Holm, 1998; Mushegian et al., 1998. The most computationally advanced and currently accepted method for establishing orthology (Tatusov et al., 1997), for example, relies on the simple pairwise similarity scores found in genome-wide BLAST searches (Altschul et al., 1990). Why would informaticians rely on methods that systematists view as outdated and inappropriate? In my view, two reasons provide partial explanations: first, in handling huge amounts of data, fast methods that give apparently unambiguous results have great appeal, and second, few molecular or computational biologists have been trained in or otherwise exposed to the principles and techniques of phylogenetic systematics.

Although they have been less widely used, phylogenetic methods—cladistic parsimony in particular—offer compelling advantages for studies of gene family evolution (see chapters by Wenzel and by Siddall in this volume). The central assumption of phylogenetic systematics is no less valid for genes in a superfamily than it is for species in a genus: if genes have evolved by duplication and divergence from common ancestors, the genes will exist in a nested hierarchy of relatedness, and these relations will be manifest in a hierarchical distribution of shared derived characters (synapomorphies) in the gene sequences (Hennig, 1963; Farris, 1983). On this theoretical foundation, the most parsimonious gene family tree—the one with the fewest parallel and reverse character changes—is the phylogeny that best explains the distribution of shared character states among extant genes as the result of descent from common ancestral genes. With the possible exception of trees with certain combinations of grossly unequal branch lengths (Felsenstein, 1978), parsimony methods thus provide a phylogenetic technique that can be applied in a wide variety of circumstances with a minimum of assumptions.

Building gene family trees

Techniques for analyzing gene family relationships with parsimony are almost identical to those used for inferring taxonomic relationships from gene sequences. First, the sequences to be analyzed must be selected. In principle, nucleotide or amino acid sequences—or both (Agosti et al., 1996)—may be used; in practice, amino acids are more often analyzed, because their signal-to-noise ratio is usually more appropriate for analyzing gene families that diversified hundreds of millions of years ago. Members of a given gene family in the published databases can be obtained by using BLAST searches (Altschul et al., 1990, available at www.ncbi.nlm.nih.gov/blast); the position-specific iterated BLAST approach is particularly useful for finding distantly related members of a family that may be missed by single-pass similarity searches (Altschul et al., 1997). For large, well-studied gene families there are often hundreds of sequences available; to use all would be extremely demanding of time and computer resources. In these cases, it is necessary to use only some of the many orthologous sequences, and taxa should be sampled as broadly as is feasible. When orthologs are highly conserved among closely related species (as is the case in many families of genes—like transcription factors and other kinds of developmental regulators—that are essential to organismal function), little phylogenetic information is lost by including just one or a few orthologs from the same taxonomic order (i.e., rodents, primates, cichlids, etc.).

Sequences must then be aligned to produce a data matrix. Numerous methods and programs for sequence alignment are available; the most theoretically justifiable are those that perform multiple alignments in a phylogenetic context, such as Clustal (Thompson et al., 1994), or better, the parsimony-based

TreeAlign (Hein, 1990) or Malign (Wheeler and Gladstein, 1995). All alignment methods should be used with attention to the sensitivity of alignments to user-specified gap-change ratios and other parameters. Paralogous groups of genes in a family are often significantly diverged from each other, and the alignment of less constrained regions often varies with the parameters. One technique to avoid the arbitrary selection of one of many plausible alignments is the "cull" procedure, in which alignment-ambiguous positions are omitted altogether (Gatesy et al., 1993). These sites often contain useful phylogenetic information, so their omission may reduce phylogenetic resolution and/or support. To avoid this problem, the "elision" procedure can be used to assemble numerous plausible alignments in a master data matrix—an approach that effectively gives higher weight to alignment-consistent positions without losing the information that is present in alignment-ambiguous sites (Wheeler et al., 1995).

Phylogenetic analysis of the aligned sequences proceeds as it does with organismal phylogeny. The diversity of most gene families usually requires heuristic search strategies, such as those available in PAUP* (Swofford et al., 1999), to seek the most parsimonious tree(s) (MPT). Confidence in the MPT can be evaluated by calculating Bremer supports (Bremer, 1994), which express the relative character support for each node in the MPT as the number of extra steps required for each node not to appear in the MPT. Bremer supports can be calculated automatically using Auto-Decay (Eriksson, 1996). Bootstrapping—the assembly of a new data matrix the same size as the original by randomly sampling sequence positions, with replacement—is also frequently used to assess confidence in a phylogeny (Felsenstein, 1985a). Because this approach in essence measures the effect of random weighting of characters, it reveals only the degree to which phylogenetic signal is uniformly distributed throughout the data set, not statistical confidence in the MPT or the extent to which the data support that tree. In gene families, it is not uncommon for many sequence sites to be completely conserved and others to be highly diverged, with only a portion of sites exhibiting the intermediate degree of variability that makes them phylogenetically informative. Furthermore, the sites that are informative at deep levels of the tree are often different from those that support resolution nearer the tips. High bootstrap values for many nodes in a gene family tree are thus not expected, even when there is substantial support for the MPT. (For other problems with bootstrapping as a measure of tree confidence, see Sanderson, 1995.)

Interpreting tree topology

Embedded trees

A gene family phylogeny is less straightforward to interpret than an organismal tree. An accurate tree tracks recency of common ancestry among sequences in a gene family, but splitting and divergence of gene lineages can be caused either by duplication of genes within a genome (producing paralogs) or by the splitting of the lineage of the organisms that contain those genes (producing orthologs). When a gene family tree includes sequences from taxa whose most recent common ancestors existed before some but not all of the duplication events that produced the paralogs in the family, then orthologs and paralogs will be interleaved on the phylogeny in a complex pattern that hierarchically reflects the order in which gene duplication and cladogenetic events occurred.

Consider the example shown in Figure 1. The evolutionary process shown (*a*) will yield the species tree (*b*) and the gene tree (*c*), if sequences are available for all relevant genes from all the taxa used in this analysis. Each branch of the tree that diverges from a node representing a gene duplication contains a replica of that part of the taxonomic tree produced by cladogenic events that occurred after the gene duplication. If all the gene duplication events happened before any of the taxa on the tree diverged from each other, then each branch leading to a group of orthologs will contain the entire species tree. If some

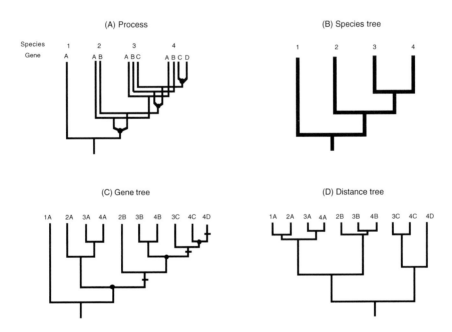

Figure 1. Cladistic and phenetic reconstructions of a gene family phylogeny. (A) Evolutionary scenario of gene duplications (marked with dark circles) and cladogenesis (unmarked nodes) that generate a hypothetical gene family. (B) Species tree for the process in A. (C) Gene tree for the process in A, correctly inferred using the parsimony criterion. Each group of shaded branches leads to a group of similar orthologs. (D) UPGMA tree for the process in A, assuming no homoplasy and ten-fold higher rates of sequence change on branches on which gene duplications lead to new paralogous genes (hashed on the gene tree cladogram in C), than on branches leading to conserved orthologs. Note that this tree fails to recover the phylogeny of the process in A.

duplications occurred before and some after the relevant cladogenic events, then subtrees of varying size, one for each paralog, will be arranged in nested fashion in the gene tree (Fig. 1c). This master phylogeny of species trees within paralog trees is analogous to other kinds of trees in which lineages duplicate at more than one level, such as biogeographic area cladograms and phylogenies of parasites from multiple host taxa (Page, 1993b).

The nesting of more recently diverged paralogs within larger groups that contain more ancient orthologs raises questions about the utility of the popular terms orthology and paralogy. Fitch (1970) defined two types of genetic homology: orthologs are genes in different genomes that have been created by the splitting of taxonomic lineages, and paralogs are genes in the same genome created by gene duplication events. But inconsistencies arise in defining orthologs when one or both gene lineages created by a cladogenetic event later undergoes gene duplications. Consider the gene tree in Figure 1c: genes C and D in the genome of species 4 are clearly paralogs of each other, and gene B in species 2, 3 and 4 are clearly orthologs of each other. But what is the relationship of 3C and 4C? These would normally be called orthologs of each other, because of their close sequence similarity and the assumption that this similarity is due to descent from a gene C in the common ancestral taxon of species 3 and 4. But homology is by definition a phylogenetic relationship, not a phenetic one. In a phylogenetic sense, 4C is no more closely related to 3C than 4D is; 4C and 4D are equally orthologous to 3C, as reflected in the common ancestry of 3C with 4C and with 4D at the same node on the tree. This problem affects not only the orthology of recent paralogs but also of ancestral genes. Gene A in the stem species 1 would normally be considered an ortholog of the other As in the tree and a paralog of all other members of the gene family; however, it is equally related phylogenetically to every other member of the gene family in the analysis. This ambiguity remains unresolved no matter how similar the sequence of 1A is to the other As and how different from all the Bs, Cs, and Ds on the tree.

Orthology as currently defined is thus a partially phenetic concept. When a gene has duplicated, choosing one member of the pair as an ortholog of a gene whose lineage branched before that duplication event can be based only on a criterion of pairwise similarity. This is not entirely unreasonable, and it is the basis for giving highly conserved genes in different species the same name, whether or not there are other family members to which they are equally related. This approach also provides a reasonable guide for functional predictions: if gene 4C is very similar in sequence to gene 3C, but its paralog 4D (which is equally orthologous to 3C) has diverged considerably, then it is likely that 4C and 3C share a conserved function (see, for example Tatusov et al., 1997; Koonin et al., 1998; Raymond et al., 1998). But phenetic similarity should not be confused with evolutionary relatedness or with homology, which is a fundamentally phylogenetic concept. For consistency, genes like these that do not form monophyletic groups of orthologs but are closest to each other in a phenetic sense should be referred to as "most similar orthologs."

Gene duplication and loss

The expected gene phylogeny allows gene duplication events to be inferred and roughly dated, given a tree that specifies the relations among the taxa used in the analysis (Goodman et al., 1979; Beintema and Neuteboom, 1983; Tsuji et al., 1994; Thornton and DeSalle, 2000). As Figure 2a shows, a gene duplication must be postulated at the base of any clade that contains a lineage whose branching order is incompatible with the taxonomic phylogeny. The location of the duplication event on the gene tree gives a lower bound of its age, because the duplication must have occurred prior to the divergence of all the lineages represented within that clade on the gene family tree. For example, Figure 2a suggests that the duplication labeled x—the event that created the paralogous gene groups A and B (and ultimately C and D, too)—occurred prior to the taxonomic divergence of the lineage that contains species 2 from that containing species 3 and 4.

This reasoning can be used even when gene sequences are not included in an analysis because they have been lost or have not been sequenced, as shown in Figure 2b. If sequences for genes 2A, 3C, and 4B are all missing, it is still necessary to postulate that duplication x occurred before the divergence of these three species, because the subtree $((2,3),4)$, is incompatible with the given taxonomic tree $(1,(2,(3,4)))$. This inference is possible despite the fact that no species is known to possess all three paralogs in its genome, as is required to justify a gene duplication inference using phenetic approaches.

Losses or incomplete sampling can be inferred in an analogous way. Once the appropriate duplications are postulated in Figure 2b, for instance, it is possible to construct a "reconciled tree" (Goodman et al., 1979; Page, 1994b; Page

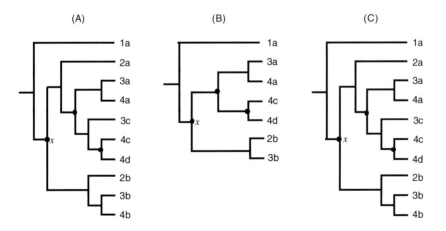

Figure 2. Inferring gene duplications and losses from a gene family tree. (A) Gene duplications must be inferred at nodes where the gene tree is incompatible with the species tree (dark circles). (B) Gene duplications can be inferred even when some sequences are lost or missing. (C) Reconciled tree for gene tree in (B), with hypothetical branches leading to lost or missing sequences shaded gray.

and Charleston, 1997) that resolves conflict between the species tree and the gene tree by including hypothetical branches for sequences that must have been lost or not yet discovered. The reconciled tree (Fig. 2c) makes clear that 2A, 3C and 4B must exist, either as genes or pseudogenes. The visual reasoning that a reconciled tree facilitates can be formalized in the following rule: each branch that diverges from a gene duplication node i must lead immediately to another node j that contains one or more gene(s) from all taxa descended from the taxonomic ancestor in which the gene duplication occurred; if they do not, then an intervening branch that leads to the lost gene or genes must be added between i *and* j. Using the reconciled tree, it also possible to predict which gene family members are likely to be found in species whose genomes have not yet been completely sequenced, providing guidance to laboratory work: given a species y known to contain gene z, all species that share a common ancestor with y more recently than the gene duplication that led to the appearance of z will also contain z (or, in the case of gene losses, a pseudogene of z).

These rules for inferring gene duplication and loss, automated in the programs Component (Page, 1993a) and GeneTree (Page, 1998a, 1998b), reconstruct the most parsimonious hypothesis of the genogenetic process, given a gene family phylogeny and a species tree. If either tree is not well supported, interpretation may be more ambiguous, requiring the investigator to weigh the parsimony criterion for inferring gene losses against the parsimony criterion by which the gene tree and species tree were constructed. For example, the alignable portions of genes in the steroid receptor family are so conserved that the taxonomic relationships of some orthologs within and among mammalian orders are unorthodox and have very low Bremer supports (Thornton and Kelley, 1998). Rather than postulate gene losses each time an anomaly like this occurs, it may be preferable to interpret the topology of the gene tree at these nodes as inaccurate. This approach can be formalized by choosing a ratio that expresses the cost of a gene loss relative to an amino acid character change and then using the method of Goodman et al. (1979) to choose the reconciled tree that minimizes the weighted sum of amino acid changes and gene losses. Determining this relative weight is ultimately a subjective and somewhat arbitrary choice (Fitch, 1979); but its use at least assures consistency and allows an analysis of the sensitivity of evolutionary reconstructions to the weight chosen for losses *versus* homoplasies. Further, for most gene families, the nodes that are interesting from an evolutionary perspective—those that reveal the timing of gene duplications and the relations among paralogs—are generally at a deeper taxonomic level, so this issue can remain unresolved without compromising the ability of the analysis to test hypotheses about gene family evolution.

Rooting

Drawing inferences about evolutionary history from a phylogeny requires that the tree be properly rooted, which is not always straightforward in gene family analyses. The only scenario in which empirical evidence would support the designation of one gene family member as an outgroup would be if a stem taxon were revealed to contain a single member of a gene family, thus designating this gene as the "ancestor." Except when whole genomes are available, the possibility that gene loss has occurred in other lineages ensures some ambiguity in rooting by this method. There is another criterion, however, by which a tree may be rooted: the degree to which it preserves the expected taxonomic structure within each group of orthologs. Each of the possible rooted trees that can be derived from an unrooted gene family tree disrupts the expected taxonomic subtrees within each group of orthologs to a varying extent, and each thus requires a different number of gene losses to be assumed. The rooted tree(s) that require the fewest assumed gene losses is the most parsimonious and therefore the preferred hypothesis of the genogenetic process. This approach is an adaptation of a method developed for rooting the deepest nodes in the tree of life using duplicated genes, such that the root is placed to preserve the expected structure among pairs of highly conserved orthologous genes (Iwabe et al., 1989).

Consider the example in Figure 3, which shows three ways of rooting an unrooted gene tree. The tree in (*b*) requires no gene losses, that in (*c*) requires one loss, and the tree in (*d*) requires four losses. Given the sequence data and the taxonomic tree, then, tree *b* is a slightly better hypothesis than *c*, and both are considerably better than *d*. To prefer a tree that is more parsimonious in terms of gene losses in this way is well justified when missing branches are due to the actual loss of a gene, since it offers the most complete explanation of the distribution of gene family members in various species as the result of inheritance from common ancestors, with a minimum of *ad hoc* hypotheses of additional gene duplications and losses. When missing branches may be due to incomplete sampling, however, the parsimony criterion is less persuasive: to suggest that there may be unidentified gene family members in the genome of a species that has not been thoroughly studied is not a burdensome *ad hoc* hypothesis. Rooting may thus remain somewhat ambiguous except in cases where complete genomes have been sequenced or exhaustive searches for gene family members have been conducted by PCR and library screening. Like the ambiguity in inferring losses, however, this problem is not an overwhelming one: even in the absence of an unambiguous root, the two plausible trees in (*b*) and (*c*) are congruent at all nodes but two, and they imply nearly identical evolutionary processes, differing only in the timing of the first gene duplication, which predates the divergence of species 2 from 3 and 4 in *(b) and* postdates it in (*c*). Rooting may thus remain incompletely determined without precluding a substantial degree of evolutionary inference.

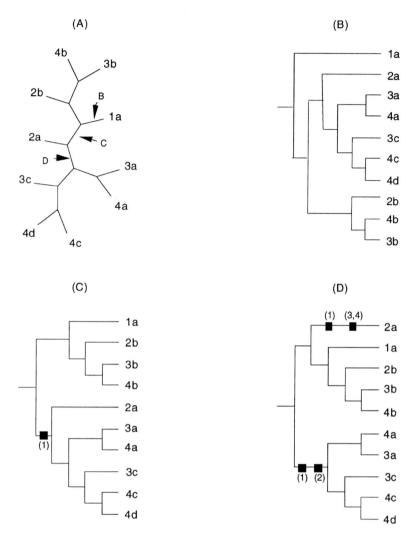

Figure 3. Inferring the root of a gene tree by minimizing *ad hoc* hypotheses of gene losses. (A) Unrooted gene family tree. (B–D) Rooted gene trees rooted at the locations labeled by the arrows in A. Gene losses that must be hypothesized are marked with dark boxes; labels show the taxa in which the gene has been lost. Tree (B) represents the most parsimoniously rooted tree.

Analyzing protein evolution in a phylogenetic context

Once a gene family phylogeny has been inferred, comparative analysis of gene and protein evolution can begin. Numerous methods are available to investigate the genetic mechanisms by which gene families diversified and took on specific functions.

Mechanisms of gene proliferation

A phylogeny provides a foundation on which to test hypotheses concerning the mechanisms by which new gene family members were created. Gene duplication can occur by tandem duplication (due to replication slippage or unequal recombination), duplications of whole genomes or large parts thereof, transposition of DNA sequences, or retrotransposition of RNA transcripts. When genomic information is available to specify the chromosomal location of genes in a superfamily, these mechanisms leave unique traces in the genome (see Wilkie et al., 1992, for example). Where tandem duplication has created paralogs, gene family members will be closely linked on a single chromosome. Large-scale duplications, such as those caused by polyploidization, result in gene family members that are scattered in the genome, and each descendant of such an event will be linked to members of the other gene families that have been duplicated in the same event. For example, the existence of numerous "tetralogous" gene groups—linked assemblages of members of several gene families repeated approximately four times throughout the genome—in mammals and other vertebrates provides evidence that two genome-wide duplications occurred early in the chordate lineage (Spring, 1997; Gibson and Spring, 1998). Gene family members created by retrotransposition can be detected by the lack of introns in their sequences; fragmented or intact retroelements may also be detectable in positions up- and downstream from genes that have been inserted by retrotransposition (Gaudieri et al., 1997). DNA transpositions have none of these characteristics, but the presence of conserved terminal sequences of mobile genetic elements in regions flanking the genes provides affirmative evidence of such a process.

New gene family members can also be created by "horizontal transfer" of genetic information between more ancient paralogous genes. Many proteins, particularly those in recognizable gene families, are composed of domains—discrete structural units with specific and often autonomous functions. Domain shuffling, which can occur by transposition of gene fragments or nonhomologous recombination, is thought to have been a major mechanism in the evolution of new proteins (Doolittle and Bork, 1993; Doolittle, 1995). Domain shuffling in the history of a gene family can be examined by analyzing separately the phylogeny of protein domains. If some members of a family have been created as evolutionary chimeras by the shuffling of domains from more ancient members, the phylogenies of the domains will be incongruent. Some incongruence is expected, however, simply as a chance result of partitioning itself, so it is necessary to ascertain whether the observed incongruence among domains is structured or random. The incongruence length difference (ILD) test of Farris and colleagues (1995) provides a non-parametric statistical test of incongruence between two data subsets; it computes the number of extra steps imposed by analyzing the subsets together as compared to their separate analysis, and it compares this degree of incongruence to that observed for a large number of random partitions in the same data set. Significant incongruence in the ILD test cannot be explained by random, non-directed homoplasy; it provides evidence of substantially conflicting phylogenetic signal between domains.

One limitation of the ILD test is that it examines the entire tree at once: significant global incongruence can be present if a single gene has been created by domain shuffling, so it would be useful to know which nodes contribute to the overall incongruence. The local incongruence length difference (LILD) test applies the ILD technique to each clade in a tree, allowing the incongruence at each node to be quantified and its statistical significance tested (Thornton and DeSalle, 2000). Other methods developed to assess recombination between alleles (reviewed in Crandall and Templeton, 1999) can be adapted for assessing the horizontal transfer of information between pairs of genes in a family. For example, Huelsenbeck and Bull's (1996) likelihood test compares the likelihood of a tree without recombination to that of a tree in which recombination —or in this case domain shuffling—occurred.

All these tests assume *a priori* knowledge of where gene sequences should be divided into subsets whose phylogenies can be separately inferred. The partition between domains can be based on structural and functional data about the gene family. Protein domains are identified biochemically with deletion experiments or the creation of chimeric proteins, and they are often separated from each other by introns or more variable coding regions. For instance, the DNA-binding and ligand-binding domains of nuclear receptors are discrete functional units that are separated by a highly diverged hinge region and at least one intron in all nuclear receptors, justifying the placement of a partition between these two regions. Alternatively, potential partitions can be inferred from the sequence data itself. Crandall and Templeton (1999) have reviewed several methods for locating recombination sites, including a phylogenetic approach based on a statistical test of the linear distribution of diagnostic characters in the sequence. With this method, the characters that support a node with potential incongruence due to domain shuffling are plotted along the length of the sequence, as are those that support alternative topologies. If domain shuffling occurred, the characters supporting the node are expected to be contiguous, with a discrete point in the linear sequence at which support for an alternative phylogeny begins to dominate. The probability that the observed clustering of synapomorphies along the sequence could have arisen by chance can be calculated by reference to the hypergeometric distribution of clustering that would be expected by chance alone.

Unlike domain shuffling, which serves as a mechanism for rapid creation of new proteins, concerted evolution tends to homogenize paralogs within a genome (Sanderson and Doyle, 1992; Dover et al., 1993). Concerted evolution can be caused by unequal recombination, gene conversion, or replication slippage. It has been important in some gene families, particularly those, like the ribosomal RNAs, which occur in tandem arrays and cause dosage repetition, a selective advantage to the organism conferred by having additional copies of a gene (reviewed in Graur and Li, 1999). Gene conversion appears to be much less frequent in families of transcription factors and other regulatory proteins, which seldom occur in tandem and for which dosage repetition is of little selective value. One study of gene families in the *C. elegans* genome, for instance, found evidence of concerted evolution for only 2 percent of genes (Semple and Wolfe, 1999), and there is no evidence of concerted evolution in the steroid receptor family (Thornton and Kelley, 1998).

A phylogeny provides a foundation on which hypotheses of concerted evolution can be tested: when pairs or groups of paralogous genes within a genome cluster together—particularly when the same pattern is repeated for the same genes in several taxa—concerted evolution is likely to be the cause. The same pattern, however, could be caused by independent duplication of the same gene late in each lineage. This ambiguity can be resolved by a detailed examination of homogeneity in different regions of the gene. The known mechanisms of concerted evolution affect DNA segments of relatively short length; outside these homogenized stretches, the sequences of family members should remain unhomogenized. In contrast, genes that are similar within a genome due to recent duplications should be relatively homogeneous along their entire lengths. Sawyer (1989) provides a technique to identify local gene conversion events in nucleotide sequences. Clades at which concerted evolution may have occurred are identified on the tree as those that unite genes from a single genome. Sites which are synonymous (do not affect the protein sequence) and contain the same nucleotide in the two candidate genes and a different one in the closest outgroup sequence may have been homogenized by concerted evolution; the clustering of such sites in contiguous regions of the gene is evidence for gene conversion rather than homoplasy. Semple and Wolfe (1999) have adapted a statistical method originally proposed to evaluate potential recombination sites for the purpose of testing whether the observed contiguity of homogenized sites is significantly greater than would have been likely to occur by chance alone.

Divergence of protein sequences and the emergence of function

The ultimate goal of gene family studies is an understanding of how duplicated genes have taken on novel biochemical and organismal functions. Domain shuffling aside, it remains a mystery how the undirected process of mutation, combined with natural selection, has resulted in the creation of thousands of new proteins with extraordinarily diverse and well-optimized functions. This problem is particularly acute in the case of tightly integrated molecular systems consisting of many interacting parts, such as ligands, receptors, and the downstream regulatory factors with which they interact. In these systems, it is not clear how a new function for any protein might be selected for unless the other members of the complex are already present, creating a molecular version of the evolutionary riddle of the chicken and the egg. Detailed studies of gene family evolution promise to shed some light on the process by which changes at the genetic level have led to the diversification of function for members of such integrated molecular systems.

To understand the evolution of structure-function relationships, we must first independently reconstruct the evolution of structure and of function. Phylogenetic methods for tracing the evolution of characters on cladograms (Williams and Fitch, 1989) and reconstructing ancestral sequences (Chang and Donoghue, 2000) are well established and have been automated in PAUP* and MacClade (Maddison and Maddison, 1997). With these techniques, it is possible to infer the evolutionary history by which various aspects of protein function—catalytic activity, expression domain, or specificity for a certain type of substrate, ligand, response element, or co-factor, to name a few—have been gained, lost, or transformed on each branch of the gene tree. In particular, we can ask whether such functions evolved consistently on a phylogeny (were created once and conserved thereafter in all descendants) or whether they have evolved in parallel, by convergence, or with reversals/loss in some lineages (see, for example, Applebury, 1994; Escriva et al., 1997; Thornton and DeSalle, 2000). When functions have evolved consistently, it is also possible to infer the functional characteristics of hypothetical ancestral proteins, even at the deepest nodes near the root of a gene family tree. Applying this approach to a tree of the transfer RNA family, for example, Fitch and Upper (1987) were able to test and corroborate the hypothesis that the genetic code evolved by a progressive reduction in ambiguity and a gradual increase in specificity of association between smaller and smaller classes of codons and amino acids.

The same techniques can be used to trace the evolution of the primary structure of proteins on a cladogram. One useful strategy is to identify specific amino acids that "diagnose" a clade of proteins, particularly a group whose members share an important functional characteristic. These synapomorphic amino acids are expected to include those that are required for the function that appeared on the same branch. Genetic and biochemical studies on structure-function relations have found that many of these phylogenetically diagnostic amino acids are indeed essential to function, validating this technique for the prediction of structure-function relationships and its utility for formulating specific hypotheses that lend focus to future structure-function studies (Thornton and Kelley, 1998).

Ancestral sequences can be reconstructed using ML methods on a tree inferred by parsimony or maximum likelihood. Yokoyama and Radlwimmer (1999), for example, reconstructed the sequence of ancestral red-green opsin proteins and predicted the color sensitivity of those sequences based on empirical evidence of the effect of specific amino acids at critical sites. They showed that mammalian color vision has evolved from an ancestral green-sensitive opsin, with rampant parallelism and reversal at both the sequence and functional levels.

Once the evolution of its individual components have been traced on a gene tree, the evolution of the relationship between structure and function can be analyzed. In particular, correlations in the evolution of structure and of function can be located on the cladogram, and the evolutionary sequence of muta-tions that led to the appearance of protein clades with unique functions can be inferred (see, for example, Kornegay et al., 1994). Ultimately, with a densely sampled gene family phylogeny, it should be possible to characterize the dynamics of the evolutionary relationship between primary structure and spe-cific aspects of protein function. Of particular interest is the extent to which functional changes take place gradually due to accumulated mutations at indi-vidual sites (Golding and Dean, 1998), or whether they are emergent proper-ties of complex combinations of sequence changes.

The role of selection in gene family evolution can be investigated by esti-mating relative and absolute rates of sequence divergence for different branch-es in the gene family tree. If the rate of mutation is more or less constant, then differences in divergence rates should indicate the strength of selection acting on any branch in the tree. Comparing the rates at which two paralogs in the same taxonomic lineage have diverged since the same cladogenetic events sug-gests the relative importance of selection in each paralog, indicating whether some proteins in a family have been more constrained by selection than others (Kissinger et al., 1997; Laudet, 1997; Thornton and Kelley, 1998). Absolute rates of divergence among orthologs can be calibrated with cladogenesis dates inferred from the fossil record. If taxa have been sampled densely enough, it should be possible to test the hypothesis that paralogs have evolved faster immediately after duplication events, followed by a slowing of sequence diver-gence thereafter (Li, 1985). Likelihood tests have also been developed that provide a more computationally sophisticated means for characterizing diver-gence rates (Sanderson, 1994).

Finally, a gene family phylogeny can be interpreted in light of "external" data to shed light on broader evolutionary issues. For example, a gene family phylogeny can be compared to the phylogeny of other gene families with which its members interact at the molecular level to understand the co-evolu-tion of interacting proteins. Fryxell (1996) has examined the evolution of pep-tide hormones, growth factors, cytokines and their receptors, using a compar-ative phylogenetic approach. These interacting gene families have diversified in a coordinated fashion, so that newly duplicated receptors gain affinity for newly duplicated ligands (see also the study of fibroblast growth factors and their receptors by Coulier et al., 1997). According to this model, simultaneous diversification of interacting protein families creates the conditions under which a duplicated receptor can take on a novel role and avoid the otherwise likely fate of transformation into a pseudogene. That is, the independent and nearly simultaneous creation of both the chicken and the egg allows both enti-ties to develop the systematic relationship with each other that leads to their perpetuation.

Some investigators have examined phylogenetic correlations between the timing of gene duplication events and major evolutionary changes in develop-mental and physiological programs. For example, the extreme diversification in the arthropod lineage of the cytochrome p450 superfamily of enzymes for oxidative detoxification appears to have occurred at about the same time that arthropods began to feed on land plants, with their wide variety of deterrent and poisonous compounds (Lewis, 1996). More ambitiously, investigators have separately proposed that the expansion of homeobox proteins, the fibroblast growth factors, or the nuclear receptors early in the chordate or vertebrate lineage caused or enabled the increased morphological and regulatory complexity of the crown vertebrates (Sidow, 1996; Coulier et al., 1997; Escriva et al., 1997). Of course, a phylogenetic correlation between the expansion of a gene family and the appearance of new organismal features does not in itself imply causality, especially since genome-wide duplications would have led to the simultaneous expansion of many gene families. The lack of support for such grand causal links is exemplified by the recent findings of large numbers of HOX clusters in fish and priapulids, neither of which is by any known measure substantially more complex than tetrapods, which have considerably fewer HOX genes (Meyer, 1998; de Rosa et al., 1999).

Towards a sampling of genomic biodiversity

The age of high-throughput sequencing promises to revolutionize gene family studies and make phylogenetic techniques and principles indispensable for genome analysis. Having complete gene sequences will allow researchers to ascertain the positive absence of a protein from a genome, removing a major source of ambiguity in the interpretation of gene family phylogenies. For a representative picture of protein evolution, however, it will be necessary to have genomic sequence data from more than the handful of organisms now being sequenced. These organisms have been chosen for their biomedical importance or their utility as model organisms for genetic and developmental analysis. Genomes from flies, worms, mice, zebrafish, yeast, *Arabidopsis*, and humans will be ultimately inadequate to inform rigorous inference about gene family evolution.

Obviously, the complete genomes of all the millions of species in nature will not be sequenced in the foreseeable future. But it is not unreasonable to hope that the choice of organisms to be studied in depth will be informed by phylogenetic relationships. From the perspective of evolutionary inference, the greatest gains will come by obtaining sequences from certain especially informative stem taxa—sister lineages to groups of major biological interest, such as the vertebrates, the chordates and bilaterians, and the metazoa. For example, the taxa critical for metazoan comparative genomics are not more rodents, more primates, more fish, or more nematodes, but the far less glamorous lamprey, hagfish, amphioxus, tunicates, echinoderms, cnidarians, sponges and choanaflagellates. No matter what taxa future sequencing projects focus on, however, this much is clear: as the genomic data come pouring in, the best way to make sense of them—at conceptual, functional and historical levels—is to begin with phylogenetics.

Acknowledgements
I thank Rob DeSalle, Darcy Kelley, and Cheryl Hayashi for helpful comments on the manuscript. This work was supported by National Science Foundation grant DEB 9870055 and by Columbia University's Center for Environmental Research and Conservation.

Molecular Systematics and Evolution: Theory and Practice
ed. by R. DeSalle, G. Giribet and W. Wheeler
© 2002 Birkhäuser Verlag/Switzerland

Evolution of spider silk proteins: insight from phylogenetic analyses

Cheryl Y. Hayashi

Department of Molecular Biology, University of Wyoming, Laramie, WY 82071, USA

Summary. Spider silks have astounding mechanical properties. In fact, dragline silk has greater tensile strength than commonly used synthetic materials such as nylon filament and capture spiral silk is among the most elastic proteins known. The recent cloning of spider silk genes has revealed that silk proteins are composed of tandem arrayed ensembles of a small number of amino-acid sequence motifs. These repetitive motifs form the structural modules within silk fibers, and are critical for determining the mechanical attributes of the silk. In this chapter, I examine the evolution of these motifs in the 11 published spider silk gene sequences. Extensive rearrangements of the motifs have occurred among the orthologous and paralogous proteins. Phylogenetic analyses suggest that numerous length mutations and recombination events have taken place in orthologous genes from closely related species and even within sets of alleles from the same species. Such genetic events appear to be critical for the homogenization of amino acid repeats within the silk proteins. The characterization of additional silk genes will clarify the relationships among novel amino acid motifs, the homogenization of motifs within a protein, and the function of silk fibers.

Spiders and silk

Silk is spun by numerous arthropods, but no group can rival the array of silk proteins produced by spiders. Shultz (1987) suggested that in primitive spiders, the primary function of silk was for reproduction. Silk is used by females to construct egg sacs and by males to form the silken platforms on which sperm is deposited for transfer to the copulatory pedipalps. Over time, spiders have greatly expanded their use of silks to perform diverse tasks (Shear, 1986). The different silk proteins are synthesized in their own specialized glands, and thin ducts connect individual silk glands to tiny spigots on abdominal spinnerets. The spinnerets of spiders are uniquely derived and are a synapomorphy for the order Araneae (Wheeler and Hayashi, 1998).

After a spider draws silk protein from the silk glands through the ducts, and out the spigots, the spider then utilizes the silk for a variety of tasks. The simplest function of silk is for the dragline, the trailing safety thread that allows a spider to retrace a path or decelerate during a fall. Silk is also critical for the protection of eggs and for the construction of retreats in the ground, on vegetation, or even underwater. Furthermore, many spiders build trip-lines, concealed trapdoors, three-dimensional obstacle courses, or aerial webs from silk in order to entrap prey. Thus, spiders have evolved suites of specialized behaviors, morphologies and proteins that are all necessary for the production and utilization of silk.

Table 1. Araneoid spiders have seven types of silk glands that lead to three pairs of spinnerets (Foelix, 1996)

Silk gland	Exit spinneret	Silk function
Major ampullate	Anterior lateral	Dragline, orb-web frame
Minor ampullate	Posterior medial	Temporary spiral
Flagelliform	Posterior lateral	Filament of capture spiral
Tubuliform	Posterior medial and lateral	Egg-sac inner wall
Pyriform	Anterior lateral	Attachment disc
Aciniform	Posterior medial and lateral	Swathing silk, egg-sac outer wall
Aggregate	Posterior lateral	Sticky droplets of capture spiral

The best-characterized silks are those spun by Araneoidea, ecribellate orb-web weaving spiders. Araneoid spiders spin seven types of silk from their spinnerets (Tab. 1). Some of these silks have attracted the attention of the military, industry, and the popular media because of the extraordinary mechanical properties of the fibers. At only one-tenth the weight of steel, dragline (major ampullate) silk, by cross-sectional area, has a tensile strength similar to Kevlar and greater than high-tensile steel (Stauffer et al., 1994; Gosline et al., 1986, 1999). Capture spiral (flagelliform) silk has a lower tensile strength than dragline silk but can be elongated well over 200% before breaking (Vollrath and Edmonds, 1989; Köhler and Vollrath, 1995). The extreme extensibility of flagelliform silk makes it one of the stretchiest proteins known.

Spider silk proteins

Until recently, spider silk proteins were characterized mainly by their amino acid compositions (e.g., Andersen, 1970; Work and Young, 1987). Liquid proteins from silk glands or spun silk fibers were hydrolyzed and the relative amounts of individual amino acids were determined by high-pressure liquid chromatography. These analyses showed that spider silks are rich in the four amino acids glycine, alanine, serine and proline (Fig. 1). They also revealed that each type of silk gland has a unique amino acid composition.

The discovery of the first spider silk cDNA was reported just 10 years ago (Xu and Lewis, 1990). This cDNA, major ampullate spidroin 1 (MaSp1), encodes a dragline protein. Since 1990, portions of other members of the spider silk gene family and a few alleles within species have been characterized (Tab. 2). These cDNA and genes were isolated from species of *Nephila* (Tetragnathidae) and *Araneus* (Araneidae). While these taxa are all within the superfamily Araneoidea, they are not closely related (Griswold et al., 1998). Thus, shared features between *Nephila* and *Araneus* silk sequences have either been maintained after divergence from a distant common araneoid ancestor at least 120 million years ago (Selden, 1989) or are the result of convergence.

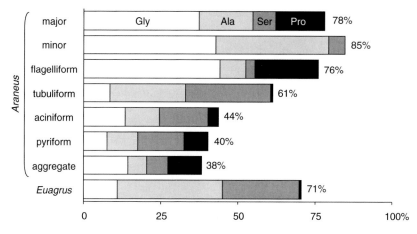

Figure 1. Four amino acids dominate the seven silks of the orb-weaver *Araneus diadematus* (Andersen, 1970) and the one silk of the mygalomorph *Euagrus* (Palmer, 1985). The percentages of glycine, alanine, serine and proline are graphed and the total percentages of these amino acids in the silk are shown.

Table 2. Genbank accession number, species, and gene name for all published spider silk cDNA and gene sequences

Accession	Species	CDNA/Gene	Reference
M37137	N.c.	MaSp1	Xu and Lewis, 1990
U03848	N.c.	MaSp1	Beckwitt and Arcidiacono, 1994
U37520	N.c.	MaSp1	Beckwitt et al., 1998
U20329	N.c.	MaSp1	Beckwitt et al., 1998
M92913	N.c.	MaSp2	Hinman and Lewis, 1992
AF027735	N.c.	MiSp1	Colgin and Lewis, 1998
AF027736	N.c.	MiSp2	Colgin and Lewis, 1998
AF027737	N.c.	MiSp2	Colgin and Lewis, 1998
AF027972	N.c.	Flag	Hayashi and Lewis, 1998
AF027973	N.c.	Flag	Hayashi and Lewis, 1998
AF218621	N.c.	Flag	Hayashi and Lewis, 2000
AF218622	N.c.	Flag	Hayashi and Lewis, 2000
AF218623	N.m.	Flag	Hayashi and Lewis, 2000
AF218624	N.m.	Flag	Hayashi and Lewis, 2000
U20328	A.b.	MaSp2	Beckwitt et al., 1998
U03847	A.b.	MaSp2	Beckwitt and Arcidiacono, 1994
U47853	A.d.	ADF-1 (= MiSp)	Guerette et al., 1996
U47854	A.d.	ADF-2	Guerette et al., 1996
U47855	A.d.	ADF-3 (= MaSp2)	Guerette et al., 1996
U47856	A.d.	ADF-4	Guerette et al., 1996

Species names are abbreviated (N.c. = *Nephila clavipes*, N.m. = *Nephila madagascariensis*, A.b. = *Araneus bicentenarius*, and A.d. = *Araneus diadematus*)

The genetic and protein data suggest that each silk gland expresses its own set of silk genes. Most of the characterized silk genes (MaSp1, MaSp2, ADF-3, ADF-4; Tab. 2) are transcribed in the major ampullate silk glands. MaSp1 and MaSp2 are two components of the dragline silk from *Nephila* while ADF-3 and ADF-4 are expressed in *Araneus*. ADF-3 appears to be an orthologue of MaSp2 and will be referred to as *A.d.* MaSp2. Minor ampullate silk genes (MiSp1, MiSp2, ADF-1; Tab. 2) have also been found in both *Nephila* and *Araneus*. ADF-1 is an orthologue of the *N.c.* MiSp genes and will be referred to as *A.d.* MiSp.

The other known silk genes are ADF-2 and Flag. ADF-2 is expressed in the tubuliform (egg-sac) glands but is likely to be only a small component of the egg sac silk proteins (Guerette et al., 1996). The amino acid composition profile of tubuliform glands shows a high level of serine and a modest amount of glycine (Fig. 1). However, the predicted product of ADF-2 is predominantly glycine with only a small fraction of serine. Because of this mismatch in amino acid profiles, ADF-2 alone cannot account for the composition of tubuliform silk. In contrast, the predicted product of Flag appears to be the dominant component of the flagelliform protein (Hayashi and Lewis, 1998).

Four amino acids dominate the silk proteins (Fig. 1). Three of these amino acids are encoded by triplets rich in guanine and cytosine (glycine = GGN; alanine = GCN; proline = CCN). Surprisingly, the nucleotide base compositions of the spider silk genes are only moderately skewed toward guanine and cytosine (55 to 61% G/C content; Fig. 2A). This composition is explained by a strong preference for codons containing adenine and thymine. The most dramatic bias is seen in the overwhelming preference of GG<u>A</u> and GG<u>T</u> codons for glycine (79 to 96%, Fig. 2B) instead of GG<u>C</u> and GG<u>G</u>. Similarly, the alanine codons GC<u>A</u> and GC<u>T</u> have a much greater frequency (65 to 94%, Fig. 2C) than GC<u>C</u> and GC<u>G</u>.

Examination of the codon usage patterns for glycine and alanine (Fig. 2B,C) shows that the silk genes have different biases even within the same species. Thus, there is no single factor, such as a GCA-rich tRNA pool in *Nephila*, that can explain the codon frequencies in the silk genes of a particular species. Perhaps gland-specific differences in tRNA availability or secondary structure requirements of individual mRNAs may be important for determining codon preferences.

Structural modules of silks

Northern blot analyses have shown that the spider silk gene mRNAs are very large. The transcripts range in size from 4.4 kilobases (kb) to 15.5 kb (Xu and Lewis, 1990; Guerette et al., 1996; Colgin and Lewis, 1998; Hayashi and Lewis, 1998). Comparison of the silk sequences (Tab. 2) reveals that all of the spider silk proteins share a common organizational scheme. The proteins are composed almost entirely of numerous repetitive elements that are flanked by

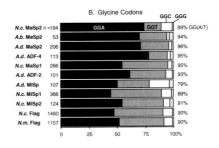

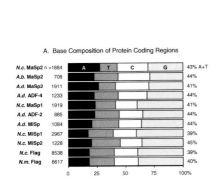

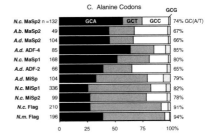

Figure 2. A. The nucleotide base compositions of the protein coding regions of spider silk genes are slightly guanine- and cytosine-rich. B. In the silk genes, codon usage for glycine is biased towards GGA and GGT. C. Codon usage for alanine is also biased toward codons ending with adenine and thymine. The number of protein coding basepairs or codons is indicated after each gene name. Species names are abbreviated as in Table 2. For *N.c.* MaSp1, the average of the nucleotide compositions/codon frequencies of three alleles (M37137, U37520, U20329) was used. *N.c.* MiSp2 was calculated using a combination of AF027736 and AF027737.

short regions of non-repetitive amino and carboxy-termini. The repetitiveness of the silk genes allows their characterization from partial gene sequences. Moreover, the repetitive portions of the 11 characterized silk proteins can be generalized as sets of consensus repeats (Fig. 3). These consensus repeats contain six types of recurrent amino acid motifs: poly-A, poly-(GA), GGX, GPGXX, GPX, and spacers. Spacers are long amino acid repeats that are atypical silk sequences because they are not glycine- or alanine-rich. The spacers in the minor ampullate silk genes are highly conserved between *N.c.* MiSp1 and *N.c.* MiSp2 (>90% identity) and have regions of high sequence identity between *Nephila* and *Araneus* (Colgin and Lewis, 1998). The spacers of Flag are nearly identical between *N. clavipes* and *N. madagascariensis* (Hayashi and Lewis, 2000). However, the MiSp spacers are not similar in sequence or length to the Flag spacers.

The repetitive amino acid motifs of the silk proteins are hypothesized to directly correspond to the mechanical properties of the silk fibers (Hayashi et al., 1999). For example, the poly-A regions and poly-(GA) sequences (Fig. 3) appear in β-sheet regions of major and minor ampullate silks (Parkhe et al., 1997; Simmons et al., 1994). Adjacent β-sheet regions are thought to link and

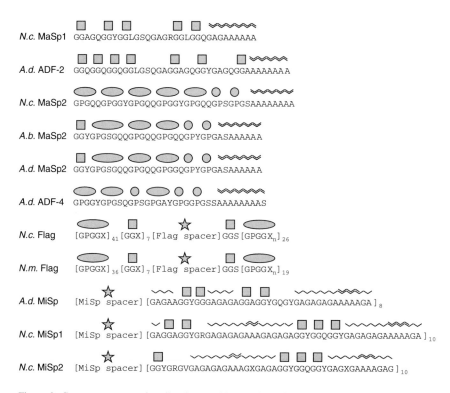

N.c. MaSp1   GGAGQGGYGGLGSQGAGRGGLGGQGAGAAAAAA

A.d. ADF-2   GGQGGQGGQGGLGSQGAGGAGQGGYGAGQGGAAAAAAAA A

N.c. MaSp2   GPGQQGPGGYGPGQQGPGGYGPGQQGPSGPGSAAAAAAAAA

A.b. MaSp2   GGYGPGSGQQGPGQQGPGQQGPYGPGASAAAAAA

A.d. MaSp2   GGYGPGSGQQGPGQQGPGGQGPYGPGASAAAAAA

A.d. ADF-4   GPGGYGPGSQGPSGPGAYGPGGPGSSAAAAAAAAS

N.c. Flag   [GPGGX]$_{41}$[GGX]$_{7}$[Flag spacer]GGS[GPGGX$_{n}$]$_{26}$

N.m. Flag   [GPGGX]$_{36}$[GGX]$_{7}$[Flag spacer]GGS[GPGGX$_{n}$]$_{19}$

A.d. MiSp   [MiSp spacer][GAGAAGGYGGGAGAGAGGAGGYGQGYGAGAGAGAAAAAGA]$_{8}$

N.c. MiSp1   [MiSp spacer][GAGGAGGYGRGAGAGAGAAAGAGAGAGGYGGQGGYGAGAGAGAAAAAGA]$_{10}$

N.c. MiSp2   [MiSp spacer][GGYGRGVGAGAGAGAAAGXGAGAGGYGGQGGYGAGXGAAAAGAG]$_{10}$

Figure 3. Consensus repeats describe the repetitive portion of each silk protein. The amino acid repeats within each translated silk gene were aligned and a consensus was determined if a majority of the repeats contained a particular amino acid. Six common motifs are depicted by symbols above each repeat: square = GGX; oval = GPGXX; circle = GPX; double zigzag = poly-A; single zigzag = poly-(GA); star = spacer. See text for definition of the spacer motifs. Subscripts denote the number of times the motif or set of motifs within parentheses is iterated within the consensus repeat. Species names are abbreviated as in Table 2.

form the crystalline areas of silk fibers. Presumably these are the regions that bind the individual protein molecules together and contribute to the extremely high tensile strength of silk fibers.

The GPGXX motif is also thought to contribute significantly to the mechanical properties of silks. Each GPGXX motif is hypothesized to form a type II β-turn, and tandem GPGXX motifs result in a series of β-turns (Hayashi and Lewis, 1998). These consecutive turns could form a structure similar to the β-spiral of elastin (Urry et al., 1995) and gluten proteins (Van Dijk et al., 1997). The β-spiral would function much like a spring, imparting elasticity to the silk fiber. Only the major ampullate and flagelliform silks are known to contain GPGXX motifs (Fig. 3), and they are the stretchiest of spider silks. As further support of the GPGXX motifs forming a spring-like structure, there is a correspondence between the number of tandem arrayed GPGXX motifs and the extensibilities of the two silks. Major ampullate silk, with up to 35% extension, has at most nine β-turns in a row, while flagelliform silk, with over 200% extension, has a minimum of 43 turns (Gosline et al., 1986; Hayashi and Lewis, 1998).

Phylogeny of silk proteins

The hypothesis that spider silk proteins are assembled from only a small set of common structural modules (Hayashi et al., 1999) will be tested as more silk genes are characterized. The diversity of spider silks produced within a spider (Tab. 1) and in different spider taxa provides an opportunity to directly relate specific DNA sequences to the unique functional properties of the silk fibers. While molecular biology techniques are used to discover the silk genes, biochemical methods can reveal protein structure, and mechanical tests are needed to assess the performance of the silk fibers, phylogenetic analyses are critical for discerning the transitions that have occurred in the evolution of silk. Cladistic methods can be applied to test hypotheses about the relationships among orthologous and paralogous silk genes and the conservation or convergence of structural modules.

Little comparative work has been done on the spider silk genes and proteins. Here I undertake a phylogenetic analysis of all the silk genes that have been reported thus far. Although the silk transcripts are several thousand nucleotide bases in length, only a fraction of the sequence information can be used in phylogenetic analyses. The primary limitation is that all of the available silk sequences (Tab. 2) represent partial-length genes. Also, because amino-terminal (leading) sequence is known for just the two Flag genes from *Nephila*, only the repetitive and carboxy-terminal (trailing) regions can be compared across members of the gene family.

Even if full-length genes were known, it would be difficult to align paralogous genes to each other because of the great differences in length of the repetitive regions. As an alternative, in this chapter, the consensus repeats (Fig. 3), that characterize entire gene arrays, were aligned to each other with the insertion of only a few gaps (Fig. 4A). This approach of using a single unit to represent many similar units is reminiscent of analyses that use one sequence to represent a multi-copy ribosomal RNA gene (e.g., Whiting et al., 1997). Although Flag is considered a member of the silk gene family, the extremely divergent Flag consensus repeats were not included in the multiple alignments. Numerous alignment ambiguities would be introduced by Flag because a single Flag repeat of ~440 amino acids (Fig. 3) is over seven times longer than the total aligned length of all the other consensus repeats (Fig. 4A).

The carboxy (C)-terminal region of the silk transcript encompasses the last ~100 amino acids before the stop codon. This region is devoid of repetitive motifs and, in general, is easy to align among the silk genes (Fig. 4B). No hypotheses have been proposed as to the possible function of the C-terminal region. However, the conservation of several sequence elements across paralogous genes suggests that the C-terminus does have an important role. Note that most of the alignment difficulties are between the Flag proteins and the other silks. Because of the extreme length of the *Nephila* Flag consensus repeats and the divergence of Flag C-termini from the other genes, the *N.c.* and *N.m.* Flag sequences were not included in cladistic analyses. The evolution of Flag, however, will be discussed in more detail below.

Phylogenetic analysis of the aligned proteins (Fig. 4) resulted in a single most parsimonious topology (Fig. 5). Because there is no outgroup, the tree was midpoint rooted. The root was placed on the node separating minor ampullate silk genes from the major ampullate and tubuliform silk genes. In future analyses, non-araneoid silk genes could be used as outgroup sequences,

A.

```
N.c. MaSp2   GPGQQGPGGYGPG--QQ---------GPGGYGPG--QQGPSGPGSAAAAAAAA--
A.b. MaSp2   -------GGYGPGSGQQ---------GPGQQGPG--QQGPYGPGASAAAAAA---
A.d. MaSp2   -------GGYGPGSGQQ---------GPGQQGPG--GQGPYGPGASAAAAAA---
A.d. ADF-4   -----GPGGYGPGS--Q---------GPS--GPG--AYGPGGPGSSAAAAAAAAS
N.c. MaSp1   --GGAGQGGYG--------------GLGSQGAG--RGGLGGQGAGAAAAAA---
A.d. ADF-2   --GGQ--GGQG-GPG----------GLGSQGAGGAGQGGYGAGQGGAAAAAAAA
A.d. MiSp    GAGAA--GGYGGGAGAGAG--------GAGGYG---QG-YGAGAGAGAAAAAGA
N.c. MiSp1   GAGGA--GGYGRGAGAGAGAAAGA--GAGAGGYG--GQGGYGAGAGAGAAAAAGA
N.c. MiSp2   -------GGYGRGVGAGAGAGAAAGXGAGAGGYG--GQGGYGAGXGAAAAGAG--
```

B.

```
N.c. MaSp2   SRLASPDSGARVASAVSNLVSSGPTSSAALSSVISNAVSQIGASNPGLSG
A.b. MaSp2   SRLSSSAASSRVSSAVSSLVSSGPTTPAALSNTISSAVSQISASNPGLSG
A.d. MaSp2   SRLSSPAASSRVSSAVSSLVSSGPTKHAALSNTISSVVSQVSASNPGLSG
A.d. ADF-4   S?LSSPAASSRVSSAVSSLVSSGPTNGAAVSGALNSLVSQISASNPGLSG
N.c. MaSp1   SRLSSPQASSRVSSAVSNLVASGPTNSAALSSTISNVVSQIGASNPGLSG
A.d. ADF-2   SRLSSPSAAARVSSAVSLVSNGGPTSPAALSSSISNVVSQISASNPGLSG
A.d. MiSp    NRLSSAGAASRVSSNVAAIASAGA---AALPNVISNIYSGVL-SS-GVSS
N.c. MiSp1   SRLSSAEASSRISSAASTLVSGGYLNTAALPSVISDLFAQVGASSPGVSD
N.c. MiSp2   SRLSSAEACARISAAASTLVS-GSLNTAALPSVISDLFAQVSASSPGVSG
N.c. Flag    SRV--PDMVNGIMSAMQGSGFNYQMFGNMLSQYSS----GSGTCNPNNVN
N.m. Flag    SRV--PDMVNGIMSAMQGSGFNYQMFGNMLSQYSS----GSGSCNPNNVN
```

```
N.c. MaSp2   CDVLIQALLEIVSACVTILSSSSIGQVNYGAASQFAQV-VGQSVLSAF-*
A.b. MaSp2   CDVLVQALLEVVSALVHILGSSSVGQINYGASAQYAQM-V?????????*
A.d. MaSp2   CDVLVQALLEVVSALVSILGSSSIGQINYGASAQYTQM-VGQSVAQALA*
A.d. ADF-4   CDALVQALLELVSALVAILSSASIGQVNVSSVSQSTQM-ISQ----ALS*
N.c. MaSp1   CDVLIQALLEVVSALIQILGSSSIGQVNYGSAGQATQI-VGQSVYQALG*
A.d. ADF-2   CDILVQALLEIISALVHILGSANIGPVNSSSAGQSASI-VGQSVYRALS*
A.d. MiSp    SEALIQALLEVISALIHVLGSASIGNVSSVGVNSALNA-VQNAVGAYAG*
N.c. MiSp1   SEVLIQVLLEIVSSLIHILSSSSVGQVDFSSVGSSAAA-VGQSMQVVMG*
N.c. MiSp2   NEVLIQVLLEIVSSLIHILSSSSVGQVDFSSVGSSAAA-VGQSMQVVMG*
N.c. Flag    --VLMDALLAALHCLSNH-GSSSFAPSPTPAAMSAYSNSVGRMF--AY-*
N.m. Flag    --VLMDALLAALHCLSNH-GSSSFAPSPTPAAMSAYSNSVGRMF--AY-*
```

Figure 4. A. The alignment of silk protein consensus repeats is shown with amino acids indicated by their one-letter abbreviations. Alternative alignments are possible but yield similar phylogenetic results. Alignment gaps are shown as dashes. B. The alignment of the carboxy-terminal regions is depicted. Species names are abbreviated as in Table 2. A minor alteration was made in the Genbank entry for *A.d.* ADF-4. See Hayashi and Lewis (1998) for an explanation of the change. Missing data are indicated by "?".

depending on the pattern of gene duplication. Assuming the rooting is correct, the tree shows that the major ampullate silk proteins form a clade with MaSp1 as the sister group to ADF-4 and the clade of MaSp2 orthologues. The minor ampullate silk proteins also form a group with the two *Nephila* MiSp sequences clustered within that group. The topology suggests that the GGX motifs and the crystalline poly-A motifs are primitive, the elasticity module GPGXX is derived within major ampullate silk proteins, and the poly-(GA) motif is restricted to minor ampullate silks (Fig. 5).

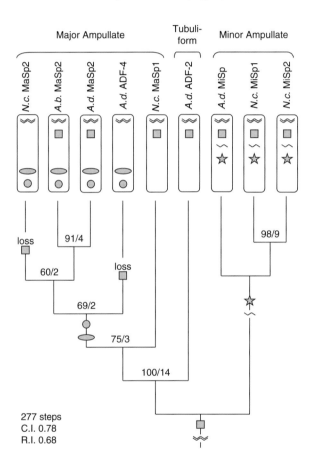

Figure 5. Parsimony analysis of the aligned silk protein amino acids (consensus repeats and carboxy-terminal regions) results in a single, most parsimonious topology. A branch and bound search was done with PAUP* (Swofford, 2000). Gaps were treated as another amino acid. Bootstrap percentages (Felsenstein, 1985a)/Bremer support (Bremer, 1994) are shown next to internodes. Bootstrap values were calculated by performing 10,000 branch and bound replicates, excluding uninformative characters. The symbols defined in Fig. 3 depict the presence or absence of common amino acid motifs in the silk sequences. The most parsimonious evolution of the common motifs is mapped onto the topology. The optimization of the GGX motif is ambiguous and one set of transformations is shown. The tree is midpoint rooted but a root has been added to indicate the possibly primitive origin of GGX and poly-A motifs. Species names are abbreviated as in Table 2.

As new members of the silk gene family are discovered and more taxa are surveyed for orthologous genes, phylogenetic analysis will be a powerful tool to develop and test hypotheses regarding the origin, conservation and convergence of structural modules. Most likely, additional structural modules will be discovered because genes are unknown for half of the araneoid silk proteins (Tab. 1) and a non-araneoid silk gene has yet to be reported. These uncharacterized silk proteins have their own suites of mechanical properties. As an extreme example of a divergent protein, aggregate silk, the adhesive glue that coats the capture spiral of an orb-web, is a mucilage rather than a fiber and, unlike all other silks, is wet and viscous. With only 38% of the aggregate silk protein being glycine, alanine, serine, or proline (Fig. 1), aggregate silk must be composed of a variety of novel structural motifs.

The characterization of additional orthologues for the known silk genes (Tab. 2) will also provide much insight into protein structure and function. Thus far, the spiders that have been studied have similar ecologies in one important aspect: they all weave orb-webs. However, one of the themes that has emerged through systematic study is that the orb web architecture has been modified, reduced, or lost many times (Griswold et al., 1998). In fact, most spiders within the superfamily of araneoid "orb-weavers" do not actually construct orb-webs. When spiders utilize different web architectures, the performance demands on their silks also change, and these changes may be reflected in the DNA sequences of the various silk protein genes. As the dataset of silk genes is expanded, combined analyses can be performed with morphological, behavioral and other molecular data. These diverse sources of evidence will allow the evolution of the silk proteins to be mapped onto the phylogeny of spiders.

Molecular evolution of MaSp1

Phylogenetic analyses can also be used to gain an understanding of the evolution and maintenance of silk protein sequences within a gene. While consensus repeats can be used to characterize silk genes (Fig. 3), the reduction of a silk gene into a single consensus repeat obscures much of the variation that is present among individual repeats within a particular gene. For example, the number of alanine residues in a poly-A region or the number of GGX motifs might vary among neighboring repeats. Despite the variability present within all the known silk genes, the sequences are remarkably well homogenized. Thus, any MaSp2 repeat from *Araneus diadematus* is always distinct from a MaSp2 repeat from *Nephila clavipes*. Exactly how sequences are maintained within a species or how all the repeats in one species diverge from the repeats in another species is not known. It may be that the molecular characteristics of the silk genes themselves affect the generation and spread of sequence variation. Factors such as biased base composition, replication slippage, gene conversion, and unequal crossing-over events could arise because silk genes are unusually long and highly internally repetitive (Hayashi and Lewis, 2000).

Beckwitt et al. (1998) compared two alleles of *N.c.* MaSp1 to the original isolate of *N.c.* MaSp1 (Xu and Lewis, 1990). Beckwitt et al. (1998) showed that most of the allelic differences are sequence rearrangements and base substitutions in the repetitive region of the genes. Very few changes were found in the C-terminal region.

Parsimony analysis provides further insight into the evolution of these alleles. If alleles diverge primarily by point mutations and localized insertion and deletion events, then it is predicted that repeats in corresponding regions of the genes would tend to be more similar to each other than to other repeats. This scenario represents the typical way that most genes evolve. However, MaSp1 has a very distinct molecular architecture compared to most genes. In essence, the individual MaSp1 repeats are serial homologues. If recombination is prevalent and rearranges elements within an array, then the repeats in different alleles are not expected to group by corresponding position. If homogenization has occurred among repeats, then all of the repeats from an individual allele would group to the exclusion of repeats from other alleles (concerted evolution, Dover et al., 1993).

I aligned and analyzed the repeats from the *N.c.* MaSp1 alleles (Fig. 6). The strict consensus of the 45 shortest trees shows that the four repeats closest to the non-repetitive C-terminal region are the only repeats to group by position among alleles. For example, the three repeats labeled "1" form a clade to the exclusion of other repeats. The fifth repeats from two of the alleles also group together. The remaining repeats are intermingled with each other on the tree. The rearrangement of repeats among the alleles suggests that shuffling has occurred, perhaps by recombination events. The pattern of repeats near the end of a tandem array being more conserved than centrally located repeats is in accord with some models of concerted evolution (e.g., Lassner and Dvorak, 1986). These models are based on the expectation that terminal regions of a repetitive locus will have a lower probability of recombination than the interior portion. This is consistent with the finding that the last four MaSp1 repeats are conserved between alleles, and it is predicted that repeats toward the 5' end of the gene will also be highly conserved.

Homogenization of Flag

Until recently, spider silk gene sequences were characterized only from cDNA and PCR-amplified DNA fragments. The first genomic DNA λ clones that encompass both 5' and 3' ends of a silk gene and substantial portions of the intervening regions were reported for Flag from two species of *Nephila* (Hayashi and Lewis, 2000). The Flag locus spans about 30 kb, half of which is exonic sequence. The coding sequence is divided into 13 exons and exons 3–13 were found to each encode a single repeat (see consensus repeats,

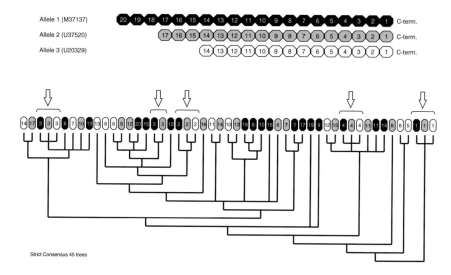

Figure 6. The strict consensus of the most parsimonious trees that result from analysis of aligned repeats from alleles of *Nephila clavipes* MaSp1. Individual repeats were numbered starting from the C-terminus. Phylogenetic analysis was performed on the inferred amino acid sequences of the aligned repeats with PAUP* (Swofford, 2000) using a heuristic search of 2000 random taxon addition replicates and tbr branch swapping. Gaps were treated as a fifth character state. There were 45 equally short trees of 314 steps. Repeats from M37137 are shown in black, repeats from U37520 are in gray, and repeats from U20329 are in white. Arrows point to the groupings of allelic repeats 1, 2, 3, 4 and 5.

Fig. 3). The repeated exons are separated by introns that also share remarkably high similarity within a species (Hayashi and Lewis, 2000). These introns give Flag an additional repetitive element beyond its encoded amino acid repeats with subrepeating structural motifs (Fig. 3).

As with the *N.c.* MaSp1 allelic repeats, phylogenetic analyses were used to test alternative hypotheses for the evolution of the repetitive Flag exons and introns. The gene elements may have independently diverged from an ancestral gene possessing the tandem array of similar exons and introns. Alternatively, the gene elements may have homogenized within each species. In the first scenario, exons and introns are expected to group by corresponding position (e.g., a pairing of exons 5 from *N. clavipes* and *N. madagascariensis*). In the second scenario, exons and introns would tend to cluster by species (e.g., a group containing exons only from *N. clavipes*) similar to the VERL repeats in abalone (Swanson and Vacquier, 1998).

The repetitive exons, of which each encodes a single repeat of ~1320 bases, were aligned and analyzed using parsimony (Fig. 7A). Between species, only exons 3 and 13 of *N. clavipes* pair with the corresponding exons of *N. madagascariensis*. In contrast, the other repeated exons tend to group by species rather than by position. An additional 605 steps (23% increase in tree length) are required to force all the exons to pair by location within the Flag gene. These analyses reveal that the coding regions of Flag have homogenized, a pattern that is indicative of concerted evolution (Dover et al., 1993). The conservation of sequence in the terminal repetitive exons (exons 3 and 13; Fig. 7A) between species is reminiscent of the conservation in terminal repeats of MaSp1 alleles (Fig. 6), and shows that homogenization of the terminal exons in silk genes is not rapid.

Analysis of the repetitive introns, each ~1420 bases in length, results in even stronger evidence for intragenic homogenization (Fig. 7B). These introns show greater similarity within species (average of 87% identity) than the repeated exons. In some cases (N.c. 5–7), the repeated introns are 99.9% similar. All of the introns within one species of *Nephila* form a clade to the exclusion of introns from the other species. To make the introns pair by corresponding positions within the Flag gene requires an additional 801 steps, a 69% increase in length from the shortest trees.

The parsimony results can be used to reject the hypothesis that the Flag gene elements are evolving in a typical pattern of independent divergences from a common ancestor of *N.c.* and *N.m.* Instead, the parsimony analyses indicate that sequence homogenization has occurred within Flag (Fig. 7). Because the sequences are homogenizing within a species, the parsimony trees cannot be interpreted as representing the phylogeny of the exons or introns. Rather, the parsimony analyses support the hypothesis that the molecular architecture of the Flag gene promotes within-gene concerted evolution. Perhaps through some combination of gene conversion and unequal crossing-

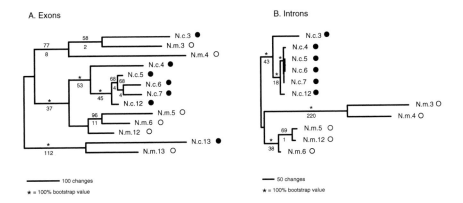

Figure 7. A. The aligned nucleotides of the repeats encoded by the Flag exons (Hayashi and Lewis, 2000) were analyzed using the branch and bound parsimony search of PAUP* (Swofford, 2000). This is one of two shortest trees (2609 steps) that was chosen after successive approximations (Farris, 1969). The other tree differed by the placement of N.m.12 as the sister group of ((N.c.4–N.c.7, N.c.12), (N.m.5, N.m.6)). B. Parsimony analysis of the aligned introns resulted in three trees (1158 steps) that differed only in the resolution of N.c.5, N.c.6, and N.c.7. Gaps were treated as a fifth character state. Species names are abbreviated as in Table 2 and the exon or intron number follows each species designation. Bootstrap values (Felsenstein, 1985a) are shown above and Bremer support (Bremer, 1994) below the internodes. Bremer support was calculated using equal weighting of characters. Branch lengths are proportional to the number of character changes.

over at the internally repetitive exons, the Flag protein remains fairly homogenized. However, the exons, with an exorbitant amount of repetitive sequence, contain potential hotspots for recombination and replication slippage (Mita et al., 1994; Beckwitt et al., 1998). An extreme mutation rate within individual exons may prevent the repeated exons in Flag from completely homogenizing (Fig. 7A). The introns, which are not internally repetitive, lack the mutational hotspots present in the coding sequence and are more efficiently homogenized than the exons (Fig. 7B).

It could be argued that the exons and introns within a species are homogenized through differential selection between species. However, several lines of evidence are inconsistent with this scenario. First, the repetitive exons in the two species contain the same motifs and have nearly identical consensus repeat patterns (Fig. 3). Second, because both *Nephila* species use flagelliform silk for the construction of symmetrical, planar orb-webs, it is difficult to envision differential selection pressures on their silk proteins. Third, the introns within a species share as much as 99.9% similarity. To argue for selection alone homogenizing the introns requires the convergent evolution of hundreds of nucleotide changes among the introns of the two species. Instead, recombination and/or gene conversion events are more likely to have homogenized the coding and non-coding sequences within a species.

The work on Flag suggests that there may be an unusual interaction of forces operating on the silk genes of spiders. The organization of the silk proteins into sets of iterated consensus repeats (Fig. 3) is evidence that the proteins may require modularity to self-assemble into fibers and to endow the fibers with specific mechanical properties (Hayashi et al., 1999). In addition, the available cDNA and genomic DNA data (Tab. 2) show that the genes encoding these modular proteins are also modular with repeating and subrepeating sets of nucleotides. Given that the silk gene DNA sequences exhibit a pattern of concerted evolution (Hayashi and Lewis, 2000), a beneficial variant repeat that arises could spread rapidly throughout the gene and result in enhanced fiber performance. Alternatively, natural selection on silk fibers could be hindered by the extreme mutation rate of the highly repetitive silk coding sequences. Spider silk genes could be viewed as a genetic battlefield where pressure to maintain functional silk fibers is opposed by frequent mutations, replication errors, and unequal crossing-over events in the highly repetitive DNA sequences encoding the modular proteins.

Future directions

With the discovery of the first spider silk gene (Xu and Lewis, 1990) and the subsequent publication of ten other silk genes (Tab. 2), the silk protein family has become increasingly amenable to comparative studies. Cladistic methods can be used to investigate the evolution of repetitive sequences within a silk gene, as shown with MaSp1 (Fig. 6) and Flag (Fig. 7). At a higher level, phylogenetic analyses can discern evolutionary links among paralogous silk proteins that perform different ecological tasks (Fig. 5). But these analyses provide just a preliminary glimpse at the evolution of spider silks. Future research will pursue three major objectives. First, silk genes need to be cloned from other species of spiders. Currently, orthologous silk gene sequences are known from at most three species, a scant fraction of spider diversity. Second, new members of the silk gene family are awaiting discovery. Less than half of the araneoid silk proteins (Tab. 1) have been well characterized by genetic data, and nothing is known about the DNA sequences that encode silk proteins spun by non-araneoid spiders. Finally, more silks need to be described by mechanical tests and biophysical studies. Mechanical test results reflect the effects of changes in protein sequence on fiber performance, and biophysical data reveal the conformation of the repetitive amino acid motifs in the silk fiber. All of these data are essential for correlating the evolutionary diversification of silk protein sequences with changes in the function of silk fibers.

Acknowledgements
I thank A. de Queiroz, J. Gatesy, R. Lewis and J. Thornton for thoughtful comments that improved the manuscript. This work was supported by grants from NSF (MCB-9806999) and ARO (DAAG55-98-1-0262).

Molecular Systematics and Evolution: Theory and Practice
ed. by R. DeSalle, G. Giribet and W. Wheeler
© 2002 Birkhäuser Verlag/Switzerland

Comparative methods and evolution

Alfried P. Vogler[1,2] and Andy Purvis[2]

[1] *Department of Entomology, The Natural History Museum, London SW7 5BD, UK*
[2] *Department of Biology, Imperial College at Silwood Park, Ascot, Berkshire SL5 7PY, UK*

Summary. Comparative biology provides a framework for the study of evolution, by seeking answers to the question of *why* traits evolved. However, the difficulties of making inferences about the biological causes of trait variation and covariation resulted in the development of very different approaches to comparative analyses, which are outlined and contrasted here. The Homology approach differs from all other approaches as it attempts to explain historical uniques, whereas the various homoplasy approaches draw inferences from repeated origination of correlated traits. A distinction is made between those approaches deriving information from correlated transitions of characters on a clado-gram (Homoplasy I) and those that simply extract correlated differences between pairs of phyloge-netically independent taxa (Homoplasy II). The latter approach has been implemented in well-devel-oped analytical procedures ("comparative methods") which, however, are principally non-historical and thus their inferences about evolution are indirect. In contrast, the study of correlation of character transitions seeks to explain trait variation as the outcome of evolution, but is hampered by the diffi-culty of reconstructing ancestral character states, a prerequisite for the analysis of correlated transi-tions. Model-based techniques for ancestral state reconstruction are being developed, but the lack of data contained in single characters limits the biological reality of parameter-intensive models.

Introduction

Molecular systematics data form the basis for modern comparative analysis. Whereas phylogenies depict the hierarchical relationships of groups, compar-ative biology goes beyond the descriptive summary of these hypotheses of relationships and attempts to understand *why* evolution took a certain path. The possibility to gain causal explanations of evolution has frequently been a pri-mary motivation for the application of molecular systematics, and therefore tight links between this discipline and comparative analysis have developed. "Comparative methods" are the explicit analytical procedures to implement the principles of comparative biology in our quest to learn about evolution. While the principal aims of comparative biology in elucidating the traits of organisms and their function are obvious and universally accepted, the prem-ises and scope of comparative approaches are highly contentious. This primer attempts to illustrate the different methodologies and their underlying assump-tions, possibilities and limitations.

The kinds of questions

Comparative approaches are the basis for "understanding" biological phenomena. Singular observations about the living world are largely meaningless if they cannot be replicated and manipulated in an experimental setting or, more pertinent to the discussion here, if they cannot be compared to independent observations made on other organisms. While comparisons are fundamental to learning about why the organisms are the way they are, they suffer from the problem of being non-historical, i.e., the explanations for our observations are not sought in the context of the evolutionary history during which the traits in extant species were acquired. However, if we ask why particular organisms are performing in a certain way, and why they have particular traits that others don't have, it would be shortsighted to simply explain the observations in the light of current function only (Coddington, 1988). Comparative approaches to biology have therefore extensively incorporated historical (phylogenetic) information.

Explaining the traits and their function therefore requires, first, to answer the question of "what happened?" This perspective makes phylogenetic trees the centerpiece of comparative methods. These trees show the relationships among organisms and—leaving aside the uncertainties of their estimation from living or fossil specimens—can be used to read off the presumed cladogenetic events in the history of a lineage. But as much as the traits of living organisms are the primary data for inferring these consecutive events of lineage separation, they are themselves the object of historical reconstruction (Coddington, 1988; Greene, 1986). In the tree, all taxa are connected to the branching points which define where trait variation begins an independent trajectory for the two sister lineages. Up to the branching point, which represents the hypothetical ancestor that gave rise to both lineages, their traits did not diverge, and comparisons between taxa should not produce any differences. Comparative approaches, therefore, concern themselves with the traits that changed after divergence, i.e., the traits uniquely derived in independent sister groups.

With the phylogenetic tree in hand we can then ask more detailed questions about the trees, the characters, the type and frequency of changes, and the correlation among these changes: which traits changed and how often did change occur? How did traits change (what are the differences between states)? In which order have traits changed? What were ancestors like? From there, it is a small step to ask the questions about the biological causes of these character changes. For example, knowing the sequence of change in two traits, it could be tested whether a particular change occurs in response to another. If there is a hypothesis that a particular change is advantageous under certain circumstances (a "selective regime"), we need to establish that this trait arose only after the lineage had been exposed to that regime, i.e., only a certain sequence of character change is consistent with the hypothesis (Baum and Larson, 1991). If we accept this view, the origin of a trait is explained as the result of an adaptation for better performance under a particular selective pressure.

Traits, in turn, affect the evolution of the lineage. The traits of organisms may determine the evolutionary success and persistence of lineages and affect the total number of species in a clade. "Key innovations" are traits which we presume to cause an increase in species numbers relative to the sister group lacking that trait (Cracraft, 1990; Heard and Hauser, 1995). While it may be difficult to establish the causal correlation of any given trait with clade size, the more general question remains: have traits shaped phylogeny, and, if so, which traits? Inevitably, the traits interact with the environment, and they determine which lineages (and their traits) proliferate or perish. On the other hand, traits are shaped by the interaction with the environment and other lineages of organisms. How have evolved traits and the environment interacted to produce ecological patterns? How have the respective changes in traits of interacting lineages produced coevolutionary patterns? Obviously, the phenomena are complex and not easily dissected to separate cause and effect. Comparative methods are an attempt to establish the macroevolutionary patterns and their causes.

The epistemological framework

Comparative methods thus can be applied to a wide range of questions about the living world but the difficulties of inference are formidable. In studies of evolution, as in any historical science, the process we attempt to assess is generally not open to direct observation. Comparative biology has produced two basic approaches to make inferences, commonly referred to as the homology approach and the homoplasy approach (Coddington, 1994). In the following, we will discuss each approach and will point out that the homoplasy approach refers to two conceptually rather different types of procedures which we will term Homoplasy I and Homoplasy II.

Homology

The homology approach is specifically concerned with "what happened" in the evolution of lineages. It relies on the study of a single inferred evolutionary event, and draws inferences from the sequence of correlated character changes. As an example we consider the evolution of phenological cycles in tiger beetles (genus *Cicindela*). All species in the temperate North American fauna can be classified into one of two types of phenologies, either exhibiting a short adult activity period in the middle of the summer, when reproduction takes place, followed by overwintering in the larval stage, or with a biphasic adult activity period in the fall and spring (overwintering as adults) and larval development in the summer (Knisley and Schultz, 1997; Vogler and Goldstein, 1997). About half of all species exhibit the Spring-Fall biphasic cycle, and these species form a well-supported monophyletic group derived with respect to the Summer cycle species. Interestingly, there is a strong correlation with their geographic distribution, as the Spring-Fall species predominate in northern latitudes and higher altitudes. So, has the Spring-Fall cycle (or adult overwintering) arisen in response to harsh climatic conditions, i.e., does it represent an adaptation to this environment? Because all species of this ecological type are a monophyletic group, the correlation of Spring-Fall cycle and northern distributional ranges can hardly be considered to be independent and thus the wide occurrence of the trait among the northern species does not constitute evidence for its adaptive value in the historical sense. Yet the trait may be adaptive in that it enhances the species's fitness and reproductive success, no matter what its historical status.

Thus, if we want to know why living things have certain traits, this question is different from the question of whether or not a trait is useful for these organisms (the trait's current utility *sensu* Baum and Larson, 1991). Under the most stringent application of the homology approach we require evidence that the trait is of current utility and that it has arisen due to the selective regime in question. This hypothesis would be corroborated by an inferred sequence of trait evolution where the transition to higher latitude precedes the origin of the Spring-Fall cycle (Fig. 1a). In a preliminary phylogenetic analysis (Vogler and Goldstein, 1997) this sequence of character transitions was in fact confirmed, consistent with the hypothesis.

If, under the homology approach, we use the historical genesis of traits as the primary evidence to answer "why" questions, however, we are left with interpreting singular events. As pointed out by Coddington (1994), homologies are logical individuals, presenting the problem of making inferences based on historical uniques. In the case of the Spring-Fall phenology, all of the 30+ species are members of a single clade and, although the character transitions may be consistent with the hypothesis in question, the corroboration would require that it holds up to a more detailed hypothesis, providing "replication" analogous to frequency in statistics. (These details could be, e.g., that the early larval stages are particularly sensitive to cold, or to a particular sequence of events in the paleontological record, etc.). However, under the paradigm of the homology approach, inferences drawn from independent origination of the variables in question, e.g., to investigate the origination of the large number of additional Spring-Fall species in the Palaearctic and the observation of the same sequence of evolutionary events, do not lend any more support (or less, if other cases are inconsistent) to the conclusions regarding a particular event. In the words of Wenzel and Carpenter (1994), "it happened", but why it happened has to be established from ancillary, circumstantial evidence surrounding this single event.

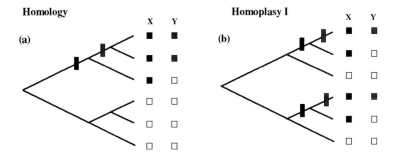

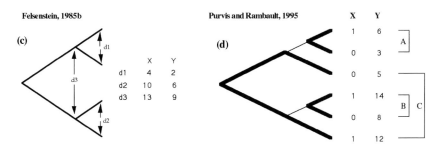

Figure 1. The principle of various approaches to comparative biology, based on hypothetical phylogenies and character distribution. (a) The homology approach implies a unique transition in trait X followed by a transition in trait Y, the sequence of which is consistent with the hypothesis. (b) Homoplasy I derives information from several correlated transitions consistent with an existing hypothesis. (c) Homoplasy II as implemented by the independent contrast methods of Felsenstein (1985b); differences (d1–d3) in (continuous) variables X and Y are determined in pair-wise comparisons indicated by the arrows. These differences can then be regressed to test for correlation. (d) Homoplasy II according to Purvis and Rambault (1995) is tested in a comparison of phylogenetically independent taxa that differ in the value for (discrete) character X, as indicated by fat lines and tested for correlation with the value in Y; in all three pair-wise comparisons possible in this example (A, B, C) the values for Y are higher when the state for X is 1.

Homoplasy I

In contrast to the homology approach, all other approaches to comparative analysis accept that evolutionary events can be assigned to logical classes (Coddington, 1994). To introduce another example, species of *Cicindela* exhibit several traits that presumably protect them from predation, such as aposematic body coloration, camouflage and chemical defenses. Traits involved in predator evasion are presumably under strong selection, as they exemplify in the most blatant way the Darwinian paradigm which would predict that organisms should invariably attempt to avoid predation. In the genus *Cicindela*, these presumably adaptive traits are highly variable. Is there any way of determining why certain species exhibit particular anti-predator traits and others don't? In this case, the multiple transitions in the relevant traits provide the opportunity for hypothesis testing from repeated occurrences of character changes, i.e., based on "homoplasy" (Fig. 1b). If we have specific hypotheses about the presence of certain traits in a set of species, e.g., particular types of coloration in a given habitat type or predator environment (the selective regime) (Pearson et al., 1988), these can be tested by predicting a sequence of character transitions consistent with this hypothesis. For example, body coloration can be implicated in camouflage to prevent detection by predators if the inferred transition in coloration from white to iridescent-green elytral coloration is preceded by a habitat shift from dune habitat to forest dwelling, but not if the transitions occur in the reverse order (Vogler and Kelley, 1996). When this approach was applied to the analysis of various aposematic and cryptic elements of the body coloration and chemical defenses, however, it did not reveal any consistent correlation of defensive-trait variation and habitat association (Vogler and Kelley, 1998), rejecting the long-held view (Pearson et al., 1988) that certain habitat types impose a consistent selective environment to which beetles adapt.

This example illustrates the repeated occurrence of particular character transitions, and thus a "homoplasy approach", whereby multiple occurrences of particular transitions might strengthen the inferences about the adaptive nature of traits. In this sense it is an extension of the homology approach that seeks the reconstruction of unique historical events (Arnold, 1994; Vogler and Kelley, 1998). In contrast to the Homology approach, which draws inferences from historical singularities, here transitions are viewed as members of classes. Importantly, this implementation of the homoplasy method attempts to draw inferences explicitly from the distribution of character transitions on the cladogram. Homoplasy approaches that rely on the analysis of phylogenetic reconstruction of character changes we term Homoplasy I, or the Transition approach.

Homoplasy II

The study of homoplastic character transitions differs from the original implementation of the homoplasy approach to comparative biology, which had been pioneered by evolutionary biologists focusing on the statistical correlation of biological traits (Felsenstein, 1985b; Harvey and Pagel, 1991). These homoplasy approaches operate on the premise that the statistical correlation of traits provides a way of testing hypotheses about their biological function. Correlation can also be used to generate hypotheses in exploratory analyses, on the assumption that the biological relevance of traits of interest may reveal itself from their statistical interdependencies. Either approach—hypothesis testing or hypothesis generation—is principally a non-historical exercise that explores the variation among organisms for such interdependencies of traits.

Rather than focus on character transitions (which requires the mapping of traits onto phylogenies, and the estimation of ancestral character states), the approach focuses on the *differences* between taxa. Replicated differences between phylogenetically independent pairs of taxa provide the data points from which the correlation is assessed: the phylogeny is used as a framework for constructing the comparisons so as to ensure their independence. This approach to use homoplastic variation based on differences between pairs of phylogenetically independent groups we call Homoplasy II, or, the Difference approach.

Phylogenetically independent comparisons were originally intended for use with continuous variables. Put simply, differences in two variables, Y and X, between sister lineages are computed and tested for whether the differences in Y are correlated with those in X. A sign test is used to assess the null hypothesis of no correlation, which predicts that positive and negative associations will be found equally (Felsenstein, 1985b). However, as Felsenstein (1985b) pointed out, it is not possible to use parametric tests in such comparisons because, although independent, the differences between pairs of taxa do not have the same expected variance given that certain pairs of taxa in the comparisons are more closely related than others. Parametric testing, and more comparisons, are possible if a model of evolution is adopted that specifies how the absolute difference between two species increases over time. For mathematical simplicity, Felsenstein (1985b) considered a Brownian motion (random walk) model of character evolution. Under this model, the variance of Y_A-Y_B is proportional to the total length of the path through phylogeny linking species A and B (Fig. 1c). This simple relationship allows the comparisons to be scaled to have a common expected variance, as required by parametric tests, and the model permits further sister-clade comparisons to be made between deeper nodes.

An alternative to this procedure is the BRUNCH algorithm (Purvis and Rambaut, 1995). It is mainly intended for continuous Y and discrete X, but it can also be useful when both variables are continuous. It produces independent contrasts between lineages differing in their value of the X variable based on the phylogeny. As in Felsenstein's (1985b) method, the goal is to produce comparisons of independent terminals that differ in one of the two variables (trait X in Fig. 1d). The maximum of such comparisons that can be obtained from a set of taxa (three in the case of Fig. 1d) is achieved if lines through the phylogeny linking the species to be compared neither touch nor cross (Burt, 1989). A sign test is used to test whether Y and X are associated, or if Y fits Brownian motion, a t-test can be used on the scaled contrasts in Y (under the null hypothesis, their mean should be zero). Not all comparisons are between sister taxa, but they are still matched pairs (the fat lines in Fig. 1d).

As pointed out above, this approach is not about historical reconstruction of traits but it merely attempts to establish if differences in traits of taxa are correlated, for pairs of phylogenetically independent taxa. The approach of analyzing differences (Homoplasy II) has attracted many criticisms (Lord et al.,

1995; Wenzel and Carpenter, 1994) which, however, in part apply to other procedures as well, or are due to misinterpretations of the methods. One set of criticisms can be refuted easily: that the results depend on indefensible estimates of ancestral states. In fact, unlike approaches based on transitions (Homoplasy I, and see below), differences methods do not require (so cannot rely upon) any estimation of ancestral states (Grafen, 1989; Grafen, 1992). The means that are calculated for superspecific clades are part of the internal working of the method: they are not estimates of ancestral states and should not be used uncritically as such. This aspect of the procedure seems to be insufficiently understood by some authors, as the contrasts are sometimes presented to produce unrealistic intermediate phenotypes on the branch points that lack the traits of either descendant (consider the drawings of intermediates of major groups of birds pictured at the nodes in Fig. 13.6 of Cotgreave and Pagel, 1997). For the purpose of the Difference (Homoplasy II) approaches, nodes have no further meaning other than to connect the pair of taxa in the comparison.

Another prevalent criticism is that the method relies on overly simple models of character change (usually Brownian motion), so rejection of the null model is not informative. However, some versions of difference methods (Burt, 1989; implemented in part in the BRUNCH algorithm) do not rely on any model of change. Other versions do (though many provide a range of alternative models: Grafen, 1989; Martins, 1997), but the adequacy of the model of character change can be tested, and remedial action taken if necessary (see below). In fact, many procedures are available for assessing the suitability of whatever evolutionary model is being used (Garland et al., 1992; Martins, 1997; Purvis and Rambaut, 1995). Standard model criticism successfully identifies most departures from model assumptions and leads to statistically satisfactory performance (Díaz-Uriarte and Garland, 1996).

Other criticisms are not specific to the difference approach. For example, that the results are likely to be very sensitive to errors in the phylogeny is one critique that applies to all phylogenetic comparative approaches. In fact, simulations demonstrate that methods based on differences are more robust to errors than are methods based on transitions, because the latter attempt more precision (Ridley and Grafen, 1996). The main discrepancy, however, between approaches based on differences and transitions is not primarily their performance and other implementation issues such as model dependency, but it is the fundamental meaning of what these different approaches attempt to deliver. Whereas transition approaches take the tree to represent a sequence of evolutionary events optimized on a diagram representing hierarchical relationships (the presumed evolutionary tree), the difference approaches simply use the tree to guide the selection of taxa to be compared, and sometimes to scale differences to have a common variance in the statistical exercise that follows. This perspective also has implications for the meaning of phylogenetic trees; in an approach that is interested in statistical correlation, the uncertainty of tree reconstruction is perceived as a statistical problem of estimating an inherently uncertain quantity. In contrast, for proponents of the transition approaches, the tree is the single best datum available, and the uncertainties about this tree are phrased as an issue of nodal support. In applications of the difference approach it is therefore becoming common practice to repeat analyses on multiple near-optimal trees, or even to use taxonomies as surrogates for phylogeny: while these are meaningless in representing a hierarchical grouping of taxa for character reconstruction, they may contain some information to guide comparisons under a difference approach, in particular because independent contrasts approaches have been designed to deal with the treatments of polytomies (and which have been shown in simulations to do so validly (Grafen, 1989; Purvis et al., 1994)).

History *versus* adaptation

The goal of all comparative approaches ultimately is to gain insight into biological causation, and as such attempts to understand the organisms' adaptations and the selective environment affecting them (although comparative analyses can also be used to study other parameters, such as geographic range size, e.g., Letcher and Harvey, 1994). But because the inference of adaptation is based on phylogenetically independent comparisons of traits (under Homoplasy II), a frequent reasoning for the application of comparative methods is to test which correlations among traits have arisen under selection, given that there is a phylogenetic history that might confound these conclusions due to the non-independence of variation in close relatives. It is therefore a widely held—but incorrect (Purvis and Webster, 1999; Vogler and Kelley, 1996)—belief that the methods remove variation that is "due to" phylogeny, and that after the removal of this phylogenetically confounded variation the effect of selection is all that remains (e.g., Lord et al., 1995).

The presence or absence of a particular trait, if it is under selection, is not an indicator of any historical processes. An organism may exhibit a trait due to selection for the expression of the trait, and the trait may have a current function; equally possible, the organism may exhibit the trait because its ancestor had the trait. Thus, the function of a trait is not necessarily correlated with a particular history of the trait. The same line of thought applies to correlated sets of traits such as those generally tested in comparative methods. Therefore the partitioning of the data into those parts that are the same in the face of a different history (and thus informative regarding selection/biological relevance) and those that are the same because of a common history (and thus uninformative regarding any biological relevance) is not meaningful. In fact, those traits that are shared in closely related organisms may be similar simply because they are under the same selective pressure. Thus, excluding those from the analysis may mean missing the biologically most informative traits.

This realization, however, also means that there is no principal primacy of historical (Homology and Homoplasy I) over non-historical (Homoplasy II) methods. Whereas the latter can newly generate hypotheses of adaptation simply from data exploration which the former can't, neither can provide positive evidence for the adaptive nature of traits. This will have to come from other sources, usually from direct observations of fitness advantages in natural situations, from functional considerations of design, or elsewhere. However, understanding the presumably selected traits in their historical context (Homology and Homoplasy I) is undoubtably more satisfactory to evolutionary biologists, whereas the direct comparison with other organisms (Homoplasy II) may appeal mostly to ecologists because this approach puts the trait variation more readily in a functional context and avoids the problems of character reconstruction. We maintain, however, that the exploration of differences will benefit from an evolutionary perspective; to go beyond the question of why a trait is present, and answering instead the question of why a trait *arose*.

The future of comparative analysis

Taking an evolutionary perspective to comparative analysis, however, is not without difficulties. The assessment of correlation of character transitions is an inherently uncertain procedure because of the difficulties with ancestral character state reconstruction. The wide use of MacClade software (Maddison and Maddison, 1992) has promoted the application of the cladistic principle to reconstructing character states for hypothetical ancestors using Fitch optimization (Farris, 1970; Fitch, 1971a). Procedures that test correlated changes in a parsimony framework are based on this type of character optimization, such as the correlated-changes test of Maddison (1990), and a recently developed test based on randomization (Barraclough et al., 1999) which may, carefully designed, permit very precise analysis of character association. The inferences about ancestral states from these optimization methods, however, have to be based on several untested assumptions, including that changes are distributed evenly throughout the cladogram, and that forward and reverse changes are occurring with equal frequency (Omland, 1999). For example, in reconstructing the loss of sexuality in lineages of aphids (Normark, 1999), standard parsimony reconstruction resulted in inferences of ancient asexual lineages which is counter to evolution-of-sex theories. However, this pattern could also be obtained if asexuals arise very frequently, as multiple independent gains of this trait would prevent the discovery of the ancestral state (presence of sex), an explanation favored on biological grounds but lacking any foundation from cladistic analysis (Normark, 1999). Attempts to determine likelihood values for these transitions are underway as a step towards more realistic estimation of character evolution. Two-parameter likelihood calculations of character states that take into account different rates of forward and reverse changes (Mooers and Schluter, 1999), for example, have achieved a better fit of the character distribution than a one-parameter model; in the case of the tiger beetle chemical defenses the improvement is by a factor of 2.77 (Mooers and Schluter, 1999).

Methods that apply the difference approach to evaluate continuous characters are well developed and well characterized. They are known to perform well, both in terms of validity and power, under a wide range of evolutionary models and scenarios in which incomplete information is available (Grafen, 1989; Purvis et al., 1994; Díaz-Uriarte and Garland, 1996). One outstanding problem is that species values are estimates made with error, and this error can greatly reduce the power of the difference approach (Purvis and Webster, 1999), though procedures are becoming available that permit the sampling error to be taken into account (Martins, 1997). The situation is less settled for discrete characters. The most valid method (Burt, 1989) has low power, while other proposed methods (Maddison, 1990; Pagel, 1994; Ridley, 1983) can be entirely invalid under very reasonable models of character change (Nee et al., 1996; Ridley and Grafen, 1996). Whereas models for continuous character evolution can be tested fairly easily, because each branch in a phylogeny can be expected to show at least some change, thus providing a large sample of differences, there may be too little information for even basic testing of the suitability of models for discrete characters. For both kinds of character, there is a trend towards developing the methods towards more complex, fully explicit, models of character change (e.g., Martins, 1997), analogous to the development of more realistic maximum likelihood models of sequence evolution in the last dozen or more years.

The development of these models, however, should not distract from the fact that the reconstruction of character change will always be fraught with uncertainty (for a salutary example, see Oakley and Cunningham, 2000). Whereas more fanciful methods of state reconstruction can possibly be developed, the principal problem will not be overcome easily. As has been pointed out (Cunningham, 1999), models will always be dependent on a single character, in contrast to likelihood approaches to the analysis of DNA sequences which can estimate models of character change from a large number of characters. This is an insurmountable problem, and we have to guard ourselves against the temptation of taking these models further than is justified given our limited knowledge of their biological reality. Oakley and Cunningham (2000) suggest that the problems of accurately reconstructing trait evolution means that historical phylogenetic analyses are perhaps better viewed as hypothesis generation than as hypothesis testing.

While this may leave us with rather dire projections for the future of comparative methods, there are also clear signs of what at first seems a startling development: not using a phylogeny for comparative analyses. Phylogenies used as frameworks for comparative analyses are hypotheses and as such are subject to corroboration and improvement. Comparative analyses, however, take them as facts. We have outlined above how some workers are ameliorating this problem by analyzing a data set on each of several possible phylogenies. A natural progression is to avoid conditioning the comparisons on any particular phylogeny. Pagel (1994) suggested that the relationship between two variables could be assessed on all possible phylogenies, weighted by their likelihood, to get an overall conclusion. In a slightly different context, studies of cospeciation have typically tried to map associations onto previously estimated phylogenies: recent approaches (Huelsenbeck, 1997; Slowinski and Page, 1999) estimated the association history simultaneously with the phylogenies. It seems likely that, as with the analysis of cospeciation, future comparative biologists will use the data from which phylogenies are estimated, rather than those estimates, as their basis for inference.

Until then, our understanding of evolutionary processes and their causes will continue to be based on improved phylogenetic hypotheses increasingly derived from multiple molecular sequences. The various methodologies for comparative analysis provide a range of options which may be more or less satisfactory for a particular purpose and a particular type of data. The sometimes vicious debate between adherents of different principles has helped to sharpen the issues, to make explicit the underlying assumptions, and to highlight flaws of methods and concepts. But the old dichotomy between the two major camps of Homoplasy and Homology approaches may not continue to be as clean as it used to be. The desire of researchers applying statistical approaches to elucidate biological causes (Pagel, 1999) and the growing trend in cladistics towards quantification (Schuh, 2000) may provide some common ground, at least on an operational level. The ever larger quantities of character information from DNA lend themselves to these quantitative approaches that were not possible previously and thus molecular systematics will be driving comparative biology and the development of "comparative methods" in the future.

Acknowledgements
Grant support for our work in comparative biology was provided by the Natural Environment Research Council (NERC) of the UK.

Molecular Systematics and Evolution: Theory and Practice
ed. by R. DeSalle, G. Giribet and W. Wheeler
© 2002 Birkhäuser Verlag/Switzerland

The use of physiological data to corroborate cospeciation events in symbiosis

Michele K. Nishiguchi

Department of Biology, New Mexico State University, MSC 3AF, Las Cruces, NM 88003, USA

Summary. The symbiotic association between sepiolid squids (Family Sepiolidae) and luminous bacteria (Genus *Vibrio*) provides an unusually tractable model to study the evolution and speciation of mutualistic partnerships. Both host and symbiont can be cultured separately, providing a new avenue to test phylogenetic congruence through molecular and physiological techniques. Combining both molecular and morphological data as well as measuring the degree of infectivity between closely related pairs can help decipher not only patterns of co-speciation between these tightly linked associations, but can also shed new light on the evolution of specificity and recognition among animal-bacterial associations.

Introduction

The presence of symbiotic associations between eukaryotic hosts and their microbial partners has long intrigued biologists who are interested in evolution and cospeciation between the individuals in the relationship. The diversity and biogeographical range of each host-symbiont pair demonstrates how each partner has co-evolved to exploit a new ecological niche, providing a new capability or function for the newly adapted symbiosis and their respective populations. A number of contemporary examples indicate that the formation of a symbiotic relationship allows the host or its symbiont to radiate into newly formed niches; these include the chemoautotrophic bacteria residing within tissues of hydrothermal vent or sewage outfall invertebrates (Cavanaugh, 1994), nitrogen-fixing bacteria in the nodules of leguminous plants (Wilkinson et al., 1996), endosymbiotic bacteria within the body cavities of aphids, termites and whiteflies (Smith and Douglass, 1987; Nardon and Grenier, 1989; Moran and Telang, 1998), and luminous bacteria within the light organs of monocentrid and anomolopid fish and sepiolid and loliginid squid (McFall-Ngai and Ruby, 1991; Haygood and Distel, 1993). Although these symbiotic partnerships may have evolved as part of an adaptation to a particular ecological niche, the question of whether the ecology or the specificity of the partnership drives the evolutionary radiation of the hosts and their symbionts has yet to be answered. Understanding how a particular symbiosis has diverged from one association into many has led to the development of several techniques that allow the elucidation between strict co-speciation or symbioses that are more promiscuous among the partners.

Studying eukaryotic-prokaryotic symbiosis has been difficult for many biologists; in many cases the symbiont cannot be cultured outside the symbiosis and therefore many of the physiological properties that are used to characterize the prokaryotic partner are not measurable or functional without the eukaryotic host. In these instances the symbiont may not retain certain morphological characters or biochemical processes that delineate it from similar free-living species. Without proper identification of specific symbionts, systematists studying co-evolving associations have had to rely on parts of the fossil record, which has provided some evidence about species interactions (Moran and Telang, 1998). Molecular techniques, which allow the amplification of specific loci from culturable and non-culturable symbionts, also allow scientists to further probe the history of symbiotic events. Using present-day specimens, one can measure characters (morphology and molecules) from both host and symbiont, and then use this information to predict how long interacting lineages have been associated. Phylogenetic analyses provide a means to assess whether symbiotic associations have evolved in parallel, or have progressively adapted so that different hosts have evolved with their symbionts, irrespective of their relationship to each other. Parallel cladogenesis would imply patterns of co-speciation, and would become more evident from the phylogenetic analyses of both partners.

Reconstructing host-symbiont phylogenies

Comparing data sets from both host and symbiont phylogenies is one method to determine if parallel cladogenesis exists between the partners. In an ideal case, unique specificity may exist among the two systems, with every single host species having one single parasite and *vice-versa*. But it is clear that other cases might exist where some hosts have no symbionts, some symbionts occur in more than one host, or hosts have more than one symbiont (Schuh, 2000). In order to test whether cospeciation is occurring among two systems, independent phylogenetic analyses can be compared using two separate phylogenies, such that cospeciation events can be inferred. In the example of many strict phylogenies, such as the aphid/*Buchnera* symbiosis, symbionts are transmitted vertically, that is, from maternal lineage to each of the offspring (Munson et al., 1991; Baumann et al., 1997). Combining molecular and morphological data of hosts and their respective symbionts, each phylogenetic tree is constructed and compared to test for phylogenetic congruence amongst the host-symbiont pairs. In species where symbiotically linked genes have been identified, those loci are ideal for further phylogenetic analyses (Lee and Ruby, 1994) and may provide greater sensitivity for developing phylogenies in parallel.

In vertically transmitted symbioses, little or no evidence exists for host-switching of partners, and the relationships tend to be strictly congruent in their phylogenetic patterns. Environmentally (or horizontally) transmitted symbioses (where hosts obtain their symbionts anew with every generation), would predict non-specific patterns of co-speciation, where symbionts from the environment will be found in hosts from the same area. Although the assumption of this pattern seems evident, it has been shown that certain environmentally transmitted symbioses still suggest strict patterns of parallel cladogenesis. In the example of the sepiolid squid-*Vibrio* mutualism, host phylogenies parallel the respective symbiont trees in a strict consensus pattern, despite the fact that the symbiosis is transmitted environmentally (Nishiguchi et al., 1998; Fig. 1). Using combined molecular data from both host and symbionts, this parallel phylogeny demonstrates that specificity can still evolve between hosts and symbionts, despite the fact that juvenile squids obtain their partners from the environment at every generation.

The program TREEMAP (Page, 1995) offers the possibility of analyzing both host and symbiont characters simultaneously and robustness of the evolutionary patterns can be tested, particularly for host-symbiont specificity and host-switching. The number of shared characters between different competing symbionts are tested, and these data are matched *a priori* for each association that is presently found in modern taxa. The number of host-switching events can then be calculated, to accommodate host-symbiont pairs that may not be strictly evolving in parallel. Thus, each node tested provides a measurement of whether host symbiont pairs are derived and co-speciated, or whether certain partners have been promiscuous in their symbiotic relationships. The robustness of the data set is then determined statistically by determining if each host-symbiont pairing is significant, compared to the other possible pairings among all the associations tested.

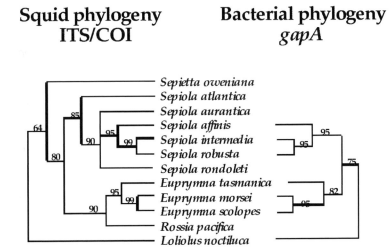

Figure 1. Phylogenetic tree of squid host species and their symbiotic bacteria isolates. Trees were inferred from unambiguously aligned positions of ITS/COI or *gapA* data sets. Numbers above nodes represent bootstrap values. For details see Nishiguchi et al. (1998).

Once parallel cladogenesis is established among the host species and their respective symbionts (using molecules and/or morphology), competition experiments between symbionts can be used to test the robustness of the prior phylogenetic analyses, which is the focus of this chapter.

Competitive hierarchy and strain specificity

In sepiolid-*Vibrio* symbiosis, newly born aposymbiotic (without symbionts) juvenile squids are infected with individual symbiont strains isolated from closely related squid taxa, and are used for measuring the individual degree of colonization and subsequent host specificity. Bacterial strains are mixed at a concentration of 1000 cells/mL seawater using one or, for competition experiments, two strains of symbiotically competent bacteria. The juvenile squids are then placed in this inoculated seawater for approximately 12 h to initiate the light organ infection (Ruby and Asato, 1993). Once infected, the juvenile squids contain enough bacteria that their luminescence can be detected with a luminometer (Turner Designs 20/20), ensuring colonization has occurred (Ruby and Asato, 1993). Single bacterial strains that are presented to the juvenile squids alone, as well as squids that are kept aposymbiotic (no symbionts), provide stable controls for the competition experiments between two different symbiont strains or species. Since bacterial cells from light-organ homogenates have approximately 100% plating efficiency, the extent of colonization can be calculated from the number of colony-forming units (CFUs) arising from diluted aliquots of whole juvenile squid homogenates that are plated on seawater tryptone agar (SWT) medium (Boettcher and Ruby, 1990; Ruby, 1996). Using two different *Vibrio* strains or species that express different luminescent phenotypes in culture, quantification is achieved by comparing the ratio of one luminescent phenotype to the other (Nishiguchi et al., 1997). When quantifying symbionts with the same luminescent phenotype, differentiation can be accomplished by using a 16S rDNA probe for a particular variable region (VA1; Fidopiastis et al., 1998) which differentiates between *V. fischeri* and *V. logei*. Using colony blot techniques, the number of colonies from each squid homogenate can be quantified for both competing symbionts (Nishiguchi, 2000).

Colonies grown from the homogenized individuals will produce various concentrations of each symbiont. The proportion of each symbiont in the juvenile squid light organ is dependent upon whether specificity exists between the symbionts and the particular squid species used in the competition (Nishiguchi et al., 1998). When two species of symbiont are presented to a hatchling juvenile, several outcomes can be expected. First, if neither symbiont is capable of infecting the host squid itself, then the competition assay between these bacterial strains will not be different from the control animals and no colonization will occur from competition (initiation, Fig. 2). If both symbiont strains are capable of infecting and colonizing the squid light organ by themselves, then

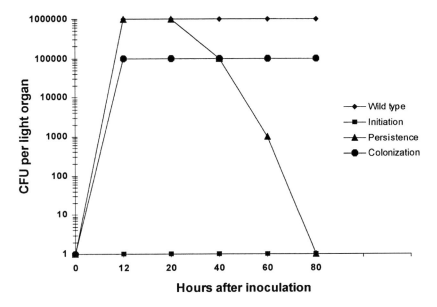

Figure 2. Classes of symbiosis-defective strains of *V. fischeri*. Schematic illustration of the colonization patterns of strains that either fail to infect ("initiation strains"), infect to a diminished extent ("colonization strains"), or are incapable of maintaining the infection ("persistence strains"), compared to
the wild-type pattern. CFU: colony-forming units.

both strains will be present in equal concentrations in the light organ during
the assay period. This would indicate that each symbiont is equally capable of
persisting inside the host light organ equally well, and neither one can be differentiated by host recognition or specificity. In contrast, if two equally competent strains of symbiont are used to infect a newly hatched juvenile, and only
one is present or dominant at the end of the assay period, then this would indicate that some preference or recognition exists for the more predominant symbiont (Nishiguchi, 2000; Nishiguchi et al., 1998).

Using specificity to delineate symbiotic relationships

Initially, if one were to compare the evolutionary history and lineages of host-
symbiont pairings, the assumption would be that each partner has shared some
common ecological or environmental past. Nealson and Hastings (1991) initially hypothesized that those associations between marine animals and luminous bacteria were initiated in a shared environment, where the ecology of the
association would predict the evolutionary history of the partners. This is the
presumed case for environmentally transmitted symbioses, where hosts must
obtain their symbionts from the surrounding seawater and bacteria are available for uptake and proliferation within the host light organ. Over evolution-

ary time, a symbiosis has the capacity to evolve species specificity, in which symbionts are more likely to be associated with a particular host species. Examples of both ecologically determined and host-specific symbioses can be observed within the sepiolid squid-luminous bacterium symbiosis (Nishiguchi et al., 1998; Nishiguchi, 2000). During the initiation of the symbiosis, juvenile squids obtain their symbionts environmentally within the first few hours after hatching (McFall-Ngai and Ruby, 1991; Nishiguchi, 2000). From this inoculum (which can be as little as 10 bacteria), the symbionts are able to successfully colonize the juvenile squids and maintain populations inside the light organ throughout the duration of the squid's life history. The occurrence of two species of *Vibrio* symbiont in one host exemplifies the notion that this environmentally transmitted symbiosis is selective for specific types of bacteria (*Vibrio*), but the population and concentration of each *Vibrio* species are affected by abiotic factors such as temperature (Nishiguchi, 2000). These subtle changes in symbiont composition may lead to the eventual differentiation and radiation of a particular host-specific association, where one host is affiliated with one type/strain of symbiont. In Mediterranean squid species of the genus *Sepiola*, the symbiosis is comprised of two species of luminous bacteria, *Vibrio fischeri* and *V. logei* (Fidopiastis et al., 1998). Both species of *Vibrio* are abundant in coastal environments where Mediterranean sepiolids are also abundant. *V. fischeri* and *V. logei* are both found in free-living populations throughout the Mediterranean, but the more psychrophilic of the two species, *V. logei*, is more common among species of sepiolids that inhabit colder temperatures (Nishiguchi, 2000). For example, in the Bay of Banyuls, France, where both symbionts reside in many of the *Sepiola* species examined, *V. logei* is the more predominant of the two bacterial species in light organs of *S. robusta*, *S. intermedia* and *S. ligulata* species. These sepiolid species are commonly found at depths below 20 m and at temperatures ranging between 10 °C to 16 °C (Nishiguchi, 2000). These squids are not affected by seasonal fluctuations in temperature due to the fact that the summer thermocline prevents changes in water temperature below 20 m. In contrast, the shallow-water *S. affinis* resides between 0–20 m in depth and is exposed to temperatures ranging between 14 °C to 24 °C throughout the year. The population of vibrios that reside in *S. affinis* changes with respect to temperature changes; in the summer, *V. fischeri* is the more predominant species in *S. affinis* light organs, where surrounding environmental temperatures are between 18–22 °C at the shallower depths (0–20 m; Tab. 1). During the winter months, when the summer thermocline disappears and temperatures drop to between 12–16 °C, *S. affinis* individuals mostly host *V. logei* in their light organs (Tab. 1). Although both species of *Vibrio* are always present in all *Sepiola* species tested thus far, there is a definitive ecological boundary that establishes whether *V. logei* or *V. fischeri* is present in higher concentrations in the light organs of sepiolid squids. *V. logei* also exhibits higher growth constants *in vitro* and in certain *Sepiola* species compared to *V. fischeri*, where depressed temperatures of the symbiotic habitat have an effect on species composition of the association (Nishiguchi, 2000).

Table 1. Total light organ concentrations of *V. fischeri* and *V. logei* from field-caught adult *Sepiola*

Squid species[*]	Depth (m)/Temp (°C)	V. fischeri[**]	V. logei[**]	Ratio (Vf:Vl)	Month caught
S. affinis (1)	15/18	6.1×10^9	2.4×10^9	72: 28	May, 1999
S. affinis (1)	20/22	2.4×10^9	6.3×10^8	79: 21	July, 1995
S. affinis (7)	10/22	$7.5 \times 10^9 \pm 6.6 \times 10^8$	$9.1 \times 10^8 \pm 2.1 \times 10^7$	89: 11	July, 1999
S. affinis (2)	20/18	$5.1 \times 10^8 \pm 4.3 \times 10^8$	$3.2 \times 10^8 \pm 1.2 \times 10^7$	61: 39	Sept, 1995
S. intermedia (2)	75/16	$4.8 \times 10^8 \pm 5.4 \times 10^7$	$8.3 \times 10^9 \pm 3.4 \times 10^9$	6: 94	July, 1999
S. intermedia (2)	60/12	$8.4 \times 10^7 \pm 4.6 \times 10^2$	$2.1 \times 10^9 \pm 1.2 \times 10^8$	4: 96	Sept, 1995
S. intermedia (1)	55/12	1.0×10^9	3.8×10^9	21: 79	Sept, 1998
S. ligulata (1)	40/16	2.1×10^{10}	7.9×10^{10}	21: 79	July, 1999
S. ligulata (3)	55/10	$7.1 \times 10^{10} \pm 3.2 \times 10^7$	$1.8 \times 10^{11} \pm 2.9 \times 10^6$	28: 72	Sept, 1998
S. robusta (1)	60/12	5.1×10^9	4.2×10^{10}	11: 89	July, 1995
S. robusta (2)	60/10	$5.4 \times 10^9 \pm 1.5 \times 10^9$	$7.1 \times 10^{10} \pm 4.5 \times 10^9$	7: 93	Sept, 1998

[*] Numbers in parentheses indicate number of adult animals used to calculate bacterial concentrations. All concentrations are in CFU (colony-forming units)/adult light organ.

[**] Values represent CFU/light organ ± standard deviation for animals with more than one representative specimen.

Vf = *V. fischeri*, Vl = *V. logei*

Table 2. Infection of juvenile *E. scolopes* with mixed inocula of squid light organ symbionts

Strains used in competition	No. of animals tested[2]	Mean proportion found in light organs at 48 h[3]
ES114 and EM17	37	98:2
ES114 and ET101	35	95:5
EM17 and ET101	17	75:25
ET101 and SA1	30	90:10
ES114 and LN101	21	100:0

[1] Bacterial isolates from squid species are as follows: ES114-*E. scolopes*, EM17 = *E. morsei*, ET101-*E. tasmanica*, SA1-*S. affinis*, *Photololigo noctiluca*.
[2] Total number of animals tested in three or four experiments for each pair of strains. All of the animals became colonized by at least one strain in these experiments.
[3] Mean averages of individual ratios of bacterial strains were arc sine transformed (Sokal and Rohlf, 1981) to determine the significance of each treatment. All mean values were significantly different to within $P < 0.05$.

Thus, ecological factors of the association can be important in establishing the symbiotic composition of the partnership, demonstrating a need for understanding abiotic factors that influence the evolution of the association.

Host Phylogeny   Strain Competition

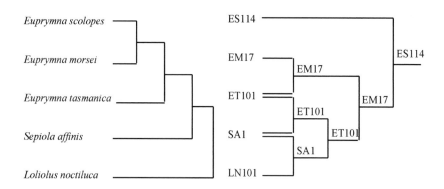

Figure 3. Comparison of species from the sepiolid host phylogeny (based on ITS and COI sequence data), to the symbiont competition experiments. According to symbiont competition experiments completed in *E. scolopes* juveniles, native ES114 (*E. scolopes* symbiont) always dominates the light organ colonization, compared to any other symbiont in a competition. EM17, the symbiont isolated from *E. morsei*, is the next best strain to dominate in a competition experiment, and complements the molecular phylogenetic data which suggest that *E. scolopes* and *E. morsei* are sister taxa. Following the phylogenetic relatedness of host squids, respective symbiont strains colonize the light organ in the same hierarchical manner as the molecular phylogenies. ES114-*Euprymna scolopes* symbiont; EM17-*Euprymna morsei* symbiont; ET101-*Euprymna tasmanica* symbiont; SA1-*Sepiola affinis* symbiont; LN101 *Photololigo noctiluca* (*Loliolus noctiluca*) symbiont.

Although no host specificity is exhibited between the *Sepiola* host and their *Vibrio* symbionts tested (Nishiguchi, 2000), a contrasting situation is present in the Indo-West Pacific genus *Euprymna*. Recent physiological data have provided evidence that strains of *V. fischeri* (the only symbiont species present in *Euprymna* species) can be independently identified and demonstrate host specificity among different species of hosts tested (Nishiguchi et al., 1998; Nishiguchi, 2001; Nishiguchi, in press). For example, all strains of *V. fischeri* isolated from either *E. scolopes* (Hawaii), *E. morsei* (Japan), or *E. tasmanica* (Australia) exhibit the same infection behavior when independently presented to newly hatched *E. scolopes* juveniles (Nishiguchi et al., 1998). But, when placed in competition with each other, native *V. fischeri* isolated from *E. scolopes* are always the dominant strain at the end of a 48-h competition with any non-native *V. fischeri* strain (Tab. 2). Even when two non-native species are present in the association, the strain of *V. fischeri* dominant in the *E. scolopes* light organs are those strains that have evolved from the more closely related host species (Nishiguchi et al., 1998). Thus, comparing both phylogenetic data and the physiological mechanisms of infection and colonization, co-speciation patterns are corroborated from both the genetic and physiological perspective (Fig. 3).

Conclusions

Defining the physiological basis of cospeciation among host–symbiont pairs and observing the degree of specificity between the partners have been difficult to achieve with experimental systems (Cavanaugh, 1994). In many animal-bacteria symbioses, the prokaryotic partner is not culturable outside of the eukaryotic host, and the symbiosis cannot be experimentally initiated in the laboratory. In the sepiolid squid-*Vibrio* mutualism, both hosts and symbionts can be cultured outside of the partnership and maintained in the laboratory, providing an experimentally tractable system to study the mechanisms which underlie cospeciation and parallel cladogenesis. Using the host's ability to differentiate between native and non-native symbionts separately or combined under varying experimental conditions, allows the investigator to measure the degree to which specificity occurs between a variety of closely related symbiotic partners, and whether their physiological behavior corroborates the phylogenetic data of host-symbiont pairs. The ability to construct and determine cospeciation patterns amongst partners in closely associated partnerships has eluded scientists for many years. The advent of molecular biology and new phylogenetic methods to analyze combined data sets have given investigators the ability to detect patterns of cospeciation and parallel evolution, and to help decipher the origin and development of symbiotic associations. Combining phylogenetic evidence with physiological assays for specificity and colonization helps to understand the extent of these complex relationships, and can increase our understanding of processes and mechanisms which underlie the basis for cospeciation.

Molecular Systematics and Evolution: Theory and Practice
ed. by R. DeSalle, G. Giribet and W. Wheeler
© 2002 Birkhäuser Verlag/Switzerland

Reexamining microbial evolution through the lens of horizontal transfer

Paul J. Planet

Department of Microbiology, Columbia University College of Physicians and Surgeons, New York, NY 10032, USA

Summary. Our ability to understand the evolution of microbial organisms revolves around a central and increasingly unsettled question: what is the nature of the mode of inheritance? The extent to which genetic information is passed vertically from parent to daughter or horizontally between distant relatives must guide reconstructions and inferences of evolutionary history, and has direct bearing on any ideas about the mechanisms of selection and diversification. Recent evidence suggests that we may have previously underestimated the contribution of horizontal gene transfer, and the dynamics and extent of this process are only beginning to be understood. The recent flood of complete genome sequences of microorganisms has already presented us with a vast array of data from which to test our hypotheses about the evolution of the entire tree of life, but what remains unclear is how we can make sense of this unwieldy data set. Analyses of this newly available data set should include explicit examinations of the contributions of both types of inheritance.

Introduction

Central to phylogenetic reconstruction is the assertion that the perceived pattern of nested hierarchy in relationships between organisms is the result of vertical transmission of genetic information (see Box 1). Phylogenetic trees are the schematic representation of this process. In this reiterative branching process, lineages that have split can never again reunite to constitute the same line of descent. Cladogenesis is a strictly committed and irreversible step.

The workings of the microbial world would seem, at first glance, to exemplify a vertical branching pattern. Organisms that reproduce asexually can create a committed lineage each time they divide or bud. This stands in contrast to sexually reproducing organisms, which must exchange genetic information in order to produce offspring.

Box. 1 definitions of modes of inheritance for asexual organisms

Vertical transfer—Inheritance of genetic material along with organismal reproduction.
Horizontal transfer—Inheritance of genetic material independent of organismal reproduction.

However, the fact that lineages of primarily asexually reproducing microorganisms can and do exchange genetic material is not a foreign concept. On the contrary, it is fundamental to our current understanding of biology. The seminal experiments that led to the discovery of DNA as the genetic material were based on the fact that traits can be inherited independently of organismal reproduction (Avery et al., 1944; Griffith, 1928). Today, mechanisms of horizontal inheritance, such as *conjugation, transformation* and *transduction*, still form the core of laboratory microbiology, because the ability to mobilize genes from one microbe to another allows us to identify mutations and make precise statements about the characteristics and functions of genes (see Box 2). Given the existence of these complex genetic systems dedicated to transfer at all, it is difficult to imagine that horizontal transfer never occurs in the environment (Syvanen, 1985). But the evolutionary importance assigned to these convenient laboratory tools depends on their frequency and extent in natural populations.

Mounting evidence suggests that many genes are, indeed, passed between microorganisms in nature, and claims of individual cases of ancient and recent horizontal transfer have flooded the literature (Kidwell, 1993; Mazodier and Davies, 1991; Smith et al., 1992; Syvanen, 1994) (Tab. 1). Historical claims have been bolstered by direct observation of horizontal transfer in the environment (Davison, 1999). Perhaps the most troubling example is the communication of antibiotic resistance in pathogenic strains of bacteria, in which conjugative plasmids have been the primary vectors (Bennett, 1987; Bennett, 1996; Datta and Hughes, 1983; Hughes and Datta, 1983). Moreover, it seems that many pathogens have acquired many of the most important genes active in host colonization and disease through horizontal transfer (Hacker et al.,

---

Box. 2 Mechanisms of horizontal gene transfer in the Archaea and Bacteria

Transformation—uptake of naked DNA from the environment. In 1944 Avery, McLeod and McCarty were able to show that DNA, taken in from the extracellular millieu, was the "transforming principle" responsible for the switch from avirulence to virulence in strains of *Streptococcus pneumoniae*, thus establishing DNA as a genetic material (Avery et al., 1944).

Conjugation—often referred to as bacterial sex, DNA is transferred by direct physical interaction between a donor and recipient (Lederberg, 1946).

Transduction—DNA transfer mediated by phage infection. In this process transferred genes are incorporated into phage genomes in which they are carried to a new cell (Zinder, 1952).

N.B. All three of these mechanisms have been shown to exist in a large spectrum of microorganisms including the Archaea as well as the Bacteria (Sowers and Schreier, 1999).

1997; Ziebuhr et al., 1999). Recent papers have suggested that horizontal transfer of DNA may, in fact, constitute the major engine driving microbial innovation and diversification (de la Cruz and Davies, 2000; Lawrence, 1997; Matic et al., 1995; Medigue et al., 1991; Ochman et al, 2000; Vulic et al., 1997; Vulic et al., 1999).

As more and more evidence accumulates, it becomes evident that ideas about the evolution of single-celled organisms must include estimations of the contribution of horizontal transfer. But how much attention should be paid to this process? Do we have the tools to address its historical importance? Can a phylogenetic model based on vertical inheritance still be used to explain a world with significant horizontal exchange?

In this chapter I begin to address some of these questions by examining some methods by which horizontally acquired genes can be detected and the dynamics of genetic exchange better understood. With these techniques in hand, I consider some major questions in microbial systematics through the lens of horizontal transfer, and discuss how emerging ideas about the relative contributions of horizontal and vertical transfer rearrange, sometimes strengthen and sometimes confuse our understanding of microbes and the earliest life forms.

Methods for detecting horizontal transfer

Two types of evidence are routinely used to identify genes that may have undergone a transfer event. The first is based on contradictions, or incongruencies, between gene relationships and organismal relationships. Horizontal transfer can explain points of incongruence between the two genealogies. A second kind of evidence relies on the identification of genes that differ in some fundamental way from the rest of the genome. It is assumed that these genes differ because they have retained a characteristic foreign signature.

Incongruence

A horizontal transfer event creates a situation in which the history of the gene diverges from that of the organism. Therefore the most direct way to identify gene transfer events that have already happened is to reconstruct and compare the two histories.

The simplest, least rigorous, and most problematic approach to comparing gene and organismal relationships is by best-hit analysis. Using pair-wise similarity, a gene can be compared to all other known genes in a range of closely and distantly related organisms. From this set, the gene with the highest amount of similarity[1] to the gene in question is identified as the best hit. The

---

[1] Or the lowest E value, which is the expected number of matches with at least this amount of similarity given a random sequence.

Table 1 Arguments for horizontal transfer events

Gene(s)	Organism(s)	Criteria used	Reference
0–17% of genes in the genome	19 Bacterial species	FS	(Ochman et al., 2000)
16.2% of the genome	*Aquifex aeolicus*	SC	(Aravind et al., 1998)
18% of the genome	*Escherichia coli*	FS	(Lawrence and Ochman, 1998) (Lawrence and Ochman, 1997)
24% of the genome	*Thermotoga maritima*	SC, PI	(Nelson et al., 1999)
F-1,6-Bisphosphatase	Bacteria to plants	PI, SC	(Smith et al., 1992) (Martin, 1998) (Martin et al., 1996)
F-ATPase	Archaea and bacteria	PI	(Sumi et al., 1992) (Hilario and Gogarten, 1993)
Fe-SOD	Prokaryotes and eukaryotes	SC, PI	(Smith et al., 1992) (Smith and Doolittle, 1992)
F-like plasmids	*Escherichia coli* and *Salmonella* spp.	SC, PI	(Boyd and Hartl, 1997)
Genes involved in replication machinery	Bacteria	PI	(Moreira 2000), (Forterre, 1999)
Glutamine Synthetase	Plants and root nodule bacteria	PI, SC, PA	(Carlson et al., 1985), (Shatters and Kahn, 1989), (Kumada et al., 1993), (Smith et al., 1992) (Brown et al., 1994) (Tiboni et al., 1993)

(continued on next page)

Table 1. (continued)

Gene(s)	Organism(s)	Criteria used	Reference
Glyceraldehyde-3-phosphate dehydrogenase (GAPDH)	Eukaryotes and prokaryotes	PI, SC	(Doolittle et al., 1990) (Martin et al., 1993) (Henze et al., 1995) (Smith et al., 1992) (Hensel et al., 1989) (Roger et al., 1996) (Martin and Schnarrenberger, 1997) (Doolittle, 1998)
Glycosyl hydrolases	*Escherichia coli* and *Bacillus subtilis*	PI, FS	(Garcia-Vallve et al., 1999)
Group I intron	α-purple bacteria	SC, PI	(Paquin et al., 1999)
inv/spa and mxi/spa operon	*Shigella flexneri* and *Salmonella enterica*	FS, PA	(Lawrence and Ochman, 1997)
Isopenicillin N-synthase	Bacteria to fungi	SC PI	(Smith et al., 1990), (Penalva et al., 1990) (Buades and Moya, 1996) (Doolittle, 1998)
naphthalene dioxygenase gene (nahAc)	*Pseudomonas* putida and naphthalene-degrading bacterium	SC, PI	(Herrick et al., 1997)
Nitrite reductase	Fungi and bacteria	SC	(Takaya, 1998)
Operational genes (as opposed to informational genes)	Bacteria and archaea	SC, PIS	(Jain et al., 1999) (Rivera et al., 1998)

(continued on next page)

Table 1. (continued)

Gene(s)	Organism(s)	Criteria used	Reference
Phosphoglucose isomerase	Plants and *E. coli*, mammals and prokaryotes	SC, PI	(Katz, 1996) (Smith et al., 1992) [Kidwell, 1993 x30 (Syvanen, 1994)
Phosphoglycerate kinase	Bacteria to plants	PI, SC	(Brinkmann and Martin, 1996) (Kaneko et al., 1996) (Martin, 1998)
Reverse gyrase (rgy)	Hyperthermophiles	PI, PA	(Forterre et al., 2000)
rpl21 and rpl22 ribosomal proteins	Endosymbiont to plants	SC, PI	(Martin et al., 1990) (Gantt et al., 1991)
rRNA operons	*Thermomonospora chromogena* to *Thermobispora bispora*	SC, PI	(Yap et al., 1999)
Rubisco genes	Plastids and proteobacteria	PI	(Delwiche and Palmer, 1996)
Transketolase	Fungi and bacteria, plants and cyanobacteria	SC, PI	(Martin, 1998)
Triosephosphate isomerase	Bacteria and eukaryotes	SC, PI	(Martin, 1998) (Lawrence, 1998)
tufA (EF-Tu)	Chloroplasts to plants	PI, SC	(Baldauf and Palmer, 1990)
Xylanase	Bacteria to fungi	SC	(Gilbert et al., 1992)

This table is not meant to be an exhaustive list of claims in the microbiology literature, but it is representative of the multiple detection criteria and the phylogenetic pervasiveness of horizontal transfer. The criteria codes are the following: PI = Phylogenetic Incongruence, SC = Best Hit or some similarity criterion, FS = Foreign signature, PA = Presence in distant relatives and/or absence in close relatives.

best-hit gene is taken to be the closest known relative of the gene. If the best hit resides in a distant relative, and close relatives also have similar genes, the inferred relationship between the gene and its best hit contradicts our preconceived notion of organismal relationships. One explanation for this contradiction is that the gene and host have distinct histories.

Best-hit analysis has been used extensively in recent whole genome publications, especially in situations where the organismal relationships are well demarcated, as is the case between the domains Archaea and Bacteria (Nelson et al., 1999; Parkhill et al., 2000a, Parkhill et al., 2000b, Tettelin et al., 2000). For instance, the presence of Archaeal best hits for Bacterial genes led Aravind et al. (1998) to suggest that nearly 10% of the genome of *Aquifex aeolicus*, a hyperthermophilic bacterium, is best described as "Archaeal" in nature and has therefore been imported from the Archaea. In the bacterium *Thermotoga maritima*, 24% of the genes in the genome give Archaeal best hits, leading to similar hypotheses (Nelson et al., 1999).

Another sequence-similarity-based technique, articulated most clearly by R.F. Doolittle (Doolittle, 1998; Doolittle et al., 1996), uses a standard range of sequence identity between organisms to identify genes that may have been horizontally transferred. Comparisons of sequence identity between cognate genes from two organisms tend to fall within a specific range (Doolittle et al., 1996). When such a comparison yields a percent identity more than a standard deviation above the mean value of the typical range, then it is possible that these genes have a special relationship—they may have shared a common ancestor more recently than their current host genomes did. Doolittle (1998) suggests that in comparisons between eukaryotes and prokaryotes a good rule of thumb for identifying horizontal transfer candidates is an identity score of more than 60%.

Though best-hit analysis and similarity-based techniques may be good first lines of attack for identifying potentially acquired genes in large and unwieldy data sets, pair-wise analysis is not a reliable way of making an inference about the phylogeny of genes (Kyrpides and Olsen, 1999; Reeck et al., 1987). Similarity-based methods do not describe the evolution of genes, stopping short at a simple statement of resemblance. Therefore, the explanatory power of the analysis is greatly reduced, denying consideration of other possible explanations for incongruent evidence (e.g., sequence saturation, duplications followed by lineage sorting, differing rates of sequence evolution, retention of ancient genes, or convergence.)

To more accurately illustrate and understand horizontal transfer we should use an approach in which explicit and well-supported phylogenetic descriptions of the evolution of genes are compared to some well-supported, standard phylogeny. Any such argument should contain each of the following steps:

(1) Construction of a gene tree
Dense sampling for potential homologs of the gene in related organisms should be followed by construction of a gene tree. Several techniques are avail-

able to infer the evolution of genes given an aligned data set of putative homologs. Although I will not discuss the relative merits of different techniques, it is worth mentioning that although maximum parsimony and maximum likelihood have not been extensively employed in microbial systematics, it is advisable to apply these techniques to derive more testable phylogenies of both organisms and genes. Distance-based methods often do not yield accurate phylogenies (Farris, 1982).

Collecting and analyzing gene family data present several specific problems. Special attention should be paid to separating orthologs from paralogs —respectively, genes related by replication at a committed lineage splitting event and duplication within a lineage. The misidentification of a paralog as an ortholog can lead to a false hypothesis of horizontal transfer (Smith et al., 1992; Syvanen, 1994). Distinguishing between paralogs and orthologs may require some method of rooting the tree. This can be extremely problematic with phylogenies of gene families, especially if the data sets are drawn from a wide array of organisms, and even more so if multiple unidentified duplications and horizontal transfers have occurred. Development of schemes for rooting gene family genealogies merits special attention in the future. Other important areas requiring methodological development include techniques that can deal with potentially enormous data sets (Goloboff, 1999; Nixon, 1999).

(2) Construction of an organismal tree
When constructing a representative organismal phylogeny one can choose several different methods. First, some gene that is thought to be refractory to horizontal transfer can be chosen to represent the organismal phylogeny (e.g.,16S, RecA etc.) However, this may not be an obvious or assumption-free choice. Another possibility is to use some method to combine several gene data sets with the expectation that the combination or consensus represents the best explanation available (for reviews on data combination see Brower et al., 1996; de Queiroz et al., 1995; Kluge, 1989; Nixon and Carpenter, 1996). A third possibility, which has only just become a realistic alternative, is to construct phylogenies from whole genomes of information. The choice between these options depends upon basic questions about the inheritance of genes, including whether or not horizontal transfer is a major evolutionary force. Taking into account the present readjustment of fundamental ideas about inheritance, it is more difficult to arrive at a logically sound, organismal phylogeny. I will revisit this issue later in this chapter.

(3) Statistical support for each tree
After reliable phylogenies have been established for both organisms and genes some criteria must be used to assess confidence in each phylogeny. Bootstrapping (Felsenstein, 1985a) is the most popular of these techniques, but other useful techniques include Jackknifing (Farris et al., 1996) and Bremer supports (Bremer, 1995).

(4) Comparison of trees

With well-supported trees for both the gene and the represented organisms, the two phylogenies can be compared. The differences between phylogenetic tree topologies can represent incongruencies between genealogies. Such historical inconsistencies may be due to horizontal transfer, and, therefore, instances of horizontal transfer can be detected and pinpointed by noting where topologies differ. However, some incongruence should be expected by chance alone, and it is extremely important to use some test of confidence. This is the step most often neglected in assessments of incongruence. Several techniques have been developed to test the statistical significance of incongruent tree topologies (Bull et al., 1993; Farris et al., 1995; Huelsenbeck and Bull, 1996; Larson, 1994; Lawrence and Hartl, 1992; Rodrigo et al., 1993). Cunningham (1997) has reviewed and tested several of these methods. Thornton and DeSalle (2000) have recently proposed a method to localize incongruence to specific nodes in a phylogeny.

It is worth noting that horizontal transfer is not the only cause of incongruence. Other possibilities such as long-branch attraction, convergent evolution, duplication and lineage sorting, and base composition bias should be ruled out before making a claim of horizontal transfer.

Traces of a foreign signature

A second way to detect horizontal transfer would be to search for sequences that have maintained some record of the places they have been. This record must be a characteristic that distinguishes the sequence from the rest of the genome.

Combined percentages of guanine and cytosine (GC) in genomic nucleic acid compositions are characters strongly correlated with the phylogeny of Bacteria and Archaea. Close relatives tend to have GC contents that are similar, and by extension GC content seems to be a relatively stable character of a genome over time (Ochman and Lawrence, 1996). However, GC content varies greatly, from 25% to 75%, throughout the Bacterial Domain and can be used to separate species and even higher-level groups. Early on, evidence suggested that GC content remains more or less homogeneous across each genome (Rolfe and Meselsohn, 1959; Sueoka, 1959). Recent whole genome sequences have confirmed that most individual stretches of sequence usually match the average genome GC percentages. (Blattner et al., 1997; Fleischmann et al., 1995; Fraser et al., 1998; Nelson et al., 1999; Parkhill et al., 2000a,b; Tettelin et al., 2000) In the aggregate, this information led Sueoka (Sueoka, 1962; Sueoka, 1988; Sueoka, 1992; Sueoka, 1993) to suggest that each organism has a "directional mutation pressure" that drives the GC content towards some characteristic value.

Other characteristic sequence biases also exist in the unicellular world. Though the genetic code of life is degenerate—i.e., several different codons specify the same amino acid—codon usage is not random. Certain species consistently prefer a specific set of codons from among the degenerate, synonymous choices (Ikemura, 1981; Ikemura, 1982; Ikemura, 1985). This preference, like GC content, unites close relatives and is more or less homogeneous throughout the genome (Sharp et al., 1988; Sharp and Matassi, 1994).

Together, GC content and codon bias analyses allow another in-road for studying horizontal gene transfer by presenting the following predictions about the evolution of a gene sequence once it has entered a foreign organism:

1. Genes that have recently moved from one genome to another should have a GC content and codon preference that mimics that of the previous host. If this donor had a sequence bias significantly different from the recipient's, then transfer can be detected by an anomalous nucleotide content—a foreign signature.
2. Once a gene has entered a new host it will be driven by directional mutation towards the new sequence environment. Lawrence and Ochman have named this predicted change in bias *amelioration* (Lawrence and Ochman, 1997; Lawrence and Ochman, 1998; Ochman and Lawrence, 1996).

Several quantitative, computer-based methods have been developed to detect significant deviation of sequence preference in data sets. Such tests include the χ^2 test of significant deviation from genomic GC contents and the Codon Adaptation Index (CAI). The χ^2 test assesses the probability that the GC content of a particular sequence would deviate from the GC content of the rest of the genome just by chance. The CAI is a convenient index of how well the codon bias of a gene matches the codon bias of a very highly expressed set of genes from a particular genome (Sharp and Li, 1987). Because highly expressed genes can be considered a standard point of reference, genes that have low CAI values deviate from the norm. The CAI has the advantage of normalizing for sequence length and is therefore a generally comparable index.

GC content and codon bias deviations have been most extensively studied in *E. coli*, from which the estimated percentages of acquired genes range from 6% to 18% of the total genome (Lawrence and Ochman, 1997; Lawrence and Ochman, 1998; Medigue et al., 1991; Ochman and Lawrence, 1996; Whittam and Ake, 1992). Other studies have used sequence bias criteria to infer recent horizontal transfer in a wide array of organisms including, among many others, *Salmonella enterica* (Ochman and Groisman, 1994) *Bacillus subtilis* (Garcia-Vallvé et al., 1999), and *Thermotoga maritima* (Nelson et al., 1999). In a recent review, Ochman et al. (2000) used sequence bias evidence from 19 fully sequenced Archaea and Bacteria to suggest that the percentage of acquired sequences is very significant and varies with genome size. Moreover, analyses of GC content have become almost requisite in identifying pathogenicity islands— alien stretches of DNA that encode virulence factors. (Hacker et al., 1997)

Identifying anomalous sequence bias has become a widespread technique for detecting foreign genes. However, this approach is limited because it relies on an indirect indication: the assumption that sequence anomaly implies horizontal transfer. Disregarding arguments based on simple random chance, it is possible that the GC and codon preferences of some genes will be "atypical" because of the workings of some currently unidentified mechanism of bias (Syvanen, 1994). Other limitations include the fact that transfer cannot be detected from organisms that have, by relation or chance, the same sequence preferences. In addition, anciently transferred sequences that have come to completely mimic the characteristic genome content of the recipient also go unnoticed.

In several recent publications, Lawrence and Ochman have discussed and developed the second prediction of characteristic sequence bias, the theory of *amelioration* (Lawrence and Ochman, 1997; Lawrence and Ochman, 1998; Ochman and Lawrence, 1996). Their work expresses Sueoka's "directional mutation pressure" in terms of the overall mutation rate of the *E. coli/Shigella/Salmonella* clade, and predicts that this rate can be used to determine the original or ancestral sequence bias of the gene before the horizontal transfer event. The crucial observation in this model, made originally by Muto and Osawa in 1987, is that the overall GC percentage and the GC frequency at each particular codon position have specific linear relationships over all bacterial genomes. Therefore, given the GC content of the organism, the "typical" GC content at each codon position can be extrapolated and *vice versa* (Fig., 1). Lawrence and Ochman (1997) suggest that certain sequences that have recently been acquired will have sequence characteristics that do not conform to the Muto and Osawa (1987) linear relationships. The relationship between overall and positional GC content will, therefore, be unlike that found in any genome; it will fall off the Muto-Osawa line. Directional mutation pressures will have acted differently at distinct codon positions. Consequently, the acquired gene will be in a state of disequilibrium somewhere on the way to demonstrating the characteristic GC environment of its new host.

Given a mathematical model of amelioration, the observed relationship between positional and overall GC content can be "back-ameliorated" until it falls somewhere on Muto and Osawa's (1987) typical, linear, genome function. This intersection is the presumed ancestral sequence bias signature.

Because Lawrence and Ochman's model of amelioration is expressed in terms of the overall mutational rate in bacteria, they suggest that by making certain assumptions about the absolute rates of sequence mutation over time, one can use back-amelioration to predict the time of entry of a foreign sequence. Using this model they recently estimated that approximately 16 kilobases of DNA have been transferred into the *E. coli* chromosome every million years (Lawrence and Ochman, 1998). This is an astoundingly high rate for a genome of 4,639,221 base pairs and implies that there has been a comparable rate of loss.

Back Amelioration

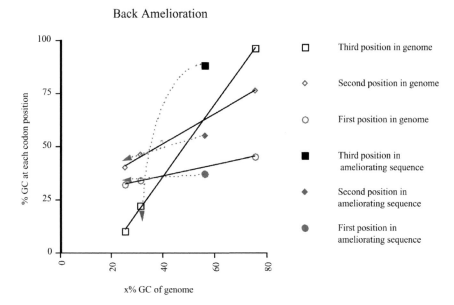

Figure 1. Amelioration. Solid lines represent the linear relationships at each codon position to overall GC content as described by Muto and Osawa (1987). Solid shapes show a hypothetical acquired sequence at disequilibrium with the Muto/Osawa relationships. Dotted lines represent mathematical models of amelioration, such that the point of intersection between dotted and solid lines is the ancestral GC state of the transferred sequence. This graph was redrawn from Lawrence and Ochman (1997).

    The amelioration model relies heavily on assumptions about Bacterial genome evolution. It assumes constant rates for overall mutation and, by extension, amelioration. If re-evaluations of rates or estimated divergence times differ, the model will have to be adjusted. In general, amelioration needs to be tested in different systems and compared to careful and robust inferences of phylogenetic incongruence. Back–amelioration, which predicts the ancestral GC state of the donor organism, seems to be particularly amenable to being tested by phylogenetic analyses in which GC content could be used as a continuous character and overlaid on a gene tree.

    Comparative phylogenetic techniques using incongruence have the advantage of directly testing the hypothesis of incongruent gene history, and allow a more rigorous examination of the hypothesis of horizontal transfer without making assumptions about the traces of sequence bias that different histories might leave on genes. But amelioration models, if the rate assumptions are accurate, may be a good independent source of information that attractively brings time into arguments of horizontal transfer events. For instance, amelioration could be used to help distinguish between conflicting equally parsimonious (or probable) phylogenetic solutions that represent different evolutionary scenarios, such as different directions of horizontal transfer or possible duplication and lineage sorting.

Horizontal transfer muddies the waters: challenges to the origin and domains of life

Since the mid-1970s, the traditional division of extant life into prokaryotes and eukaryotes has given way to the tripartite separation of the Archaea, Eucarya, and Bacteria. This partitioning has emerged largely from the work of Woese and his collaborators (Olsen et al., 1994; Woese, 1987; Woese and Fox, 1977; Woese et al., 1990). In contrast to the previous dichotomy in which prokaryotes were defined in terms of eukaryotic features they do not have (e.g., no nucleus, no membrane-bound organelles, etc.), the tripartite classification was based on sequence and structural comparisons of a common gene, the small (16S) subunit of rRNA (Woese, 1987; Woese and Fox, 1977).

At the outset, the specific relationships between the three primary groupings of life could not be described. This stemmed from the theoretical impossibility of placing the root of life in any particular branch, given only three groups and one gene (Fig. 2A). Rooting would allow both a phylogenetic description of the nature of the last universal cellular ancestor (LUCA) of all known life and a determination of the relationships between domains.

In, 1989 two groups, following the technique of Schwartz and Dayhoff (1978), used the idea of reciprocal rooting of paralogous genes to find a root (Gogarten et al., 1989; Iwabe et al., 1989). This method depends on a gene duplication in the LUCA (or some organism preceding the LUCA) and the retention of both copies of the gene in all extant organisms included in the study. The two paralogous groups can then be used to mutually root one another, so that the duplication event represents the LUCA (Fig. 2B). 16S sequences cannot be used for this purpose, so the initial studies used widely accepted paralogs—the elongation factors (EF-Tu and EF-G) and vacuolar ATPases (α and β).

Reciprocally rooted phylogenies placed the root in a somewhat unexpected location. The Archaea and Eucarya grouped together to the exclusion of the Bacteria on both sides of the duplication, implying that the root of known life lay in the Bacterial branch (Fig. 2B). This root dealt a formidable blow to the prokaryote/eukaryote distinction, and led Woese et al. (1990) to suggest that each of the three partitions should be reclassified as new primary taxonomic units called Domains. Alternative models have questioned both this rooting and the monophyly of the members of the "resolved" trichotomy (Forterre et al., 1992; Gupta, 1998a; Lake, 1985; Lake, 1988; Lake et al., 1985; Lake et al., 1984; Rivera and Lake, 1992). Others have argued that methodological shortcomings and sequence saturation may prohibit both rooting the tree and creating a reliable phylogeny in the first place (Philippe and Forterre, 1999; Teichmann and Mitchison, 1999). But in other studies of gene families the Iwabe-Gogarten root has held (Brown and Doolittle, 1995; Brown and Doolittle, 1997; Brown et al., 1997; Gribaldo and Cammarano, 1998) and, despite critiques, this fundamental reorganization has gained widespread acceptance as the natural, and paradigmatic, taxonomy of the universal tree of life.

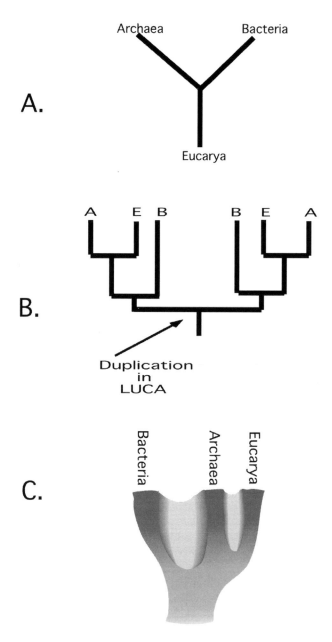

Figure 2. Progression of ideas of the Universal Tree: (A) The unrooted tripartite tree (Woese, 1987; Woese and Fox, 1977) (B) Reciprocal paralog rooting places the root in the Bacterial branch (Gogarten et al., 1989; Iwabe et al., 1989) (C) A three-domain scheme allows horizontal transfer, but is dominated by vertical transmission. Darker areas indicate more vertical lineage commitment. This particular tree is a depiction of the "genetic annealing" model proposed by Woese (1998).

Despite the tight hold of the Bacteria rooted tree, increasing appreciation of the influence of horizontal transfer in the evolution of microorganisms has resulted in notions of the LUCA, the first derivative lineages, and ultimately the reality of the tree of life being questioned. In order to maintain a natural, three-domain scheme, there must be some real difference, or vertical lineage commitment, within each major branch. A high level of horizontal transfer among the domains would compromise this and potentially obscure vertical phylogenetic signals. Many arguments that try to maintain the integrity of the universal tree in the face of horizontal transfer rest on the question of extent. For instance, while making a strong case for the exchange of certain genes between the Archaea and Bacteria, Hilario and Gogarten (1993) concluded, on the basis of the agreement of multiple data sets, that the sisterhood of the Eucarya and Archaea is probably due to vertical inheritance. A similar type of argument suggests that the amount of horizontal inheritance has become less frequent over time. In such a model, Woese postulates that the LUCA, not an organism *per se*, lived in a world in which the predominant mode of inheritance was horizontal. As genetic systems became more complex and precise, horizontal transfer was edged out by descent, and the three major vertical lineages eventually emerged—a process referred to as "genetic annealing" (Fig. 2 C) (Woese, 1998). This theory does not require the cessation of all horizontal exchange, but it predicts that after "genetic annealing", a certain portion of genes in each genome should constitute a uniquely derived and vertically inherited signature for each domain (Graham et al., 2000).

Other arguments submit that regardless of horizontal transfer in some classes of genes, a certain set of genes is so refractory to exchange that the natural tree of organismal descent can be accurately represented by the phylogeny of these genes. Is there a way to pick out these marker genes from the vast array of sequences? Several authors theorize that the probability of transfer is somehow related to the functional classification of the gene (Andrade et al., 1999; Brown and Doolittle, 1997; Jain et al., 1999; Rivera and Lake, 1992; Vellai, 1999). For instance, Jain et al. (1999) have recently theorized that genes that encode proteins involved in complex interactions are less likely to be horizontally transferred because they are less likely to function properly in a heterologous system (see also Rivera et al., 1998). From this viewpoint, the extent to which the phylogeny of the gene reflects the organismal phylogeny is more or less dependent on the gene's function, and good marker genes could be identified by functional classification. Though this is a tantalizing connection, the exact way in which function and horizontal transfer are related requires much more theoretical and experimental development.

Both of the preceding arguments maintain that, despite horizontal transfer, the tree of life can still be recovered. Other appraisals have not been so optimistic. In a recent review, R.F. Doolittle (1999) reasons that unless the effect of horizontal gene transfer can be eliminated, either by relying on its improbability or concentrating on some set of marker genes, we must abandon the idea of the universal organismal tree altogether. In its place a more natural representation might be a reticulated tree, or net, in which modern organisms are the outcomes of multiple horizontal and vertical events (Arber, 2000; Doolittle, 1999; Hilario and Gogarten, 1993; Martin, 1999). However, there is room for caution before we entirely dispose of the notion of a universal tree. As pointed out in a reply to Doolittle's paper, the demonstration of vast amounts of horizontal transfer between domains is supported, in large part, by best-hit analyses[2], and there is a need to confirm reported events with gene-by-gene, rigorous, phylogenetic and statistical testing (Stiller, 1999).

It is clear that a model that allows the blurring of boundaries should replace the previously clean divisions of the universal tree; the reality of this representation will ultimately be judged on how blurry those boundaries are. Given the eroding effects of horizontal transfer, reconstructions of an overarching organismal tree should be viewed now with a critical eye. New definitions of the tree should include explicit assessments of the detractions of horizontal transfer (and other processes such as sequence saturation) from the data. In addition, we should be aware of the pitfalls of techniques such as paralogous rooting that involve only one or two gene families.

Horizontal transfer clarifies the view: operons, species, and genomes

Because of our general ignorance of the dynamics and nature of inheritance over time, the study of microbial evolution has suffered from an inability to explain and organize certain basic characteristics of the Archaea and Bacteria. Already, evolutionary biologists have begun to reexamine certain qualities of asexually reproducing microorganisms with the assumption that horizontal transfer may be an important force molding microbial evolution. Emerging theories, combined with a rapidly expanding database, elegantly re-order the microbial world and may provide solutions to phylogenetic dilemmas.

Operons

One major difference separating the Archaea and Bacteria from the Eucarya is that prokaryotic genes involved in the same biochemical processes are more likely to be found clustered in linear arrangements, called operons, that are expressed from the same promoter (Demerec and Hartman, 1959; Jacob et al., 1960; Yanofsky and Le, 1959). Several models have emerged to explain the tendency for genes to cluster (Beckwith, 1996; Lawrence, 1999; Lawrence and Roth, 1996). Probably the most popular model suggests that it is beneficial for the organism to co-regulate genes that are needed in the same pathway.

---

[2] Several notable exceptions include Delwiche and Palmer (1996); Jain et al. (1999); Nelson et al. (1999).

A new theory proposed by Lawrence and Roth (1996) postulates that genes cluster because, by staying in close physical proximity, they are more likely to be horizontally transferred together. The reasoning for the model is as follows: when a single gene is transferred into a heterologous system, its protein product has a lower probability of making the right contacts at the right times because it is not adapted to the system. This and other possible problems decrease the chances that it will function properly. Because positive selection will only work directly on genes that function and do so advantageously, it is in the best interest of the gene to be transferred along with a system to which it is adapted. Because of the mechanics of transfer, genes that cluster have a higher probability of being transferred together. Therefore, clustering is the result of a selective pressure created by horizontal transfer; it is a method by which genes ensure their own persistence. Lawrence and Roth (1996) call this the *selfish operon* model.

One corollary of the selfish operon model is that a weakly selected set of genes, involved in a process that is only important in certain, nonessential phases of the life cycle, will tend to cluster more than a set involved in some essential process. This relies on the idea that selection will act on weakly or intermittently selected genes differently at different times. If the period during which the genes are not necessary is long enough in any particular organismal lineage, they may be doomed to eradication. Clustering, by increasing the chance that the whole, functional system will be passed into a new host, allows genes to rely on horizontal transfer events to avoid extinction. Maximizing the probability and success of horizontal transfer makes the cluster more likely to be under positive selection more of the time.

The relationship between weak or inconsistent selection and horizontal transfer provides a worthwhile vantage point from which to view the evolution of pathogenicity islands and virulence factors that reside on mobile elements such as phages or plasmids. Certain pathogens can live in the environment as well as in a host—or in multiple hosts—and therefore do not need host-specific or environment-specific genes for at least part of the time. If the pathogen remains in the environment long enough, host colonization or virulence genes will likely be lost. However, if pathogenicity genes cluster and have some means of actively transferring themselves, they have a greater chance of being in a lineage that contacts a host and initiates pathogenesis.

It has become increasingly clear that many pathogenicity determinants for major human diseases are carried in clusters on mobile, or previously mobile, genetic elements (Cheetham and Katz, 1995; Groisman and Ochman, 1994; Groisman and Ochman, 1996; Hacker et al., 1997; Ziebuhr et al., 1999). For example, both host colonization and toxin production in *Vibrio cholerae* rely on two pathogenicity islands, the TCP and CTX elements, that are transduced from one *V. cholerae* strain to another, and even between species, by filamentous phages (Boyd et al., 2000; Karaolis et al., 1999; Kimsey et al., 1998; Waldor, 1998; Waldor and Mekalanos, 1996). Other examples include the infamous Shiga toxin-bearing phage of *E. coli* strain O157:H7 (O'Brien et al.,

1984) the pathogenicity islands of pathogenic *Yersinia* (Buchrieser et al., 1998; Carniel et al., 1996; Fetherston and Perry, 1994; Fetherston et al., 1992), and the Cag Island of *Helicobacter pylori* (Censini et al., 1996). These systems make excellent models for studies of the way selection acts on horizontally transferred systems composed of intermittently selected elements.

Species, innovation and diversification

Bacterial and Archaeal systematics have had a somewhat irresolute stance on the biological meaning of species. This stems from the fundamental differences in inheritance between sexually and asexually reproducing organisms. First, the very broad range of horizontal gene transfer observable among microorganisms means that very few insurmountable barriers isolate one microbial species from another, and this makes the biological (isolation) species concept (Mayr, 1942) difficult to apply. Asexual reproduction further complicates matters because committed, clonal lineages can be formed during a reproductive event. Should these exact, or nearly exact, duplicates be designated as different species? Perhaps because of these difficulties, phenetic definitions of species have been popular in bacterial systematics (Sneath and Sokal, 1973). Because such descriptions classify organisms based on some arbitrary cut-off in similarity, the names given to species are no more than a convenient way to talk about organisms that resemble one another. If a definable entity exists that may be called a species, then a phenetic approach seems less reasonable and perhaps counterproductive. Furthermore, if we wish to include ideas about common ancestry then we should use a system that explicitly incorporates phylogeny (for example, Nixon and Wheeler, 1990).

Defining a microbial species concept rests primarily on understanding the relationship between widespread horizontal transfer and clonal (vertical) population structure. Bacterial population genetics has made an important contribution in this regard by showing that although bacterial populations are mostly clonal, recombination among genomes is not absent (Smith et al., 1991). Recombination in an asexually reproducing population must be preceded by a horizontal transfer event, and, therefore, genetic evidence for recombination is good evidence for horizontal transfer. Therefore, inheritance in populations of bacteria is mostly vertical with horizontal recombination events that interrupt what Milkman and Bridges (1990) refer to as a "clonal frame". However, the contribution of recombination varies to a high degree between species. Linkage disequilibrium studies of bacterial populations show that the prevalence of recombination varies from the almost entirely clonal to the highly sexual (Caugant et al., 1981; Ochman et al., 1983; Smith et al., 1993; Souza et al., 1992; Whittam et al., 1983a, Whittam et al., 1983b). In other words, our currently named species demonstrate a wide range of inheritance dynamics. Any definition of species that relies on sexual isolation must take this variability into account.

The presence of recombination in bacterial species has revitalized hopes of developing a microbial biological species concept. Vellai et al. (1999) suggest that species be defined by the sequences which they cannot exchange, but it is unclear that such sequences could be unambiguously identified or even that they exist at all. Dykhuizen and Green (1991) propose that species be defined by localizing incongruence between gene tree topologies. They presume that where asexual, clonal lineage splitting predominates, the phylogenies of a given set of genes will be congruent. In areas of the tree in which organisms freely exchange genes, different genes should no longer exhibit congruent phylogenies. Therefore, the nodes in the tree after which there is significant incongruence would represent speciation events. Maynard-Smith (1996) points out that, given the differences in recombination frequency in currently identified species, this method would result in a serious rearrangement and renaming process that would favor very small divisions in mostly clonal lineages (e.g., *Salmonella*) and very broad classifications in highly sexual species (e.g., *Neisseria*). Nevertheless, this reclassification system would have the advantage of normalizing the relationship between the unit species and the amount of recombination.

Another assumption of Dykhuizen and Green's (1991) sexual isolation model is that at one definable point in time—the speciation event—a subgroup of organisms will begin to exchange genes freely only among themselves. This may be untenable in light of recent suggestions that horizontal transfer is rampant between very distantly related organisms. Such exchange would make a "species" a meaninglessly large group. In addition, if the amount of exchange does not increase at one moment in time but increases (with phylogenetic proximity) more gradually, then all incongruencies may not localize to one specific node, making the boundary of the species ambiguous.

It may not be realistic to expect that one universal species concept applies to all organisms especially given very distinct dynamics of inheritance. The fact that genetic exchange and reproduction are decoupled in the Bacteria and Archaea distinguishes these organisms from sexually reproducing species in that they have very different courses of diversification. Ochman and Groisman (1994) point out that most of the variation in observable phenotypic characters is due to sequences that are specific to one genome or the other rather than changes in common genes. Many of these "species-specific" sequences have anomalous GC contents, suggesting that they were acquired through horizontal transfer with distant relatives. They go on to propose that acquired, novel sequences allow species to occupy some previously unavailable ecological niche or carry out some novel activity. Similarly, loss of specific systems or genes could also define a new lineage with different potentialities (Lawrence, 1998). This model, in which diversification is determined in a clonal lineage by the acquisition or loss of genes, has been proposed or discussed by several authors (de la Cruz and Davies 2000, Lawrence, 1997; Martin, 1998; Matic et al., 1995; Medigue et al., 1991; Ochman et al., 2000; Vulic et al., 1997; Vulic et al., 1999). It should be noted that novel acquisition does not necessarily include all recombinational events such as those between closely related homologous (orthologous) sequences (Ochman et al., 2000).

Could novel acquisition or loss define a species concept for asexually reproducing microorganisms? Despite the fact that such a definition would not be comparable to the biological species concept, it is useful in terms of the phenotypic information that we might want to know about a microbial species. In the words of Maynard-Smith (1996), "One cannot decide on a sensible method of naming and classifying unless one has some idea of what one wants to do with the system." The novel acquisition/loss model is especially convenient if one is interested in ecological niche or pathogenesis. Because many transferred genetic elements encode novel pathways or virulence determinants, novel acquisitions of such systems instantly endow a lineage with a novel phenotype or ability (Groisman and Ochman, 1996). A gene loss may either free a lineage from some constraint or limit some ability.

This model is nicely illustrated by the relationship between *Yersinia pseudotuberculosis*, which causes an enteric human disease with low mortality, and *Yersinia pestis,* the etiologic agent of bubonic and pneumonic plague. Nucleotide sequences from *Y. pestis* are often one hundred percent identical to *Y. pseudotuberculosis,* which has led some researchers, arguing within a phenetic framework, to propose that the two species be lumped together—reclassified as subspecies (Bercovier et al. 1980). More recent work has reiterated the very close relationship between *Y. pseudotuberculosis* and *Y. pestis*: "[I]f not for tradition, the results would justify changing the name of *Y. pestis* to reflect the fact that it is not an independent species"(Achtman et al., 1999). Independent of tradition, the fact remains that *Y. pseudotuberculosis* and *Y. pestis* cause two very different diseases and have very different life cycles, which may be a good reason to make a formal biological distinction between the two. Acquisitions or losses that allow *Y. pestis* or *Y. pseudotuberculosis* to colonize and persist in their respective niches could describe such a distinction. *Y. pestis* has two "*pestis*-specific" plasmids (Brubaker, 1991) and several chromosomal, *Y. pseudotuberculosis*-like genes that have been inactivated by insertion elements (Carnoy et al., 2000; Simonet et al., 1996). Each whole gene difference could be seen as a synapomorphy uniting the monophyletic species *Y. pestis*, making the novel gene acquisition/loss framework amenable to cladistic analysis and a phylogenetic classification scheme. On the other hand, if *only* a whole gene loss or acquisition can define a new species, then this method runs a significant risk of creating paraphyletic or polyphyletic species, and any acquisition/loss argument should be tested with multiple other sources of phylogenetic information to insure species monophyly.

The novel acquisition/loss model may also be supported by ideas emerging from molecular biology. Several recent studies have proposed that bacterial speciation might involve a molecular mechanism that increases the amount of recombination leading to the fixation of acquired genes. Experimental evidence indicates that stressed organisms up-regulate the SOS system and down-regulate mismatch repair systems, thus increasing recombination rates (Matic et al., 1995; Vulic et al., 1997; Vulic et al., 1999). Given that recombination seems to be the major obstacle to genetic exchange, this shift allows newly acquired genes to be incorporated into the genome (Majewski et al., 2000; Matic et al., 1995; Roberts and Cohan, 1993; Vulic et al., 1997; Vulic et al., 1999). In this way, bacterial diversification (i.e., speciation) may involve a period of relatively increased acquisition via horizontal transfer followed by the re-instatement of the previous steeper, but not insurmountable, boundary to exchange.

Genomes *versus* genes

At the time that this chapter is written, 31 Archaeal and Bacterial genomes are fully sequenced and annotated, another 17 are sequenced but not yet annotated, and 69 more are in progress (www.nlm.nih.gov). This means that within the next few years we should have at our disposal the fully annotated sequences of at least 117 Archaea and Bacteria. This gives microbial systematists an unparalleled opportunity to understand the evolution of genomes and organisms. A data set of this magnitude and breadth will not be available to eukaryotic systematists for some time.

But how should we evaluate the immense amount of data in whole genomes, and how will this vast data set inform our ideas about microbial organismal evolution? It may be that comparative whole-genome analyses, because of their completeness, can resolve some of the obstacles to phylogenetic reconstruction posed by horizontal gene transfer. But first we must understand how the possibility of widespread and frequent horizontal transfer affects whole-genome phylogenies.

At the heart of the dilemma is the hotly debated topic of whether or when potentially incongruent data sets should be combined to reconstruct organismal phylogeny (for reviews see Brower et al., 1996; de Queiroz et al., 1995; Kluge, 1989; Nixon and Carpenter, 1996). In a genome in which some or all of the component genes may have experienced horizontal transfer events at some point in the past, should all of the genes be analyzed separately or combined into a single data set?

Several recent attempts have been made to reconstruct the phylogeny of the universal tree based on data sets drawn from whole genomes of information (Gupta, 1998b; Huynen et al., 1998; Huynen and Bork, 1998; Huynen and Snel 2000, Snel et al., 1999). The power of these analyses rests as much on knowing all of the genes or sequences present in a given organism as on the ability to say precisely which are absent. Moreover, other characters and structural features, such as gene arrangement and position, codon bias, and GC content, can be included along with sequence comparisons in the analysis. In general, whole-genome phylogenies have the advantage of not taking the phylogeny of a single gene to represent the phylogeny of organisms or genomes. But these analyses assume that the depth and breadth of the vertical data should not be masked by anomalous historical signals coming from genes that have been horizontally transferred. If this assumption is true, then the best technique for obtaining a hypothesis of prokaryotic organismal phylogeny from whole-genome analysis would be to include as much extractable data as possible and carry out a simultaneous phylogenetic analysis (Brower et al., 1996; Nixon and Carpenter, 1996). Of course, the computational power required for such an analysis would be immense.

In any estimation, vertical inheritance is a much more frequent occurrence than horizontal transfer, and assuming that vertical signal will dominate the analysis over short evolutionary periods is very reasonable. However, long-term accumulation of horizontal transfers may affect reconstructions of deep phylogeny. Given recent estimations of the frequency of horizontal transfer in the Archaea and Bacteria, we may have no hope of recovering a universal organismal tree with any data set.

The entire question rests on the extent of horizontal gene transfer over time, and the best way to estimate this is by testing phylogenetic incongruence between many individual gene histories. A remarkable point made by Woese (1998) in his discussion of the universal ancestor is that we should "let molecular phylogenetic trees represent exactly what they in the first instance do represent, histories of individual genes or gene groupings". Gene families, barring intragenic recombination, can still be analyzed with a vertical phylogenetic model.

One way to represent the frequency of horizontal transfer in gene phylogeny would be to treat genes as parasites in host genomes and apply techniques used for reconciling host-parasite trees (Hafner and Page, 1995; Page, 1994b; Page, 1993b; Page, 1994a). Given a well-supported parasite (gene) tree that is significantly incongruent with the host (genome) tree, the two trees could be reconciled by minimizing, but allowing, some set of *ad hoc* hypotheses such as gene duplications followed by lineage sorting, convergence, and horizontal transfer. This can be done with computer programs such as TreeMap or GeneTree (Page, 1996; Page, 1998b). If all the genes in a genome were compared and reconciled in this manner, one could build the reconciled, reticulate tree of life, allowing a rigorous assessment of how much horizontal transfer is represented in microbial evolution.

However, it is not obvious which phylogeny should be used as the genome (host) phylogeny. A genome tree produced by simultaneous analysis of whole-genome data would represent the best-possible, vertical, phylogenetic explanation of the data. Such a whole-genome host phylogeny has the advantage of giving a much stronger test of incongruence. However, because we do not yet know what effect horizontal transfer would have on this phylogeny, it is not clear what this standard would actually represent.

It may be more logically consistent to abandon an organismal hypothesis altogether when it comes to deep phylogeny, and limit our gene-by-gene comparisons and reconciliations to individual gene trees. To this end we may want to choose some gold standard, gene phylogeny as an arbitrary basis for comparison—a phylogenetic scaffold upon which all other gene trees could be hung. However, this technique has methodological limitations. If we accept the hypothesis that any gene can undergo a horizontal transfer event, we must always be aware that the standard gene scaffold does not necessarily represent organismal or genome evolution. In fact, *any* scaffold that does not entirely represent organismal evolution will actually describe horizontal transfer events that did not happen (Fig. 3). Because of this, whether we choose a whole-

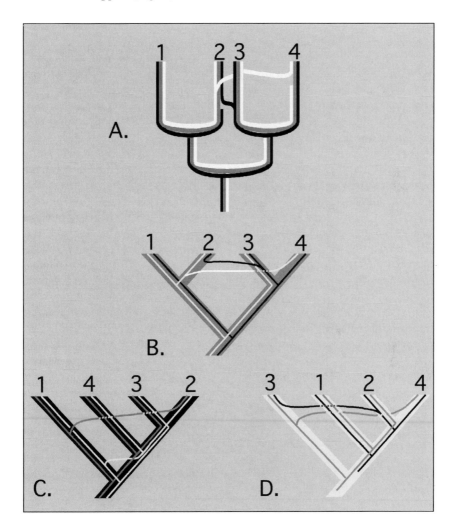

Figure 3. Problems with a standard gene scaffold. (A) Represents the true phylogeny. Each different shade represents one gene phylogeny. B–D are reconstructed phylogenies allowing up to one transfer event for each tree reconciliation. (B) Using the gray gene phylogeny as a scaffold, the reconciled phylogeny accurately describes the evolutionary scenario (transfers from taxon 3 to 2 and 2 to 4) because it accurately represents organismal phylogeny. In (C) and (D) using either the white or black gene as the scaffold reconstructs a history of events that did not happen.

genome or single-gene scaffold, the outcome of the exercise would simply be a representation of the number of significant incongruencies and possible transfer events with respect to the scaffold. Completing the exercise would allow us to assess how much confidence we should give to a vertical organismal tree for microorganisms, and an idea of the extent to which the tree of life might be better described as a web.

Conclusion

The existence of horizontal transfer on a large scale represents a major threat to phylogenies of microorganisms that rely on the assumption of the exclusivity of vertical transfer. A strictly bifurcating organismal tree is not an appropriate representation of a world in which both horizontal and vertical transfer play significant roles in inheritance. Furthermore, if horizontal transfer is as widespread and frequent as we suspect it is, then it is certain to have been a potent force during evolution affecting innovation, selection and diversification in microbial life. However, vertical phylogenies may still be the best way to understand the evolution of genes, and, by extension, to understand the exact way that horizontal inheritance has influenced life. For the moment, we may want to heed Woese's (1998) advice to "release this notion of organismal lineages altogether and see where that leaves us." To truly understand horizontal transfer in the net of life, rigorous gene-by-gene phylogenies should be compared and reconciled to each other or with some agreed-upon standard scaffold for comparison. Constructing a reticulated phylogeny should allow an appraisal of how well our current ideas of inheritance can explain, among other things, the formation of operons, the reality of species, genome evolution, the nature of the last universal common ancestor, and the reality of a universal tree of life.

References

Abouheif, E. (1997) Developmental genetics and homology: a hierarchical approach. *TREE* 12: 405–408.

Abouheif, E., Akam, M., Dickinson, W.J., Holland, P.W.H., Meyer, A., Patel, N.H., Raff, R.A., Roth, V.L. and Wray, G.A. (1997) Homology and developmental genes. *Trends Ecol Evol* 13: 432–433.

Abouheif, E., Zardoya, R. and Meyer, A. (1998) Limitations of metazoan 18S rRNA sequence data: implications for reconstructing a phylogeny of the animal kingdom and inferring the reality of the Cambrian explosion. *J Mol Evol* 47: 394–405.

Achtelig, M. (1975) Die Abdomenbasis der Neuropteroidea (Insecta, Holometabola). *Zoomorphologie* 82: 201–242.

Achtman, M., Zurth, K., Morelli, G., Torrea, G., Guiyoule, A. and Carniel, E. (1999) *Yersinia pestis*, the cause of plague, is a recently emerged clone of *Yersinia pseudotuberculosis*. *Proc Natl Acad Sci USA* 96: 14,043–14,048.

Agosti, D., Jacobs, D. and DeSalle, R. (1996) On combining protein sequences and nucleic acid sequences in phylogenetic analysis: the homeobox protein case. *Cladistics* 12: 65–82.

Aguinaldo, A.M.A., Turbeville, J.M., Lindford, L.S., Rivera, M.C., Garey, J.R., Raff, R.A. and Lake, J.A. (1997) Evidence for a clade of nematodes, arthropods and other moulting animals. *Nature* 387: 489–493.

Aguinaldo, A.M. and Lake, J.A. (1998) Evolution of the multicellular animals. *Amer Zool* 38: 878–887.

Akam, M. (1998) The Yin and Yang of Evo-Devo. *Cell* 92: 153.

Allard, M.W. and Carpenter, J.M. (1996) On weighting and congruence. *Cladistics* 12: 183–198.

Allard, M.W., Farris, J.S. and Carpenter, J.M. (1999) Congruence among mammalian mitochondrial genes. *Cladistics* 15: 75–84.

Altschul, S.F., Gish, W., Miller, W., Myers, E.W. and Lipman, D.J. (1990) Basic local alignment search tool. *J Mol Biol* 215: 403–410.

Altschul, S.F., Madden, T.L., Schaffer, A.A., Zhang, J., Zhang, Z., Miller, W. and Lipman, D.J. (1997) Gapped BLAST and PSI-BLAST: a new generation of protein database search programs. *Nucl Acid Res* 25: 3389–3402.

Álvarez, Y.J., Juste, B., Tabares, E., Garrido-Pertierra, A., Ibáñez, C. and Bautista, J.M. (1999) Molecular phylogeny and morphological homoplasy in fruitbats. *Mol Biol Evol* 16: 1061–1067.

Amadon, D. (1949) The seventy-five percent rule for subspecies. *Condor* 51: 250–258.

Amero, S.A., Kretsinger, R.H., Moncrief, N.D., Yamamoto, K.R. and Pearson, W.R. (1992) The origin of nuclear receptor proteins: A single precursor distinct from other transcription factors. *Mol Endocrinol* 6: 3–7.

Andersen, S. (1970) Amino acid composition of spider silks. *Comp Biochem Physiol* 35: 705–711.

Anderson, S., de Bruijn MH, Coulson AR, Eperon IC, Sanger, F. and Young IG (1982) Complete sequence of bovine mitochondrial DNA: Conserved features of the mammalian mitochondrial genome. *J Mol Biol* 156: 683–717.

Andrade, M.A., Ouzounis, C., Sander, C., Tamames, J. and Valencia, A. (1999) Functional classes in the three domains of life. *J Mol Evol* 49: 551–7.

Applebury, M.L. (1994) Relationships of G-protein-coupled receptors: A survey with the photoreceptor opsin subfamily. *In*: D.M. Fambrough (ed.): *Molecular Evolution of Physiological Processes*. Rockefeller University Press, New York, pp. 235–248.

Aravind, L., Tatusov, R.L., Wolf, Y.I., Walker, D.R. and Koonin, E.V. (1998) Evidence for massive gene exchange between archaeal and bacterial hyperthermophiles. *Trends Ecol Evol* 14: 442–444.

Arber, W. (2000) Genetic variation: molecular mechanisms and impact on microbial evolution. *FEMS Microbiol Rev* 24: 1–7.

Archie, J.W. and Felsenstein, J. (1993) The number of evolutionary steps on random and minimum length trees for random evolutionary data. *Theor Pop Biol* 43: 52–79.

Arctander, P., Johansen, C. and Coutellec-Vreto, M. (1999) Phylogeography of three closely related African bovids (Tribe Alcelaphini). *Mol Biol Evol* 16: 1724–1739.

Aristotle (1991) *Historia animalium*. (D.M. Balme, ed. and Trans.). Harvard University Press, Cambridge.

Arnason, U., Gullberg, A. and Widegren, B. (1991) The complete nucleotide sequence of the mitochondrial DNA of the fin whale, *Balaenoptera physalus*. *J Mol Evol* 33: 556–568.

Arnold, E.N. (1994) Investigating the evolutionary effects of one feature on another: Does muscle spread suppress caudal autonomy in lizards? *J Zool, Lond* 232: 505–523.

Avery, O.T., MacLeod, C.M. and McCarty, M. (1944) Studies on the chemical nature of the substance inducing transformation in pneumococcal types. *J Exp Med* 79: 137–59.

Avise, J.C. (2000) *Phylogeography: The history and formation of species.* Harvard University Press, Cambridge.

Avise, J.C., Arnold, J., Ball, R.M., Bermingham, E., Lamb, T., Neigel, J.E., Reeb, C.A. and Saunders, N.C. (1987) Intraspecific phylogeography: the mitochondrial DNA bridge between population genetics and systematics. *Annu Rev Ecol Syst* 18: 489–522.

Avise, J.C., Nelson, W.S. and Sibley, C.G. (1994a) DNA sequence support for a close phylogenetic relationship between some storks and New World vultures. *Proc Natl Acad Sci USA* 91: 5173–5177.

Avise, J.C., Nelson, W.S. and Sibley, C.G. (1994b) Why one-kilobase sequences from mitochondrial DNA fail to solve the hoatzin phylogenetic enigma. *Mol Phylogenet Evol* 3: 175–187.

Ayala, F.J. and Rzhetsky, A. (1998) Origin of the metazoan phyla: molecular clocks confirm paleontological estimates. *Proc Natl Acad Sci USA* 95: 606–611.

Baker, R.H. and DeSalle, R. (1997) Multiple sources of character information and the phylogeny of Hawaiian drosophilids. *Syst Biol* 46: 654–673.

Baker, R.H., Wilkinson, G.S. and DeSalle, R. (2001) The phylogenetic utility of different types of molecular data used to infer evolutionary relationships among stalk-eyed flies (Diopsidae). *Syst Biol* 50: 87–105.

Baker, R.H., Yu, X. and DeSalle, R. (1998) Assessing the relative contribution of molecular and morphological characters in simultaneous analysis trees. *Mol Phylogenet Evol* 9: 427–436.

Baldauf, S.L. and Palmer, J.D. (1990) Evolutionary transfer of the chloroplast tufA gene to the nucleus. *Nature* 344: 262–265.

Bang, R., DeSalle, R. and Wheeler, W. (2000) Transformationalism, taxism and systematics. *Syst Biol* 49: 19–27.

Barraclough, T.G., Hogan, J.E. and Vogler, A.P. (1999) Testing whether ecological factors promote cladogenesis in a group of tiger beetles (Coleoptera: Cicindelidae). *Proc R Soc Lond B* 266: 1061–1067.

Barrett, M., Donoghue, M.J. and Sober, E. (1991) Against consensus. *Syst Zool* 40: 486–493.

Barrowclough, G.F. (1992) Biochemical studies of the higher level systematics of birds. *Bull Brit Ornithol Club Centenary Suppl* 1992: 39–52.

Barry, D. and Hartigan, J.A. (1987) Statistical analysis of hominoid molecular evolution. *Stat Sci* 2: 191–210.

Baum, D.A. (1992) Phylogenetic species concepts. *Trends Ecol Evol* 7: 1–2.

Baum, D.A. and Donoghue, M.J. (1995) Choosing among alternative "phylogenetic" species concepts. *Syst Bot* 20: 560–573.

Baum, D.A. and Larson, A. (1991) Adaptation reviewed: A phylogenetic methodology for studying character macroevolution. *Syst Zool* 40: 1–18.

Baum, D.A. and Shaw, K.L. (1995) Genealogical perspectives on the species problem. *In*: P.C. Hoch and G.D. Stephenson (eds): *Experimental and molecular approaches to plant biosystematics.* Missouri Botanical Garden, St. Louis, pp. 289–303.

Baumann, P., Moran, N.A. and Baumann, L. (1997) The evolution and genetics of aphid endosymbionts. *Bioscience* 47: 12–20.

Baumann, P., Munson, M.A., Lai, C.-Y., Clark, M.A., Baumann, L., Moran, N.A. and Campbell, B.C. (1993) Origin and properties of bacterial endosymbionts of aphids, whiteflies, and mealybugs. *ASM News* 59: 21–24.

Baverstock, P.R. and Moritz, C. (1990) Sampling Design. *In*: D.M. Hillis and C. Moritz (eds): *Molecular Systematics.* Sinauer Associates, Sunderland, pp. 13–24.

Baverstock, P.R. and Moritz, C. (1996) Project design. *In*: D.M. Hillis, C. Moritz and B.K. Mable (eds): *Molecular Systematics.* Sinauer Associates, Sunderland, pp. 17–28.

Beckwith, J. (1996) The operon: an historical account. *In*: F.C. Neidhardt (ed.): *Escherichia coli and Salmonella: Cellular and molecular biology.* ASM Press, Washington, D.C., pp. 1439–1443.

Beckwitt, R. and Arcidiacono, S. (1994) Sequence conservation in the C-terminal region of spider silk proteins (spidroin) from *Nephila clavipes* (Tetragnathidae) and *Araneus bicentenarius* (Araneidae). *J Biol Chem* 269: 6661–6663.

Beckwitt, R., Arcidiacono, S. and Stote, R. (1998) Evolution of repetitive proteins: spider silks from

Nephila clavipes (Tetragnathidae) and *Araneus bicentenarius* (Araneidae). *Insect Biochem Molec Biol* 28: 121–130.

Beintema, J.J. and Neuteboom, B. (1983) Origin of the duplicated ribonuclease gene in guinea-pig: Comparison of the amino acid sequences with those of two close relatives: *Capybara* and cuis ribonuclease. *J Mol Evol* 19: 145–52.

Bennett, P.M. (1987) Genetic basis of the spread of antibiotic resistance genes. *Ann Ist Super Sanita* 23: 819–25.

Bennett, P.M. (1996) The Spread of Drug Resistance. *In*: D.M. Roberts (ed.): *Evolution of microbial life: Fifty-fourth Symposium of the Society for General Microbiology held at the University of Warwick, March 1996*, Cambridge University Press, Cambridge, pp. 239–250.

Benton, M.J. (1993) *The Fossil Record, Volume 2*. Chapman and Hall, New York.

Benton, M.J. (1999) Early evolution of modern birds and mammals: Molecules *versus* morphology. *BioEssays* 21: 1043–1051.

Bercovier, H.H., Alonso, J.M., Brault, J., Fanning, G.R., Steigerwalt, A.G. and Brenner, D.J. (1980) Intra- and interspecies relatedness of *Yersinia pestis* by DNA hybridization and its relationship to *Yersinia pseudotuberculosis*. *Curr Microbiol* 4: 225–229.

Beverly, S.M. and Wilson, A.C. (1985) Ancient origin for Hawaiian Drosophilidae inferred from protein comparisons. *Proc Natl Acad Sci USA* 82: 4753–4757.

Bilinski, S., Bünnig, J. and Simiczyjew, B. (1998) The ovaries of Mecoptera: Basic similarities and one exception to the rule. *Folia Histochem Cytobiol* 36: 189–195.

Bjorklund, M. (1999) Are third positions really that bad? A test using vertebrate cytochrome b. *Cladistics* 15: 191–197.

Blanchette, M., Kunisawa, T. and Sankoff, D. (1999) Gene order breakpoint evidence in animal mitochondrial phylogeny. *J Mol Evol* 49: 193–203.

Blattner, F.R., Plunkett, G., III, Bloch, C.A., Perna, N.T., Burland, V., Riley, M., Collado-Vides, J., Glasner, J.D., Rode, C.K., Mayhew et al. (1997) The complete genome sequence of *Escherichia coli* K-12. *Science* 277: 1453–1474.

Bock, W.J. (1974) Philosophical foundations of classical evolutionary classification. *Syst Zool* 22: 375–392.

Boettcher, K.J. and Ruby, E.G. (1990) Depressed light emission by symbiotic *Vibrio fischeri* of the sepiolid squid *Euprymna scolopes*. *J Bacteriol* 172: 3701–3706.

Bolker, J.A. and Raff, R.A. (1996) Developmental genetics and traditional homology. *Bioessays* 18: 489–494.

Bonci, A., Chiesurin, A., Muscas, P. and Rossolini, G.M. (1997) Relatedness and phylogeny within the family of periplasmic chaperones involved in the assembly of pilli or capsule-like structures of gram-negative bacteria. *J Mol Evol* 44: 299–309.

Borchiellini, C., Boury-Esnault, N., Vacelet, J. and Le Parco, Y. (1998) Phylogenetic analysis of the Hsp70 sequences reveals the monophyly of Metazoa and specific phylogenetic relationships between animals and fungi. *Mol Biol Evol* 15: 647–655.

Boyd, E.F. and Hartl, D.L. (1997) Recent horizontal transmission of plasmids between natural populations of *Escherichia coli* and *Salmonella enterica*. *J Bacteriol* 179: 1622–1627.

Boyd, E.F., Moyer, K.E., Shi, L. and Waldor, M.K. (2000) Infectious CTXPhi and the *Vibrio* pathogenicity island prophage in *Vibrio mimicus*: Evidence for recent horizontal transfer between *V. mimicus* and *V. cholerae*. *Infect Immunity* 68: 1507–13.

Brady, R.H. (1985) On the independence of systematics. *Cladistics* 1: 113–126.

Bremer, K. (1988) The limits of amino acid sequence data in angiosperm phylogenetic reconstruction. *Evolution* 42: 795–803.

Bremer, K. (1995) Branch support and tree stability. *Cladistics* 10: 295–304.

Brendel, V., Brocchieri, L., Sandler, S.J., Clark, A.J. and Karlin, S. (1997) Evolutionary comparisons of RecA-like proteins across all major kingdoms of living organisms. *J Mol Evol* 44: 528–541.

Brinkmann, H. and Martin, W. (1996) Higher-plant chloroplast and cytosolic 3-phosphoglycerate kinases: A case of endosymbiotic gene replacement. *Plant Mol Biol* 30: 65–75.

Brochu, C. (1997) Morphology, fossils, divergence timing, and the phylogenetic relationships of *Gavialis*. *Syst Biol* 46: 479–522.

Brower, A.V.Z. (1994) Rapid morphological radiation and convergence among races of the butterfly *Heliconius erato* inferred from patterns of mitochondrial DNA evolution. *Proc Natl Acad Sci USA* 91: 6491–6495.

Brower, A.V.Z. (1996a) A new mimetic species of *Heliconius* (Lepidoptera: Nymphalidae), from southeastern Colombia, as revealed by cladistic analysis of mitochondrial DNA sequences. *Zool J Linn Soc* 116: 317–332.

Brower, A.V.Z. (1996b) Combining data in phylogenetic analysis (a response to Huelsenbeck et al., 1996) *Trends Ecol Evol* 11: 334–335.

Brower, A.V.Z. (1999) Delimitation of phylogenetic species with DNA sequences: a critique of Davis and Nixon's population aggregation analysis. *Syst Biol* 48: 199–213.

Brower, A.V.Z. (2000a) Homology and the inference of systematic relationships: some historical and philosophical perspectives. *In:* R.W. Scotland and T. Pennington (ed.): *Homology and Systematics.* Taylor and Francis, London, pp. 10–21.

Brower, A.V.Z. (2000b) On the validity of *Heliconius tristero* and *Heliconius melpomene mocoa* Brower, with notes on species concepts in *Heliconius* Kluk (Lepidoptera: Nymphalidae). *Proc Entomol Soc Wash* 103: 678–687.

Brower, A.V.Z. (2000c) Evolution is not an assumption of cladistics. *Cladistics* 16: 143–154.

Brower, A. and DeSalle, R. (1994) Practical and theoretical considerations for choice of DNA sequence region in insect molecular systematics, with a short review of published studies using nuclear gene regions. *Annu Rev Entomol* 87: 702–716.

Brower, A.V.Z., DeSalle, R. and Vogler, A. (1996) Gene trees, species trees, and systematics: A cladistic perspective. *Annu Rev Ecol Syst* 27: 423–50.

Brower, A.V.Z. and Egan, M.G. (1997) Cladistics of *Heliconius* butterflies and relatives (Nymphalidae: Heliconiiti): the phylogenetic position of Eueides based on sequences from mtDNA and a nuclear gene. *Proc R Soc Lond B* 264: 969–977.

Brower, A.V.Z. and Schawaroch, V. (1996) Three steps of homology assessment. *Cladistics* 12: 265–272.

Brown, J.R. and Doolittle, W.F. (1995) Root of the universal tree of life based on ancient aminoacyl-tRNA synthetase gene duplications. *Proc Natl Acad Sci USA* 92: 2441–2445.

Brown, J.R. and Doolittle, W.F. (1997) Archaea and the prokaryote-to-eukaryote transition. *Microbiol Mol Biol Rev* 61: 456–502.

Brown, J.R., Masuchi, Y., Robb, F.T. and Doolittle, W.F. (1994) Evolutionary relationships of bacterial and archaeal glutamine synthetase genes. *J Mol Evol* 38: 566–76.

Brown, J.R., Robb, F.T., Weiss, R. and Doolittle, W.F. (1997) Evidence for the early divergence of tryptophanyl- and tyrosyl-tRNA synthetases. *J Mol Evol* 45: 9–16.

Brown, S., Rouse, G.W., Hutchings, P. and Colgan, D.J. (1999) Assessing the usefulness of histone H3, U2 snRNA and 28S rDNA in analyses of polychaete relationships. *Aust J Zool* 47: 499–516.

Brubaker, R.R. (1991) Factors promoting acute and chronic diseases caused by *Yersiniae*. *Clin Microbiol Rev* 4: 309–324.

Brundin, L. (1972) Evolution, causal biology, and classification. *Zool Scr* 1: 107–120.

Buades, C. and Moya, A. (1996) Phylogenetic analysis of the isopenicillin-N-synthetase horizontal gene transfer. *J Mol Evol* 42: 537–542.

Buchrieser, C., Brosch, R., Bach, S., Guiyoule, A. and Carniel, E. (1998) The high-pathogenicity island of *Yersinia pseudotuberculosis* can be inserted into any of the three chromosomal asn tRNA genes. *Mol Microbiol* 30: 965–978.

Bull, J.J., Badgett, M.R., Wichman, H.A., Huelsenbeck, J.P., Hillis, D.M., Gulati, A., Ho, C. and Molineux, J. (1997) Exceptional convergent evolution in a virus. *Genetics* 147: 1497–1507.

Bull, J.J., Huelsenbeck, J.P., Cunningham, C.W., Swofford, D.L. and Waddell, P.J. (1993) Partitioning and combining data in phylogenetic analysis. *Syst Biol* 42: 384–397.

Burns, K.J. (1998) Molecular systematics of tanagers (Thraupinae): Evolution and biogeography of a diverse radiation of neotropical birds. *Mol Phylogenet Evol* 8: 334–348.

Burt, A. (1989) Comparative methods using phylogenetically independent contrasts. *Oxf Surv Evol Biol* 6: 33–53.

Campbell, B.C., Steffen-Campbell, J.D., Sorensen, J.T. and Gill, R.J. (1995) Paraphyly of Homoptera and Auchenorrhyncha inferred from 18S rDNA nucleotide sequences. *Syst Entomol* 20: 175–194.

Cao, Y., Janke, A., Waddell, P.J., Westerman, M., Takenaka, O., Murata, S., Okada, N., Pääbo, S. and Hasegawa, M. (1998) Conflict among individual mitochondrial proteins in resolving the phylogeny of eutherian orders. *J Mol Evol* 47: 307–322.

Carlson, T.A., Guerinot, M.L. and Chelm, B.K. (1985) Characterization of the gene encoding glutamine synthetase I (glnA) from *Bradyrhizobium japonicum*. *J Bacteriol* 162: 698–703.

Carmean, D. and Crespi, B. (1995) Do long branches attract flies? *Nature* 373: 666.

Carmean, D., Kimsey, L.S. and Berbee, M.L. (1992) 18S rDNA sequences and holometabolous insects. *Mol Phylogenet Evol* 1: 270–278.

Carniel, E., Guilvout, I. and Prentice, M. (1996) Characterization of a large chromosomal "high-pathogenicity island" in biotype 1B *Yersinia enterocolitica*. *J Bacteriol* 178: 6743–6751.

Carnoy, C., Mullet, C., Muller-Alouf, H., Leteurtre, E. and Simonet, M. (2000) Superantigen YPMa exacerbates the virulence of *Yersinia pseudotuberculosis* in mice. *Infect Immunity* 68: 2553–2559.

Carpenter, J.M. (1988) Choosing among multiple equally parsimonious cladograms. *Cladistics* 4: 291–296.

Carpenter, J.M. (1992) Random cladistics. *Cladistics* 8: 147–153.

Carranza, S., Baguñà, J. and Riutort, M. (1997) Are the Platyhelminthes a monophyletic primitive group? An assessment using 18S rDNA sequences. *Mol Biol Evol* 14: 485–497.

Carroll, R. (1988) *Vertebrate paleontology and evolution*. Freeman, New York.

Carson, H.L. (1971) *The ecology of* Drosophila *breeding sites*. Harold L. Lyon Arboretum Lecture Number Two. University of Hawaii Press, Honolulu.

Carson, H.L. (1987) Tracing ancestry with chromosomal sequences. *Trends Ecol Evol* 2: 203–207.

Carson, H.L. and Kaneshiro, K.Y. (1976) *Drosophila* of Hawaii: Systematics and ecological genetics. *Annu Rev Ecol Syst* 7: 311–345.

Carson, H.L., Hardy, D.E., Spieth, H.T. and Stone, W.S. (1970) The evolutionary biology of the Hawaiian Drosophilidae. *In*: M.K. Hecht and W.C. Steere (eds): *Essays in Evolution and Genetics in Honor of Theodosius Dobzhansky*. Appleton Century Crofts, New York, pp. 437–543.

Carter, M., Hendy, M.D., Penny, D. Székely, L.A. and Wormald, N.C. (1990) On the distribution of lengths of evolutionary trees. *SIAM J Discr Math* 3: 38–47.

Caugant, D.A., Levin, B.R. and Selander, R.K. (1981) Genetic diversity and temporal variation in the *E. coli* population of a human host. *Genetics* 98: 467–490.

Cavalli-Sforza, L.L. and Edwards, A.F.W. (1967) Phylogenetic analysis: models and estimation procedures. *Am J Hum Genet* 19: 233–257.

Cavalier-Smith, T., Allsopp, M.T.E.P, Chao, E.E., Boury-Esnault, N. and Vacelet, J. (1996) Sponge phylogeny, animal monophyly, and the origin of the nervous system: 18S rRNA evidence. *Can J Zool* 74: 2031–2045.

Cavanaugh, C.M. (1994) Microbial symbiosis: Patterns of diversity in the marine environment. *Amer Zool* 34: 79–89.

Cavender, J.A. (1978) Taxonomy with confidence. *Math Biosci* 40: 271–280.

Censini, S., Lange, C., Xiang, Z., Crabtree, J.E., Ghiara, P., Borodovsky, M., Rappuoli, R. and Covacci, A. (1996) cag, a pathogenicity island of *Helicobacter pylori*, encodes type I- specific and disease-associated virulence factors. *Proc Natl Acad Sci USA* 93: 14,648–14,653.

Cerchio, S. and Tucker, P. (1998) Influence of alignment on the mtDNA phylogeny of cetacea: questionable support for a Mysticeti/Physeteroidea clade. *Syst Biol* 47: 336–344.

Chalwatzis, N., Hauf, J., van de Peer, Y., Kinzelbach, R. and Zimmerman FK (1996) 18S ribosomal RNA genes of insects: Primary structure of the genes and molecular phylogeny of the Holometabola. An. Entomol. Soc. 89: 788–803.

Chang, B.S.W. and Donoghue, M.J. (2000) Recreating ancestral proteins. *Trends Ecol Evol* 15: 109–114.

Chang, J. (1996a). Inconsistency of evolutionary tree topology reconstruction methods when substitution rates vary across characters. *Math Biosci* 134: 189–215.

Chang, J. (1996b) Full reconstruction of Markov models on evolutionary trees: Identifiability and consistency. *Math Biosci* 137: 51–73.

Charleston, M.A., Hendy, M.D. and Penny, D. (1994) The effects of sequence length, tree topology and number of taxa on the performance of phylogenetic methods. *J Comp Biol* 1: 133–151.

Charleston, M. and Steel, M.A. (1995) Five surprising properties of parsimoniously colored trees. *Bull Math Biol* 57: 367–375.

Cheetham, B.F. and Katz, M.E. (1995) A role for bacteriophages in the evolution and transfer of bacterial virulence determinants. *Mol Microbiol* 18: 201–208.

Chervitz, S.A., Aravind, L., Sherlock, G., Ball, C.A., Koonin, E.V., Dwight, S.S., Harris, M.A., Dolinski, K., Mohr, S., Smith, T. et al. (1998) Comparison of the complete protein sets of worm and yeast: Orthology and divergence. *Science* 282: 2022–2028.

Chiappe, L.M. (1995) The first 85 million years of avian evolution. *Nature* 378: 349–355.

Chippindale, P.T. and Wiens, J.J. (1994) Weighting, partitioning, and combining characters in phylogenetic analysis. *Syst Biol* 43: 278–287.

Chiu, J., DeSalle, R., Lam, H.M., Meisel, L. and Coruzzi, G. (1999) Molecular evolution of glutamate receptors: A primitive signaling mechanism that existed before plants and animals diverged. *Mol Biol Evol* 16: 826–838.

Coddington, J.A. (1988) Cladistic tests of adaptational hypotheses. *Cladistics* 4: 3–22.

Coddington, J.A. (1994) The roles of homology and convergence in studies of adaptation. *In*: P. Eggleton and R. Vane-Wright (eds): *Phylogenetics and ecology*. Academic Press, London, pp. 53–78.

Coddington, J.A. and Scharff, N. (1996) Problems with "soft" polytomies. *Cladistics* 12: 139–145.

Cohen, B.L., Gawthrop, A.B. and Cavalier-Smith, T. (1998) Molecular phylogeny of brachiopods and phoronids based on nuclear-encoded small subunit ribosomal RNA gene sequences. *Phil Trans R Soc Lond B* 353: 2039–2061.

Colgan, D.J., McLauchlan, A., Wilson, G.D.F., Livingston, S.P., Edgecombe, G.D., Macaranas, J., Cassis, G. and Gray, M.R. (1998) Histone H3 and U2 snRNA DNA sequences and arthropod molecular evolution. *Aust J Zool* 46: 419–437.

Colgin, M. and Lewis, R. (1998) Spider minor ampullate silk proteins contain new repetitive sequences and highly conserved non-silk-like "spacer regions." *Protein Sci* 7: 667–672.

Collazo, A. and Fraser, S.E. (1996) Integrating cellular and molecular approaches into studies of development and evolution: The issue of morphological homology. *Aliso* 14: 237–262.

Collins, T., Wimberger, P. and Naylor, G. (1994) Compositional bias, character-state bias, and character-state reconstruction using parsimony. *Syst Biol* 43: 482–496.

Cooper, A. and Penny, D. (1997) Mass survival of birds across the Cretaceous-Tertiary boundary: Molecular evidence. *Science* 275: 1109–1113.

Cotgreave, P. and Pagel, M.D. (1997) Predicting and understanding rarity: The comparative approach. *In*: W.E. Kunin and K.J. Gaston (eds): *The biology of rarity*. Chapman and Hall, London, pp. 237–261.

Coulier, F., Pontarotti, P., Roubin, R., Hartun, G., Goldfarb, M. and Birnbaum, D. (1997) Of worms and men: An evolutionary perspective on the fibroblast growth factor (FGF) and FGF receptor families. *J Mol Evol* 44: 43–56.

Coyne, J.A. (1994) Ernst Mayr and the origin of species. *Evolution* 48: 19–30.

Cracraft, J. (1981) Pattern and process in paleobiology: The role of cladistic analysis in systematic paleontology. *Paleobiology* 7: 456–78.

Cracraft, J. (1983) Species concepts and speciation analysis. *In*: R.F. Johnston (ed.): *Current Ornithology*. Plenum Press, New York-London, pp. 159–187.

Cracraft, J. (1986) The origin and early diversification of birds. *Paleobiology* 12: 383–399.

Cracraft, J. (1987a) DNA hybridization and avian phylogenetics. *Evol Biol* 21: 47–96.

Cracraft, J. (1987b) Species concepts and the ontology of evolution. *Biol Philosoph* 2: 329–346.

Cracraft, J. (1989) Speciation and its ontology: the empirical consequences of alternative species concepts for understanding patterns and processes of differentiation. *In*: D. Otte and J.A. Endler (eds): *Speciation and its consequences*. Sinauer Associates, Sunderland, pp. 29–59.

Cracraft, J. (1990) The origin of evolutionary novelties: Pattern and process at different hierarchical levels. *In*: M.H. Nitecki (ed.): *Evolutionary Innovations*. University of Chicago Press, Chicago, pp. 21–44.

Cracraft, J. (1997) Species concepts in systematics and conservation biology—an ornithological viewpoint. *In*: M.F. Claridge, H.A. Dawach and M.R. Wilson (eds): *Species. The units of biodiversity*. The Systematics Association Special Volume Series 54, Chapman and Hall, London, pp. 325–339.

Cracraft, J. and Clarke, J. The basal clades of modern birds. Proceedings of the J. H. Ostrom Symposium. *Bull Peabody Mus*; *in press*.

Craddock, E.M. and Kambysellis, M.P. (1997) Adaptive radiation in the Hawaiian *Drosophila* (Diptera: Drosophilidae): Ecological and reproductive character analyses. *Pac. Sci.* 51: 475–489.

Crandall, K.A., Kelsey, C.R., Imamichi, H., Lane, H.C. and Salzman, N.P. (1999) Parallel evolution of drug resistance in HIV: Failure of nonsynonymous/synonymous substitution rate ratio to detect selection. *Mol Biol Evol* 16: 372–382.

Crandall, K.A. and Templeton, A.R. (1999) Statistical approaches to detecting recombination. *In*: K.A. Crandall (ed.): *The Evolution of HIV*. John Hopkins University Press, Baltimore, pp. 153–176.

Crandall, K.A., Templeton, A.R. and Sing, C.F. (1994) Intraspecific phylogenies: Critique of Davis and Nixon's population aggregation analysis. *Syst Biol* 48: 199–213.

Cronin, M., Stuart, R., Pierson, B. and J. Patton (1996) k-casein gene phylogeny of higher ruminants (Pecora, Artiodactyla). *Mol Phylogenet Evol* 6: 295–311.

Cunningham, C.W. (1997) Can three incongruence tests predict when data should be combined? *Mol Biol Evol* 14: 733–740.

Cunningham, C.W. (1999) Some limitations of ancestral character-state reconstruction when testing evolutionary hypotheses. *Syst Biol* 48: 665–674.

Cunningham, C.W., Zhu, H. and Hillis, D.M. (1998) Best-fit maximum-likelihood models for phylogenetic inference: Best empirical tests with known phylogenies. *Evolution* 54: 978–987.

Cummings, M.P., Otto, S.P. and Wakeley, J. (1995) Sampling properties of DNA sequence data in phylogenetic analysis. *Mol Biol Evol* 12: 814–822.

Darwin, C. (1859) *On the origin of species by means of natural selection or the preservation of favored races in the struggle for life.* John Murray, London.

Datta, N. and Hughes, V.M. (1983) Plasmids of the same Inc groups in Enterobacteria before and after the medical use of antibiotics. *Nature* 306: 616–617.

Davis, J.I. and Nixon, K.C. (1992) Populations, genetic variation, and the delimitation of phylogenetic species. *Syst Biol* 41: 421–435.

Davison, J. (1999) Genetic exchange between bacteria in the environment. *Plasmid* 42: 73–91.

de la Cruz, I. and Davies, I. (2000) Horizontal gene transfer and the origin of species: Lessons from bacteria. *Trends Microbiol* 8: 128–133.

De Laet, J. and Smets, E. (1999) Data decisiveness, missing entries, and the DD index. *Cladistics* 15: 25–37.

de Queiroz, K. (1998) The general lineage concept of species, species criteria, and the process of speciation: A conceptual unification and terminological recommendations. *In*: D.J. Howard and S.H. Berlocher (eds): *Endless Forms: Species and Speciation.* Oxford University Press, New York, pp. 57–78.

de Queiroz, K. and Donoghue MJ (1988) Phylogenetic systematics and the species problem. *Cladistics* 4: 317–338.

de Queiroz, K. and Donoghue MJ (1990) Phylogenetic systematics of Nelson's version of cladistics? *Cladistics* 6: 61–75.

de Queiroz, A. and Donoghue MJ, Kim, J. (1995) Separate *versus* combined analysis of phylogenetic evidence. *Annu Rev Ecol Syst* 26: 657–681.

de Queiroz, K. and Gauthier, J. (1992) Phylogenetic taxonomy. *Annu Rev Ecol Syst* 23: 449–480.

de Rosa, R., Grenier JK, Andreeva, T., Cook CE, Adoutte, A., Akam, M. and Carroll SB, Balavoine, G. (1999) Hox genes in brachiopods and priapulids and protostome evolution. *Nature* 399: 772–776.

DeBeer, G.R. (1971) *Homology, an unsolved problem.* Oxford University Press, Oxford.

Delwiche, C.F. and Palmer, J.D. (1996) Rampant horizontal transfer and duplication of rubisco genes in eubacteria and plastids. *Mol Biol Evol* 13: 873–882.

Demerec, M. and Hartman, P.E. (1959) Complex loci in microorganisms. *Annu Rev Microbiol* 13: 377–406.

dePinna, M. (1991) Concepts and tests of homology in the cladistic paradigm. *Cladistics* 7: 367–394.

DeSalle, R. (1992) The origin and possible time of divergence of the Hawaiian Drosophilidae: Evidence from DNA sequences. *Mol Biol Evol* 9: 905–916.

DeSalle, R. and Brower, A.V.Z. (1997) Process partitions, congruence, and the independence of characters: Inferring relationships among closely related Hawaiian *Drosophila* from multiple gene regions. *Syst Biol* 46: 751–764.

DeSalle, R. and Giddings, L.V. (1986) Disconcordance of nuclear and mitochondrial DNA phylogenies in Hawaiian *Drosophila. Proc Natl Acad Sci USA* 83: 6902–6906.

DeSalle, R. and Grimaldi, D.A. (1991) Morphological and molecular systematics of the Drosophilidae. *Annu Rev Ecol Syst* 22: 447–475.

DeSalle, R. and Grimaldi, D.A. (1992) Characters and the systematics of Drosophilidae. *J Hered* 83: 182–188.

DeSalle, R., Wray, C. and Absher, R. (1994) Computational problems in molecular systematics. *In*: B. Schierwater, B. Streit, G.P. Wagner and R. DeSalle (eds): *Molecular Ecology and Evolution: Approaches and Applications.* Birkhäuser, Basel, pp. 353–370.

Díaz-Uriarte, R. and Garland, T.J. (1996) Testing hypotheses of correlated evolution using phylogenetically independent contrasts: sensitivity to deviations from Brownian motion. *Syst Biol* 45: 27–47.

Donoghue, M.J. (1985) A critique of the biological species concept and recommendations for a phylogenetic alternative. *Bryologist* 88: 172–181.

Donoghue, M.J. (1994) Progress and prospects in reconstructing plant phylogeny. *Ann MO Bot Gard* 81: 405–418.

Donoghue, M.J. and Sanderson, M.J. (1992) The suitability of molecular and morphological evidence in reconstructing plant phylogeny. *In*: P.S. Soltis, D.E. Soltis and J.J. Doyle (eds): *Molecular systematics of plants*. Chapman and Hall, New York, pp. 340–368.

Donoghue, M.J. and Mathews, S. (1998) Duplicate genes and the root of the angiosperms, with an example using phytochrome sequences. *Mol Phylogenet Evol* 9: 489–500.

Doolittle, R.F. (1995) The multiplicity of domains in proteins. *Annu Rev Biochem* 64: 287–314.

Doolittle, R.F. (1998) The case for gene transfer between very distantly related organisms. *In*: M. Syvanen and C.I. Kado (eds): *Horizontal gene transfer*. Chapman and Hall, London-New York, pp. 311–320.

Doolittle, R.F. (1999) Phylogenetic classification and the universal tree. *Science* 284: 2124–2129.

Doolittle, R.F. and Bork, P. (1993) Evolutionarily mobile modules in proteins. *Sci Amer* 269/4: 50–56.

Doolittle, R.F., Feng, D.F., Anderson, K.L. and Alberro, M.R. (1990) A naturally occurring horizontal gene transfer from a eukaryote to a prokaryote. *J Mol Evol* 31: 383–388.

Doolittle, R.F., Feng, D.F., Tsang, S., Cho, G. and Little, E. (1996) Determining divergence times of the major kingdoms of living organisms with a protein clock. *Science* 271: 470–477.

Dover, G.A., Ruiz Linares, A., Bowen, T. and Hancock, H.M. (1993) Detection and quantification of concerted evolution and molecular drive. *Meth Enzymol* 224: 525–541.

Doyle, J.J. (1992) Gene trees and species trees: molecular systematics as one-character taxonomy. *Syst Bot* 17: 144–163.

Doyle, J.J. (1995) The irrelevance of allele tree topologies for species delimitation, and a non-topological alternative. *Syst Bot* 20: 574–588.

Dress, A., Huson, D. and Moulton, V. (1996) Analysing and visualizing sequence and distance data using SPLITSTREE. *Discrete Appl Math* 71: 95–109.

Dubuisson, J. and-Y. (1997) Systematic relationships within the genus *Trichomanes sensu lato* (Hymenophyllaceae, Filicopsida): Cladistic analysis based on anatomical and morphological data. *Bot J Linn Soc* 123: 265–296.

Dubuisson, J.-Y., Hebant-Mauri, R. and Galtier, J. (1998) Molecules and morphology: Conflicts and congruence within the Fern genus *Trichomanes* (Hymenophyllaceae). *Mol Phylogenet Evol* 9: 390–397.

Durando, C.M., Baker, R.H., Etges, W.J., Heed, W.B., Wasserman, M. and DeSalle, R. (2000) Phylogenetic analysis of the *repleta* species group of the genus *Drosophila* using multiple sources of characters. *Mol Phylogenet Evol* 16: 296–307.

Dykhuizen, D.E. and Green, L. (1991) Recombination in *Escherichia coli* and the definition of biological species. *J Bacteriol* 173: 7257–7268.

Easteal, S. (1999) Molecular evidence for the early divergence of placental mammals. *BioEssays* 21: 1052–1058.

Edgecombe, G.D., Giribet, G. and Wheeler, W.C. (1999) Filogenia de Chilopoda: combinando secuencias de los genes ribosómicos 18S y 28S y morfología [Phylogeny of Chilopoda: combining 18S and 28S rRNA sequences and morphology]. *In*: A. Melic, J.J. DE Haro, M. Mendez and I. Ribera (eds): *Filogenia y Evolución de Arthropoda. Bol Soc Entomol Aragonesa* 26: 293–331.

Edwards, A.W.F. (1972) *Likelihood*. Cambridge University Press, Cambridge.

Edwards, A.W.F. (1996) The origin and early development of the method of minimum evolution for the reconstruction of phylogenetic trees. *Syst Biol* 45: 79–91.

Edwards, A.W.F. and Cavalli-Sforza, L.L. (1963) The reconstruction of evolution, Hereditary 18: 533, Ann. *Hum Genet* 27: 104–105.

Eernisse, D.J. (1998) Arthropod and annelid relationships re-examined. *In*: Fortey, RA and Thomas, RH (eds): *Arthropod Relantionships*. Chapman and Hall, London, pp. 43–56.

Eernisse, D. and Kluge, A. (1993) Taxonomic congruence *versus* total evidence, and amniote phylogeny inferred from fossils, molecules, and morphology. *Mol Biol Evol* 10: 1170–1195.

Eldredge, N. (1985) *Unfinished Synthesis*. Oxford University Press, Oxford.

Eldredge, N. and Cracraft, J. (1980) *Phylogenetic patterns and the evolutionary process*. Columbia University Press, New York.

Ellegren, H. and Fridolfsson, A. and-K. (1997) Male-driven evolution of DNA sequences in birds. *Nat Genet* 17: 182–184.

Erdös, P.L. and Székely, L. (1992) Evolutionary trees: an integer multicommodity max-flow-min-cut theorem. *Adv Appl Math* 13: 375–389.

Erdös, P.L., Steel, M.A., Székely, L.A. and Warnow, T. (1999) A few logs suffice to build (almost) all trees (Part 1). *Random Struct Algorithm* 14: 153–184.

Erdös, P.L. and Székely, L.A. (1993) Counting bichromatic evolutionary trees. *Discrete Appl Math* 47: 1–8.

Eriksson, T. (1996) Auto-Decay, version 2.9.5 (software and documentation). Stockholm University, Stockholm.

Escriva, H., Safi, R., Hanni, C., Langlois, M.C., Saumitou-Laprade, P., Stehelin, D., Capron, A., Pierce, R. and Laudet, V. (1997) Ligand binding was acquired during evolution of nuclear receptors. *Proc Natl Acad Sci USA* 94: 6803–6808.

Farrell, B.D. (1998) "Inordinate fondness" explained: Why are there so many beetles? *Science* 281: 555–558.

Farris, J.S. (1969) A successive approximations approach to character weighting. *Syst Zool* 18: 374–385.

Farris, J.S. (1970) Methods for computing Wagner trees. *Syst Zool* 19: 83–92.

Farris, J.S. (1971) The hypothesis of nonspecificity and taxonomic congruence. *Annu Rev Ecol Syst* 2: 277–302.

Farris, J.S. (1973) A probability model for inferring evolutionary trees. *Syst Zool* 22: 250–256.

Farris, J.S. (1982) Distance data in phylogenetic analysis. *Adv Cladist* 1: 3–23.

Farris, J.S. (1983) The logical basis of phylogenetic analysis. *In*: N.I. Platnick and V.A. Funk (eds): *Advances in Cladistics, Volume 2: Proceedings of the Second Willi Hennig Society*. Columbia University Press, New York, pp. 7–36.

Farris, J.S. (1989) The retention index and the rescaled consistency index. *Cladistics* 5: 417–419.

Farris, J.S. (1999) Likelihood and inconsistency. *Cladistics* 15: 199–204.

Farris, J.S., Albert, V.A., Kälersjö, M., Lipscomb, D. and Kluge, A.G. (1996) Parsimony jackknifing outperforms Neighbor-joining. *Cladistics* 12: 99–124.

Farris, J.S., Källersjö, M., Kluge, A.G. and Bult, C. (1994) Testing significance of congruence. *Cladistics* 10: 315–320.

Farris, J.S., Källersjö, M., Kluge, A.G. and Bult, C. (1995) Constructing a significance test for incongruence. *Syst Biol* 44: 570–572.

Farris, J. and Kluge, A. (1985) Parsimony, synapomorphy, and explanatory power: a reply to Duncan. *Taxon* 34: 130–135.

Farris, J., Kluge, A. and Eckardt, M.J. (1970) A numerical approach to phylogenetic systematics. *Syst Zool* 19: 172–189.

Feduccia, A. (1995) Explosive evolution in Tertiary birds and mammals. *Science* 267: 637–638.

Feduccia, A. (1996) *The Origin and Evolution of Birds*. Yale University Press, New Haven.

Felsenstein, J. (1973) Maximum likelihood and minimum-steps method for estimating evolutionary trees from data on discrete characters. *Syst Zool* 22: 240–249.

Felsenstein, J. (1978) Cases in which parsimony and compatibility methods will be positively misleading. *Syst Zool* 27: 401–410.

Felsenstein, J. (1985a) Confidence limits on phylogenies: An approach using the bootstrap. *Evolution* 39: 783–791.

Felsenstein, J. (1985b) Phylogenies and the comparative method. *Amer Naturalist* 125: 1–15.

Felsenstein, J. and Sober, E. (1986) Parsimony and likelihood: An exchange. *Syst Zool* 35: 617–626.

Feng, D.F. and Doolittle, R.F. (1987) Progressive sequence alignment as a prerequisite to correct phylogenetic trees. *J Mol Evol* 25: 351–360.

Feng, D.F. and Doolittle, R.F. (1990) Progressive alignment and phylogenetic tree construction of protein sequences. *Methods Enzymol* 183: 375–387.

Fetherston, J.D. and Perry, R.D. (1994) The pigmentation locus of *Yersinia pestis* KIM6+ is flanked by an insertion sequence and includes the structural genes for pesticin sensitivity and HMWP2. *Mol Microbiol* 13: 697–708.

Fetherston, J.D., Schuetze, P. and Perry, R.D. (1992) Loss of the pigmentation phenotype in *Yersinia pestis* is due to the spontaneous deletion of 102 kb of chromosomal DNA which is flanked by a repetitive element. *Mol Microbiol* 6: 2693–2704.

Fidopiastis, P.M., Boletzky, S. v. and Ruby, E.G. (1998) A new niche from *Vibrio logei*, the predominant light organ symbiont of squids in the genus *Sepiola*. *J Bacteriol* 180: 59–64.

Fiely, S.G., Olsen, G.J., Lane, D.J., Giovannoni, S.J., Ghiselin, M.T., Raff, E.C., Pace, N.R. and Raff, R.A. (1988) Molecular phylogeny of the animal kingdom. *Science* 239: 748–753.

Fitch, W.M. (1970) Distinguishing homologous from analogous proteins. *Syst Zool* 19: 99–113.

Fitch, W.M. (1971a) Toward defining the course of evolution: minimal change for a specific tree topology. *Syst Zool* 20: 406–416.

Fitch, W.M. (1971b) Rate of change of concomitantly variable codons. *J Mol Evol* 1: 84–96.

Fitch, W.M. (1979) Cautionary remarks on using gene expression events in parsimony procedures. *Syst Zool* 28: 375–379.

Fitch, W.M. and Smith, T.F. (1983) Optimal sequence alignments. *Proc Natl Acad Sci USA* 80: 1382–1386.

Fitch, W.M. and Upper, K. (1987) The phylogeny of tRNA sequences provides evidence for ambiguity reduction in the origin of the genetic code. *Cold Spring Harbor Sym Quant Biol* 52: 759–767.

Fleischmann, R.D., Adams, M.D., White, O., Clayton, R.A., Kirkness, E.F., Kerlavage, A.R., Bult, C.J., Tomb, J.F., Dougherty, B.A. and Merrick, J.M. (1995) Whole-genome random sequencing and assembly of *Haemophilus influenzae* Rd. *Science* 269: 496–512.

Foelix, R. (1996) *Biology of spiders*, 2nd edition. Oxford University Press, New York.

Folmer, O., Black, M., Hoeh, W., Lutz, R. and Vrijenhoek, R.C. (1994) DNA primers for amplification of mitochondrial cytochrome c oxidase subunit I from diverse metazoan invertebrates. *Mol Mar Biol Biotechnol* 3: 294–299.

Forterre, P. (1999) Displacement of cellular proteins by functional analogues from plasmids or viruses could explain puzzling phylogenies of many DNA informational proteins. *Mol Microbiol* 33: 457–465.

Forterre, P., Benachenhou-Lahfa, N., Confalonieri, F., Duguet, M., Elie, C. and Labedan, B. (1992) The nature of the last universal ancestor and the root of the tree of life, still open questions. *Biosystems* 28: 15–32.

Forterre, P., Bouthier De La Tour, C., Philippe, H. and Duguet, M. (2000) Reverse gyrase from hyperthermophiles: probable transfer of a thermoadaptation trait from archaea to bacteria. *Trends Genet* 16: 152–154.

Fraser, C.M., Norris, S.J., Weinstock, G.M., White, O., Sutton, G.G., Dodson, R., Gwinn, M., Hickey, E.K., Clayton, R., Ketchum, K.A. et al. (1998) Complete genome sequence of *Treponema pallidum*, the syphilis spirochete. *Science* 281: 375–388.

Friedlander, T.P., Regier, J.C. and Mitter, C. (1994) Phylogenetic information content of five nuclear gene sequences in animals: initial assessment of character sets from concordance and divergence studies. *Syst Biol* 43: 511–525.

Frost, D.R. and Hillis, D.M. (1990) Species in concept and practice: Herpetological applications. *Herpetologica* 46: 87–104.

Frost, D.R. and Kluge, A.G. (1994) A consideration of epistemology in systematic biology, with special reference to species. *Cladistics* 10: 259–294.

Fryxell, K.J. (1996) The coevolution of gene family trees. *Trends Genet* 12: 364–369.

Gantt, J.S., Baldauf, S.L., Calie, P.J., Weeden, N.F. and Palmer, J.D. (1991) Transfer of rpl22 to the nucleus greatly preceded its loss from the chloroplast and involved the gain of an intron. *EMBO J* 10: 3073–3078.

García-Vallvé, S., Palau, J. and Romeu, A. (1999) Horizontal gene transfer in glycosyl hydrolases inferred from codon usage in *Escherichia coli* and *Bacillus subtilis*. *Mol Biol Evol* 16: 1125–34.

Garey, J.R., Krotec, M., Nelson, D.R. and Brooks, J. (1996) Molecular analysis supports a tardigrade-arthropod association. *Invertebr Biol* 115: 79–88.

Garey, J.R., Mackey, L.Y., Brooks, J.M., Winnepenninckx, B. and Backeljau, T. (1995) Animal phylogeny: Ribosomal RNA studies of aschelminthes. *J Cell Biochem Suppl* 345–345.

Garland, T., Harvey, P.H. and Ives, A.R. (1992) Procedures for the analysis of comparative data using phylogenetically independent contrasts. *Syst Biol* 41: 18–32.

Gatesy, J. (1997) More support for a Cetacea/Hippopotamidae clade: The blood-clotting protein gene g-fibrinogen. *Mol Biol Evol* 14: 537–543.

Gatesy, J. (1998) Molecular evidence for the phylogenetic affinities of Cetacea. *In*: Thewissen, J. (ed.): *Advances in Vertebrate Paleobiology*. Plenum Press, New York, pp. 63–111.

Gatesy, J. and Arctander, P. (2000) Hidden morphological support for the phylogenetic placement of *Pseudoryx nghetinhensis* with bovine bovids: A combined analysis of gross anatomical evidence and DNA sequences from five genes. *Syst Biol* 49: 515–538.

Gatesy, J., DeSalle, R. and Wheeler, W.C. (1993) Alignment-ambiguous nucleotide sites and the exclusion of systematic data. *Mol Phylogenet Evol* 2: 152–157.

Gatesy, J., Milinkovitch, M., Waddell, V. and Stanhope, M. (1999a) Stability of cladistic relationships between Cetacea and higher-level artiodactyl taxa. *Syst Biol* 48: 6–20.

Gatesy, J., O'Grady, P. and Baker, R.H. (1999b) Corroboration among data sets in simultaneous analysis: Hidden support for phylogenetic relationships among higher level artiodactyl taxa. *Cladistics* 15: 271–313.

Gaudieri, S., Leelayuwat, C., Towsend, D.C., Kulski, J.K. and Dawkins, R.L. (1997) Genomic characterization of the region between HLA-B and TNF: Implications for the evolution of multicopy gene families. *J Mol Evol* 44 (suppl. 1): S147-S154.

Gauthier, J., Kluge, A. and Rowe, T. (1988) Amniote phylogeny and the importance of fossils. *Cladistics* 4: 105–209.

Gee, H. (2000) Homegrown computer roots out phylogenetic networks. *Nature* 404: 214.

Geisler, J. and Luo, Z. (1998) Relationships of Cetacea to terrestrial ungulates and the evolution of cranial vasculature. *In*: Thewissen, J. (ed.): *The emergence of whales, advances in vertebrate paleobiology*. Plenum Press, New York, pp. 163–212.

Genecodes (1999) Sequencher, version 3.1.1. Genecodes Co.

Gentry, A. (1992) The subfamilies and tribes of the family Bovidae. *Mammal Rev* 22: 1–32.

Geoffroy Saint-Hilaire, E. (1824) Note complémentaire sur les prétendus osselets de l'ouie des poissons. *Mém Mus Nat Hist, Paris* 11: 253–260.

Georges, A., Birrell, J., Saint, K.M., McCord, W. and Donnellan, S.C. (1998) A phylogeny for the side-necked turtles (Chelonia: Pleurodira) based on mitochondrial and nuclear gene sequence variation. *Biol J Linn Soc* 67: 213–246.

Ghiselin, M.T. (1966) An application of the theory of definitions to taxonomic principles. *Syst Zool* 15: 127–130.

Ghiselin, M.T. (1976) The nomenclature of correspondence: A new look at "Homology" and "Analogy". *In*: R.B. Masterton, W. Hodos and H. Jerison (eds): *Evolution, brain, and behavior: Persistent problems*. Lawrence Erlbaum, Hillsdale, pp. 129–142.

Ghiselin, M.T. (1997) *Metaphysics and the origin of species*. State University of New York Press, Albany.

Gibson, T.J. and Spring, J. (1998) Genetic redundancy in vertebrates: Polyploidy and persistence of genes encoding multidomain proteins. *Trends Genet* 14: 46–49.

Gilbert, D. (1992) SeqApp Version 1.9a. Indiana University, Bloomington.

Gilbert, H.J., Hazlewood, G.P., Laurie, J.I., Orpin, C.G. and Xue, G.P. (1992) Homologous catalytic domains in a rumen fungal xylanase: Evidence for gene duplication and prokaryotic origin. *Mol Microbiol* 6: 2065–2072.

Gilbert, S.F., Opitz, J.M. and Raff, R.A. (1996) Resynthesizing evolutionary and developmental biology. *Develop Biol* 173: 357–372.

Giribet, G. (1999) Ecdysozoa *versus* Articulata, dos hipótesis alternativas sobre la posición de los Artrópodos en el reino Animal. *In*: A. Melic, J.J. DE Haro, M. Mendez and I. Ribera (eds): *Filogenia y Evolución de Arthropoda. Bol Soc Entomol Aragonesa* 26: 145–160.

Giribet, G. (2001) Exploring the behavior of POY, a program for direct optimization of molecular data. *Cladistics* 17: S60–S70.

Giribet, G., Carranza, S., Baguñà, J., Riutort, M. and Ribera, C. (1996) First molecular evidence for the existence of a Tardigrada + Arthropoda clade. *Mol Biol Evol* 13: 76–84.

Giribet, G., Distel, D.L., Polz, M., Sterrer, W. and Wheeler, W.C. (2000) Triploblastic relationships with emphasis on the acoelomates, and the position of Gnathostomulida, Cycliophora, Plathelminthes and Chaetognatha; a combined approach of 18S rDNA sequences and morphology. *Syst Biol* 49: 539–562.

Giribet, G., Rambla, M., Carranza, S., Baguñà, J., Riutort, M. and Ribera, C. (1999) Internal phylogeny of the order Opiliones (Arthropoda, Arachnida) based in a combined approach of complete 18S rDNA, partial 28S rDNA sequences and morphology. *Mol Phylogenet Evol* 11: 296–307.

Giribet, G. and Ribera, C. (1998) The position of arthropods in the animal kingdom: a search for a reliable outgroup for internal arthropod phylogeny. *Mol Phylogenet Evol* 9: 481–488.

Giribet, G. and Ribera, C. (2000) A review of arthropod phylogeny: new data based on ribosomal DNA sequences and direct character optimization. *Cladistics* 16: 204–231.

Giribet, G. and Wheeler, W.C. (1999a) The position of arthropods in the animal kingdom: Ecdysozoa, islands, trees, and the "parsimony ratchet". *Mol Phylogenet Evol* 13: 619–623.

Giribet, G. and Wheeler, W.C. (1999b) On gaps. *Mol Phylogenet Evol* 13: 132–143.

Givnish, T.J. and Sytsma, K.J. (1997) Consistency, characters, and the likelihood of correct phylogenetic inference. *Mol Phylogenet Evol* 7: 320–330.

Givnish, T.J. and Sytsma, K.J. (1998) Homoplasy in molecular *versus* morphological data: The likelihood of correct phylogenetic inference. *In*: T.J. Givnish and K.J. Sytsma (eds): *Molecular evolution and adaptive radiation*. Cambridge University Press, Cambridge, pp. 55–101.

Givnish, T.J., Evans, T.M., Pires, J.C. and Sytsma, K.J. (1999) Polyphyly and convergent morphological evolution in Commelinales and Commelinidae: Evidence from *rbcL* sequence data. *Mol Phylogenet Evol* 12: 360–385.

Gladstein, D.L. and Wheeler, W.C. (1997) POY. Software for direct optimization of DNA and other data. American Museum of Natural History, New York, ftp://ftp.amnh.org/pub/molecular/poy.

Gogarten, J.P., Kibak, H., Dittrich, P., Taiz, L., Bowman, E.J., Bowman, B.J., Manolson, M.F., Poole, R.J., Date, T. and Oshima, T. (1989) Evolution of the vacuolar H$^+$-ATPase: implications for the origin of eukaryotes. *Proc Natl Acad Sci USA* 86: 6661–6665.

Golding, G.B. and Dean, A.M. (1998) The structural basis of molecular adaptation. *Mol Biol Evol* 15: 355–369.

Goldman, N. (1990) Maximum likelihood inference of phylogenetic trees, with special reference to a Poisson process model of DNA substitution and to parsimony analysis. *Syst Zool* 39: 345–361.

Goldman, N. (1993) Statistical tests of models of DNA substitution. *J Mol Evol* 36: 182–198.

Goldstein, P.Z. and DeSalle, R. (2000) Phylogenetic species, nested hierarchies, and character fixation. *Cladistics* 16: 364–384.

Goldstein, P.Z., DeSalle, R., Amato, G. and Vogler, A.P. (2000) Conservation genetics at the species boundary. *Conserv Biol* 14: 120–131.

Goloboff, P.A. (1991) Homoplasy and the choice among cladograms. *Cladistics* 7: 215–232.

Goloboff, P.A. (1993) Estimating character weights during tree search. *Cladistics* 9: 83–91.

Goloboff, P.A. (1994) NONA: a tree searching program. Program and documentation. Fundacion e Instituto Miguel Lillo.

Goloboff, P.A. (1999) Analyzing large data sets in reasonable times: Solutions for composite optima. *Cladistics* 15: 415–428.

Goodman, M., Czelusniak, J., Moore, G.W. and Matsuda, G. (1979) Fitting the gene lineage into its species lineage: a parsimony strategy illustrated by cladograms constructed from globin sequences. *Syst Zool* 28: 132–163.

Gosline, J., DeMont, M. and Denny, M. (1986) The structure and properties of spider silk. *Endeavour* 10: 37–43.

Gosline, J., Guerette, P., Ortlepp, C. and Savage, K. (1999) The mechanical design of spider silks: from fibroin sequence to mechanical function. *J Exp Biol* 202: 3295–3303.

Gould, S.J. (1977) *Ontogeny and Phylogeny*. Harvard University Press, Cambridge.

Grafen, A. (1989) The phylogenetic regression. *Phil Trans R Soc* 326: 119–157.

Grafen, A. (1992) The uniqueness of the phylogenetic regression. *J Theor Biol* 156: 405–423.

Graham, D.E., Overbeek, R., Olsen, G.J. and Woese, C.R. (2000) An archaeal genomic signature. *Proc Natl Acad Sci USA* 97: 3304–3308.

Graham, S.W., Kohn, J.R., Morton, B.R., Eckenwalder, J.E. and Barrett, S.C.H. (1998) Phylogenetic congruence and discordance among one morphological and three molecular data sets from Pontederiaceae. *Syst Biol* 47: 545–567.

Graur, D. and Li WH (1999) *Fundamentals of Molecular Evolution. Second Edition*. Sinauer, Sunderland.

Graybeal, A. (1998) Is it better to add taxa or characters to a difficult phylogenetic problem? *Syst Biol* 47: 9–17.

Greene, H.W. (1986) Diet and arboreality in the emerald monitor, *Varanus prasinus*, with comments on the study of adaptation. *Fieldiana Zool N Ser.* 31: 1–12.

Gribaldo, S. and Cammarano, P. (1998) The root of the universal tree of life inferred from anciently duplicated genes encoding components of the protein-targeting machinery. *J Mol Evol* 47: 508–516.

Griffith, F. (1928) The significance of pneumococcal types. *J Hygien* 27: 113–159.

Grimaldi, D. (1990) A phylogenetic, revised classification of genera in the Drosophilidae (Diptera). *Bull Am Mus Nat Hist* 197: 1–139.

Griswold, C., Coddington, J., Hormiga, G. and Scharff, N. (1998) Phylogeny of the orb-web building spiders (Araneae, Orbiculariae: Deinopoidea, Araneoidea). *Zool J Linn Soc* 123: 1–99.

Groisman, E.A. and Ochman, H. (1994) How to become a pathogen. *Trends Microbiol* 2: 289–294.

Groisman, E.A. and Ochman, H. (1996) Pathogenicity islands: Bacterial evolution in quantum leaps. *Cell* 87: 791–794.

Groth, J.G. (1998) Molecular phylogenetics of finches and sparrows: consequences of character state removal in cytochrome *b* sequences. *Mol Phylogenet Evol* 10: 377–390.

Groth, J.G. and Barrowclough, G.F. (1999) Basal divergences in birds and the phylogenetic utility of the nuclear RAG-1 gene. *Mol Phylogenet Evol* 12: 115–123.

Gu, X. (1998) Early metazoan divergence was about 830 million years ago. *J Mol Evol* 47: 369–371.

Guerette, P., Ginzinger, D., Weber, B. and Gosline, J. (1996) Silk properties determined by gland-specific expression of a spider fibroin gene family. *Science* 272: 112–115.

Gupta, R.S. (1998a) Life's third domain (Archaea): An established fact or an endangered paradigm? *Theor Pop Biol* 54: 91–104.

Gupta, R.S. (1998b) Protein phylogenies and signature sequences: A reappraisal of evolutionary relationships among archaebacteria, eubacteria, and eukaryotes. *Microbiol Mol Biol Rev* 62: 1435–1491.

Hacker, J., Blum-Oehler, G., Muhldorfer, I. and Tschape, H. (1997) Pathogenicity islands of virulent bacteria: Structure, function and impact on microbial evolution. *Mol Microbiol* 23: 1089–1097.

Hackett, S.J., Griffiths, C.S., Bates, J.M. and Klein, N.K. (1995) A commentary on the use of sequence data for phylogeny construction. *Mol Phylogenet Evol* 4: 350–356.

Hafner, M.S. and Page, R.D. (1995) Molecular phylogenies and host-parasite cospeciation: Gophers and lice as a model system. *Phil Trans R Soc Lond B* 349: 77–83.

Halanych, K.M., Bacheller, J.D., Aguinaldo, A.M., Liva, S.M., Hillis, D.M. and Lake, J.A. (1995) Evidence from 18S ribosomal DNA that the lophophorates are protostome animals [published erratum appears in *Science* 1995 Apr 28, 268(5210): 485]. *Science* 267: 1641–1643.

Halanych, K.M. (1996) Testing hypotheses of Chaetognath origins: Long branches revealed by 18S ribosomal DNA. *Syst Biol* 45: 223–246.

Halanych, K.M. (1998) Lagomorphs misplaced by more characters and fewer taxa. *Syst Biol* 47: 138–146.

Halanych, K.M., Bacheller, J.D., Aguinaldo, A.M., Liva, S.M., Hillis, D.M. and Lake, J.A. (1996) Lophophorate phylogeny. *Science* 272: 283.

Hamel, A.M. and Steel, M.A. (1997) The length of a leaf coloration on a random binary tree. *SIAM J Discr Math* 10: 359–372.

Hardy, D.E. (1965) Diptera: Cyclorrhaphia II, Series Schizophora, Section Acalypterae I, Family Drosophilidae. *Insects of Hawaii* 12: 1–814.

Hardy, D.E. and Kaneshiro, K.Y. (1981) Drosophilidae of Pacific Oceania. *In*: M. Ashburner, H.L. Carson and J.J.N. Thompson (eds): *The Genetics and Biology of* Drosophila. Academic Press, New York, pp. 309–348.

Harris, E.E. and Disotell, T.R. (1998) Nuclear gene trees and the phylogenetic relationships of the *Mangabeys* (Primates: Papionini). *Mol Biol Evol* 15: 892–900.

Harrison, R.G. (1989) Animal mitochondrial DNA as a genetic marker in population and evolutionary biology. *TREE* 4: 6–11.

Harrison, M.K. and Crespi, B.J. (1999) Phylogenetics of *Cancer* crabs (Crustacea: Decapoda: Brachyura). *Mol Phylogenet Evol* 12: 186–199.

Harrison, R.G. (1998) Linking evolutionary pattern and process: the relevance of species concepts for the study of speciation. *In*: D.J. Howard and S.H. Berlocher (eds): *Endless forms: Species and speciation*. Oxford University Press, New York, pp. 19–31.

Harry, M., Solignac, M. and Lachaise, D. (1998) Molecular evidence for parallel evolution of adaptive syndromes in fig-breeding *Lissocephala* (Drosophilidae). *Mol Phylogenet Evol* 9: 542–551.

Harvey, P.H., Brown, A.J.L., Smith, J.M. and Nee, S. (1996) *New Uses for New Phylogenies*. Oxford University Press, Oxford.

Harvey, P.H. and Pagel, M.D. (1991) The comparative method in evolutionary biology. *In*: M. May and P.H. Harvey (eds): *Oxford Series in Ecology and Evolution*. Oxford University Press, Oxford.

Hayashi, C. and Lewis, R. (1998) Evidence from flagelliform silk cDNA for the structural basis of elasticity and modular nature of spider silks. *J Mol Biol* 275: 773–784.

Hayashi, C. and Lewis, R. (2000) Molecular architecture and evolution of a modular spider silk protein gene. *Science* 287: 1477–1479.

Hayashi, C., Shipley, N. and Lewis, R. (1999) Hypotheses that correlate the sequence, structure, and mechanical properties of spider silks. *Int J Biol Macromol* 24: 271–275.

Haygood, M.G. and Distel, D.L. (1993) Bioluminescent symbionts of flashlight fishes and deep-sea anglerfishes from unique lineages related to Vibrios. *Nature* 363: 154–156.

Heard, S.B. and Hauser, D.L. (1995) Key evolutionary innovations and their ecological mechnisms. *Historical Biol* 10: 151–173.

Hedges, S.B.C. and G. Sibley (1994) Molecules *versus* morphology in avian evolution: The case of the "pelecaniform" birds. *Proc Natl Acad Sci USA* 91: 9861–9865.

Hedges, S.B. and Maxson, L. (1996) Re: Molecules and morphology in amniote phylogeny. *Mol Phylogenet Evol* 6: 312–314.

Hedges, S.B., Parker, P.H., Sibley, C.G. and Kumar, S. (1996) Continental breakup and the ordinal diversification of birds and mammals. *Nature* 381: 226–229.

Hedges, S.B., Simmons, M.D., van Dijk MAM, Caspers, G.,-J. and de Jong WW, Sibley CG (1995) Phylogenetic relationships of the Hoatzin, an enigmatic South American bird. *Proc Natl Acad Sci USA* 92: 11,662–11,665.

Heed, W.B. (1968) Ecology of the Hawaiian Drosophilidae. *Univ TX Publ* 6818: 388–419.

Heed, W.B. (1971) Host plant specificity and speciation in Hawaiian *Drosophila*. *Taxon* 20: 115–121.

Hein, J. (1989) A new method that simultaneously aligns and reconstructs ancestral sequences for any number of homologous sequences, when the phylogeny is given. *Mol Biol Evol* 6: 649–668.

Hein, J. (1990) Unified approach to alignment and phylogenies. *Meth Enzymol* 183: 626–644.

Hennig, W. (1950) *Grundzüge einer Theorie der phylogenetischen Systematik.* Deutscher Zentralverlag, Berlin.

Hennig, W. (1963) Phylogenetic systematics. *Annu Rev Entomol* 10: 97–116.

Hennig, W. (1966) *Phylogenetic Systematics.* University of Illinois Press, Urbana.

Hennig, W. (1981) *Insect Phylogeny.* Academic Press, New York.

Hensel, R., Zwickl, P., Fabry, S., Lang, J. and Palm, P. (1989) Sequence comparison of glyceraldehyde-3-phosphate dehydrogenases from the three urkingdoms: Evolutionary implication. *Can J Microbiol* 35: 81–85.

Henze, K., Badr, A., Wettern, M., Cerff, R. and Martin, W. (1995) A nuclear gene of eubacterial origin in *Euglena gracilis* reflects cryptic endosymbioses during protist evolution. *Proc Natl Acad Sci USA* 92: 9122–9126.

Herrick, J.B., Stuart-Keil, K.G., Ghiorse, W.C. and Madsen, E.L. (1997) Natural horizontal transfer of a naphthalene dioxygenase gene between bacteria native to a coal tar-contaminated field site. *Appl Environ Microbiol* 63: 2330–2337.

Higgins, D.G. (1994) CLUSTAL V: Multiple alignment of DNA and protein sequences. *Methods Mol Biol* 25: 307–318.

Higgins, D.G., Bleasby, A.J. and Fuchs, R. (1992) CLUSTAL V: Improved software for multiple sequence alignment. *Comput Appl Biosci* 8: 189–191.

Higgins, D.G. and Sharp, P.M. (1989) Fast and sensitive multiple sequence alignments on a microcomputer. *Comput Appl Biosci* 5: 151–153.

Higgins, D.G. and Sharp, P.M. (1988) CLUSTAL: A package for performing multiple sequence alignment on a microcomputer. *Gene* 73: 237–244.

Higgins, D.G., Thompson, J.D. and Gibson, T.J. (1996) Using CLUSTAL for multiple sequence alignments. *Methods Enzymol* 266: 383–402.

Hilario, E. and Gogarten, J.P. (1993) Horizontal transfer of ATPase genes—the tree of life becomes a net of life. *Biosystems* 31: 111–119.

Hillis, D.M. (1987) Molecular *versus* morphological approaches to systematics. *Annu Rev Ecol Syst* 18: 23–42.

Hillis, D.M. (1995) Approaches for assessing phylogenetic accuracy. *Syst Biol* 44: 3–16.

Hillis, D.M. (1996) Inferring complex phylogenies. *Nature* 383: 130–131.

Hillis, D.M. (1998) Taxonomic sampling, phylogenetic accuracy, and investigator bias. *Syst Biol* 47: 3–8.

Hillis, D., Bull, J., White, M., Badgett, M. and Molineux, I. (1992) Experimental phylogenetics: Generation of a known phylogeny. *Science* 255: 589–592.

Hillis, D., Huelsenbeck, J. and Cunningham, C. (1994) Application and accuracy of molecular phylogenies. *Trends Ecol Evol* 264: 671–677.

Hillis, D.M., Moritz, C. and Mable, B. (1996) *Molecular Systematics,* 2nd edition. Sinauer, Sunderland.

Hinman, M. and Lewis, R. (1992) Isolation of a clone encoding a second dragline silk fibroin. *J Biol Chem* 267: 19,320–19,324.

Hoelzer, G.A. and Melnick, D.J. (1994) Patterns of speciation and limits to phylogenetic resolution. *Trends Ecol Evol* 9: 104–107.

Holland, N.D. (1996) Homology, homeobox genes, and the early evolution of the vertebrates. *In*: M.T. Ghiselin and G. Pinna (eds): *New Perspectives on the History of Life*. California Academy of

Sciences, San Francisco, pp. 63–70.

Holm, L. (1998) Unification of protein families. *Curr Opin Struct Biol* 8: 372–379.

Holsinger, K.E. (1984) The nature of biological species. *Phil Sci* 51: 293–307.

Honeycutt, R., Nedbal, M., Adkins, R. and Janecek, L. (1995) Mammalian mitochondrial DNA evolution: A comparison of the cytochrome *b* and cytochrome *c* oxidase II genes. *J Mol Evol* 40: 260–272.

Houde, P. (1987) Critical evaluation of DNA hybridization studies in avian systematics. *Auk* 103: 17–32.

Huelsenbeck, J. (1991) When are fossils better than extant taxa in phylogenetic analysis? *Syst Biol* 40: 458–469.

Huelsenbeck, J. (1997) Is the Felsenstein zone a fly trap? *Syst Biol* 46: 69–74.

Huelsenbeck, J.P. (1998) Systematic bias in phylogenetic analysis: is the Strepsiptera problem solved? *Syst Biol* 47: 519–537.

Huelsenbeck, J.P. and Bull, J.J. (1996) A likelihood ratio test for detection of conflicting phylogenetic signal. *Syst Biol* 42: 247–264.

Huelsenbeck, J.P. and Hillis, D.M. (1993) Success of phylogenetic methods in the four-taxon case. *Syst Biol* 42: 247–264.

Huelsenbeck, J.P. and Rannala, B. (1997) Phylogenetic methods come of age: Testing hypotheses in an evolutionary context. *Science* 276: 227–232.

Huelsenbeck, J.P. and Crandall, K.A. (1997) Phylogeny estimation and hypothesis testing using maximum likelihood. *Annu Rev Ecol Syst* 28: 437–466.

Hughes, V.M. and Datta, N. (1983) Conjugative plasmids in bacteria of the 'pre-antibiotic' era. *Nature* 302: 725–726.

Hull, D.L. (1988) *Science as a Process: An Evolutionary Account of the Social and Conceptual Development of Science*. University of Chicago Press, Chicago.

Hunter, I.J. (1964) Paralogy, a concept complementary to homology and analogy. *Nature* 204: 604.

Huynen, M.A. and Bork, P. (1998) Measuring genome evolution. *Proc Natl Acad Sci USA* 95: 5849–5856.

Huynen, M., Dandekar, T. and Bork, P. (1998) Differential genome analysis applied to the species-specific features of *Helicobacter pylori*. *FEBS Lett* 426: 1–5.

Huynen, M.A. and Snel, B. (2000) Gene and context: integrative approaches to genome analysis. *Adv Protein Chem* 54: 345–379.

Hwang, U., Kim, W., Tautz, D. and Friedrich, M. (1998) Molecular phylogenetics at the Felsenstein zone: Approaching the Strepsiptera problem using 5.8S and 28S rDNA sequences. *Mol Phylogenet Evol* 9: 470–480.

Ikemura, T. (1981) Correlation between the abundance of *Escherichia coli* transfer RNAs and the occurrence of the respective codons in its protein genes: a proposal for a synonymous codon choice that is optimal for the *E. coli* translational system. *J Mol Biol* 151: 389–409.

Ikemura, T. (1982) Correlation between the abundance of yeast transfer RNAs and the occurrence of the respective codons in protein genes. Differences in synonymous codon choice patterns of yeast and *Escherichia coli* with reference to the abundance of isoaccepting transfer RNAs. *J Mol Biol* 158: 573–597.

Ikemura, T. (1985) Codon usage and tRNA content in unicellular and multicellular organisms. *Mol Biol Evol* 2: 13–34.

International Commission for Zoological Nomenclature. (1985) *International Code of Zoological Nomenclature*. International Trust for Zoological Nomenclature and the British Museum (Natural History), London.

Iwabe, N., Kuma, K., Hasegawa, M., Osawa, S. and Miyata, T. (1989) Evolutionary relationship of archaebacteria, eubacteria, and eukaryotes inferred from phylogenetic trees of duplicated genes. *Proc Natl Acad Sci USA* 86: 9355–9359.

Jacob, F., Perrin, F., Sanchez, C. and Monod, J. (1960) L'opéron: groupe de gènes à expression coordonnée par un opérateur. *C R Acad Sci* 250: 1727–1729.

Jain, R., Rivera, M.C. and Lake, J.A. (1999) Horizontal gene transfer among genomes: The complexity hypothesis. *Proc Natl Acad Sci USA* 96: 3801–3806.

Janies, D. (2001) Phylogenetic relationships of extant echinoderm classes. *Can J Zool* 79: 1232–1250.

Janies, D. and DeSalle, R. (1999) Development, evolution and corroboration. *Anat Rec* 257: 6–14.

Janies, D. and Wheeler, W.C. (2001) Efficiency of parallel direct optimization. *Cladistics* 17: S17–S82.

Jeanmougin, F., Thompson, J.D., Gouy, M., Higgins, D.G. and Gibson, T.J. (1998) Multiple sequence alignment with Clustal X. *Trends Biochem Sci* 23: 403–405.

Jiggins, C.D., McMillan, W.O., King, P. and Mallet, J. (1997) The maintenance of species differences across a *Heliconius* hybrid zone. *Heredity* 79: 495–505.

Jiggins, C.D., McMillan, W.O., Neukirchen, W. and Mallet, J. (1996) What can hybrid zones tell us about speciation? The case of *Heliconius erato* and *H. himera* (Lepidoptera: Nymphalidae). *Biol J Linn Soc* 59: 221–242.

Kahn, N.W. and Quinn, T.W. (1999) Male driven evolution among Eoaves? A test of the replicative division hypothesis in a heterogametic female (ZW) system. *J Mol Evol* 49: 750–759.

Källersjö, M., Farris, J.S., Kluge, A. and Bult, C. (1992) Skewness and permutation. *Cladistics* 8: 275–287.

Källersjö, M., Albert, V.A. and Farris, J.S. (1999) Homoplasy increases phylogenetic signal. *Cladistics* 8: 275–287.

Kambysellis, M.P., Ho, K.,-F., Craddock EM, Piano, F., Parisi, M. and Cohen, J. (1995) Pattern of ecological shifts in the diversification of Hawaiian *Drosophila* inferred from a molecular phylogeny. *Curr Biol* 5: 1129–1139.

Kaneko, T., Sato, S., Kotani, H., Tanaka, A., Asamizu, E., Nakamura, Y., Miyajima, N., Hirosawa, M., Sugiura, M. and Sasamoto et al. (1996) Sequence analysis of the genome of the unicellular cyanobacterium *Synechocystis* sp. strain PCC6803. II. Sequence determination of the entire genome and assignment of potential protein-coding regions (supplement). *DNA Res* 3: 185–209.

Kaneshiro, K.Y. (1976) A revision of generic concepts in the biosystematics of Hawaiian Drosophilidae. *Proc Hawaii Entomol Soc* 22: 255–278.

Kaneshiro, K.Y. and Boake, C.R.B. (1987) Sexual selection and speciation: Issues raised by Hawaiian *Drosophila*. *Trends Ecol Evol* 2: 207–212.

Kaneshiro, K.Y., Gillespie, R.G. and Carson, H.L. (1995) Chromosomes and male genitalia of Hawaiian *Drosophila*. *In*: W.L. Wagner and V.A. Funk (eds): *Hawaiian Biogeography: Evolution on a Hot Spot Archipelago*. Smithsonian Institition Press, Washington and London, pp. 57–71.

Karaolis, D.K., Somara, S. and Maneval, D.R., Jr., Johnson, J.A. and Kaper, J.B. (1999) A bacteriophage encoding a pathogenicity island, a type-IV pilus and a phage receptor in cholera bacteria. *Nature* 399: 375–379.

Katz, L.A. (1996) Transkingdom transfer of the phosphoglucose isomerase gene. *J Mol Evol* 43: 453–459.

Kellogg, E.A. and Campbell, C.S. (1987) Phylogenetic analysis of the Graminae. *In*: T. Soderstrom, K. Hilu, C. Campbell and M. Barkworth (eds): *Grass Systematics and Evolution*. Smithsonian Institition Press, Washington and London, pp. 310–322.

Kidwell, M.G. (1993) Lateral transfer in natural populations of eukaryotes. *Annu Rev Genet* 27: 235–256.

Kim, J. (1996) General inconsistency conditions for maximum parsimony: Effects of branch lengths and increasing numbers of taxa. *Syst Biol* 45: 363–374.

Kim, J., Kim, W. and Cunningham, C.W. (1999) A new perspective on lower metazoan relationships from 18S rDNA sequences. *Mol Biol Evol* 16: 423–427.

Kimsey, H.H., Nair, G.B., Ghosh, A. and Waldor, M.K. (1998) Diverse CTXphis and evolution of new pathogenic *Vibrio cholerae*. *Lancet* 352: 457–458.

Kimura, M. (1980) A simple method for estimating evolutionary rates of base substitutions through comparative studies of nucleotide sequences. *J Mol Evol* 16: 111–120.

Kishino, H. and Hasegawa, M. (1989) Evaluation of the maximum likelihood estimate of the evolutionary tree topologies from DNA sequence data, and the branching order of the Hominoidea. *J Mol Evol* 29: 170–179.

Kissinger, J.C., Hahn, J.H. and Raff, R.A. (1997) Rapid evolution in a conserved gene family: Evolution of the acting gene family in the sea urchin genus *Heliocidaris* and related genera. *Mol Biol Evol* 14: 654–665.

Kjer, K.M. (1995) Use of rRNA secondary structure in phylogenetic studies to identify homologous positions: An example of alignment and data presentation from the frogs. *Mol Phylogenet Evol* 4: 314–330.

Kjer, K.M., Baldridge, G.D. and Fallon, A.M. (1994) Mosquito large subunit ribosomal RNA: simultaneous alignment of primary and secondary structure. *Biochim Biophys Acta* 1217: 147–155.

Kluge, A.G. (1989) A concern for evidence and a phylogenetic hypothesis of relationships among Epicrates (Boidae, Serpentes). *Syst Zool* 38: 7–25.

Kluge, A.G. (1997) Testability and the refutation and corroboration of cladistic hypotheses. *Cladistics* 13: 81–96.

Kluge, A. and Farris, J.S. (1969) Quantitative phyletics and the evolution of anurans. *Syst Zool* 18: 1–32.

Kluge, A.G. and Wolf, A.J. (1993) Cladistics: What's in a word? *Cladistics* 9: 183–199.

Knisley, C.B. and Schultz, T.D. (1997) *The biology of tiger beetles*. Virginia Museum of Natural History, Martinsville.

Kobayashi, M., Wada, H. and Satoh, N. (1994) Phylogenetic position of diploblasts inferred from amino acid sequences of elongation factor 1-alpha. *Zool Sci* 11: 36–36.

Kobayashi, M., Wada, H. and Satoh, N. (1995) Deuterostome phylogeny inferred from amino acid sequence of elongation factor 1 alpha. *Zool Sci* 12: 34–34.

Kobayashi, M., Wada, H. and Satoh, N. (1996) Early evolution of the Metazoa and phylogenetic status of diploblasts as inferred from amino acid sequence of elongation factor-1-alpha. *Mol Phylogenet Evol* 5: 414–422.

Köhler, T. and Vollrath, F. (1995) Thread biomechanics in the two orb-weaving spiders *Araneus diadematus* (Araneae, Araneidae) and *Uloborus walckenaerius* (Araneae, Uloboridae). *J Exp Zool* 271: 1–17.

Kojima, S. (1998) Paraphyletic status of Polychaeta suggested by phylogenetic analysis based on the amino acid sequences of elongation factor-1 alpha. *Mol Phylogenet Evol* 9: 255–261.

Kojima, S., Hashimoto, T., Hasegawa, M., Murata, S., Ohta, S., Seki, H. and Okada, N. (1993) Close phylogenetic relationship between Vestimentifera (tube worms) and Annelida revealed by the amino acid sequence of elongation factor-1 alpha. *J Mol Evol* 37: 66–70.

Koonin, E.V., Tatusov, R.L. and Galperin, M.Y. (1998) Beyond complete genomes: From sequence to structure and function. *Curr Opin Struct Biol* 8: 355–363.

Koonin, E.V., Tatusov, R.L. and Rudd, K.E. (1996) Protein sequence comparison at genome scale. *Meth Enzymol* 266: 295–323.

Kornegay, J.R., Schilling, J.W. and Wilson, A.C. (1994) Molecular adaptation of a leaf-eating bird: stomach lysozyme of the hoatzin. *Mol Biol Evol* 11: 921–928.

Koshi, J.M. and Goldstein, R.A. (1996) Probabilistic reconstruction of ancestral protein sequences. *J Mol Evol* 42: 313–320.

Kraus, F., Jarecki, L., Miyamoto, M., Tanhauser, S. and Laipis, P. (1992) Mispairing and compensational changes during the evolution of mitochondrial ribosomal RNA. *Mol Biol Evol* 9: 770–774.

Kristensen, N.P. (1991) Phylogeny of extant hexapods. *In*: P.B.C.I. D. Naumann, J.F. Lawrence, E.S. Nielsen, J.P. Spradberry, R.W. Taylor, M.J. Whitten and M.J. Littlejohn (eds): *The insects of Australia: A textbook for students and research workers*, 2nd edition. CSIRO, Melbourne Univ. Press, Melbourne, pp. 125–140.

Kristensen, N.P. (1995) Fourty [sic] years' insect phylogenetic systematics. *Zool Beitr N F* 36: 83–124.

Kumada, Y., Benson, D.R., Hillemann, D., Hosted, T.J., Rochefort, D.A., Thompson, C.J., Wohlleben, W. and Tateno, Y. (1993) Evolution of the glutamine synthetase gene, one of the oldest existing and functioning genes. *Proc Natl Acad Sci USA* 90: 3009–3013.

Kumar, S. and Hedges, S.B. (1998) A molecular timescale for vertebrate evolution. *Nature* 392: 917–920.

Kyrpides, N.C. and Olsen, G.J. (1999) Archaeal and bacterial hyperthermophiles: horizontal gene exchange or common ancestry? *Trends Genet* 15: 298–299.

Lake, J.A. (1985) Evolving ribosome structure: Domains in archaebacteria, eubacteria, eocytes and eukaryotes. *Annu Rev Biochem* 54: 507–530.

Lake, J.A. (1988) Origin of the eukaryotic nucleus determined by rate-invariant analysis of rRNA sequences. *Nature* 331: 184–186.

Lake, J.A. (1989) Origin of the multicellular animals. *In*: B. Fernhölm, K. Bremer and H. Jurnvall (eds): *The hierarchy of life. Molecules and morphology in phylogenetic analysis*. Excerpta Medica, Amsterdam-New York-Oxford, pp. 273–278.

Lake, J.A. (1990) Origin of the Metazoa. *Proc Natl Acad Sci USA* 87: 763–766.

Lake, J.A., Clark, M.W., Henderson, E., Fay, S.P., Oakes, M., Scheinman, A., Thornber, J.P. and Mah, R.A. (1985) Eubacteria, halobacteria, and the origin of photosynthesis: the photocytes. *Proc Natl Acad Sci USA* 82: 3716–3720.

Lake, J.A., Henderson, E., Oakes, M. and Clark, M.W. (1984) Eocytes: A new ribosome structure indicates a kingdom with a close relationship to eukaryotes. *Proc Natl Acad Sci USA* 81: 3786–3790.

Lamboy, W.F. (1994) The accuracy of the maximum parsimony method for phylogenetic reconstruction with morphological characters. *Syst Biol* 19: 489–505.

Lanyon, S.M. and Omland, K.E. (1999) A molecular phylogeny of the blackbirds (Icteridae): Five lineages revealed by cytochrome *b* sequence data. *Auk* 116: 629–639.

Larson, A. (1994) The comparison of morphological and molecular data in phylogenetic systematics. *In*: B. Schierwater, B. Streit, G.P. Wagner and R. DeSalle (eds): *Molecular Ecology and Evolution: Approaches and Applications*. Birkhäuser, Basel, pp. 371–390.

Lassner, M. and Dvorak, J. (1986) Preferential homogenization between adjacent and alternate subrepeats in wheat rDNA. *Nucl Acid Res* 14: 5499–5512.

Laudet, V. (1997) Evolution of the nuclear receptor superfamily: early diversification from an ancestral orphan receptor. *J Mol Endocrinol* 19: 207–226.

Laudet, V., Hanni, C., Coll, J., Catzeflis, F. and Stehelin, D. (1992) Evolution of the nuclear receptor gene superfamily. *EMBO J* 11: 1003–1013.

Lawrence, J.G. (1997) Selfish operons and speciation by gene transfer. *Trends Microbiol* 5: 355–359.

Lawrence, J.R. (1998) Roles of Horizontal Transfer in Bacterial Evolution. *In*: M. Syvanen and C.I. Kado (eds): *Horizontal gene transfer*. Chapman and Hall, London-New York, pp. 474–484.

Lawrence, J. (1999) Selfish operons: The evolutionary impact of gene clustering in prokaryotes and eukaryotes. *Curr Opin Genet Develop* 9: 642–648.

Lawrence, J.G. and Hartl, D.L. (1992) Inference of horizontal genetic transfer from molecular data: An approach using the bootstrap. *Genetics* 131: 753–760.

Lawrence, J.G. and Ochman, H. (1997) Amelioration of bacterial genomes: Rates of change and exchange. *J Mol Evol* 44: 383–397.

Lawrence, J.G. and Ochman, H. (1998) Molecular archaeology of the *Escherichia coli* genome. *Proc Natl Acad Sci USA* 95: 9413–9417.

Lawrence, J.G. and Roth, J.R. (1996) Selfish operons: horizontal transfer may drive the evolution of gene clusters. *Genetics* 143: 1843–1860.

Lecointre, G., Philippe, H., Van Le H.L. and Le Guyader, H. (1993) Species sampling has a major impact on phylogenetic inference. *Mol Phylogenet Evol* 2: 205–224.

Lederberg, J.T.E. (1946) Gene recombination in *E. coli*. *Nature* 158: 558.

Lee, K., Feinstein, J. and Cracraft, J. (1997) The phylogeny of ratite birds: Resolving conflicts between molecular and morphological data sets. *In*: D.P. Mindell (ed.): *Avian Molecular Evolution and Systematics*. Academic Press, New York, pp. 173–211.

Lee, K.-H. and Ruby, E.G. (1994) Competition between *Vibrio fischeri* strains during the initiation and maintenance of a light organ symbiosis. *J Bacteriol* 176: 1985–1991.

Letcher, A.J. and Harvey, P.H. (1994) Variation in geographical range size among mammals of the Palearctic. *Amer Naturalist* 144: 30–42.

Lewis, D.F.V. (1996) *Cytochromes P450: Structure, Function and Mechanism*. Taylor and Francis, London.

Lewis, R.E. and Lewis, J.H. (1985) Notes on the geographical distribution and host preferences in the order Siphonaptera. *J Med Entomol* 22: 134–152.

Li WH (1985) Accelerated evolution following gene duplication and its implications for the neutralist- selection controversy. *In*: T. Ohta (ed.): *Population Genetics and Molecular Evolution*. Springer-Verlag, Berlin, pp. 335–352.

Littlewood, D.T.J. and Rohde, K. and Clough, K.A. (1999) The interrelationships of all major groups of Platyhelminthes: phylogenetic evidence from morphology and molecules. *Biol J Linn Soc* 66: 75–114.

Littlewood, D.T.J., Telford, M.J. and Clough, K.A. and Rohde, K. (1998) Gnathostomulida—an enigmatic metazoan phylum from both morphological and molecular perspectives. *Mol Phylogenet Evol* 9: 72–79.

Liu, F.-G., R. and Miyamoto, M.M. (1999) Phylogenetic assessment of molecular and morphological data for eutherian mammals. *Syst Biol* 48: 54–64.

Lockhart, P.J., Steel, M.A., Penny, D. and Hendy, M.D. (1994) Recovering evolutionary trees under a more realistic model of sequence evolution. *Mol Biol Evol* 11: 605–612.

Lord, J.M., Westoby, M. and Leishman, M. (1995) Seed size and phylogeny in six temperate floras: constraints, niche conservatism, and adaptation. *Amer Naturalist* 146: 349–364.

Lowe, C.J. and Wray, G.A. (1997) Radical alterations in the roles of homeobox genes during echinoderm evolution. *Nature* 389: 718–721.

Luckett, W. and Hong, N. (1998) Phylogenetic relationships between the orders Artiodactyla and

Cetacea: A combined assessment of morphological and molecular evidence. *J Mammal Evol* 5: 127–182.

Luckow, M. (1995) Species concepts: Assumptions, methods, and applications. *Syst Bot* 20: 589–605.

Lundberg, J.G. (1972) Wagner networks and ancestors. *Syst Zool* 21: 398–413.

Maddison, D.R., Baker, M.D. and Ober, K.A. (1999) Phylogeny of Carabid beetles as inferred from 18S ribosomal DNA (Coleoptera: *Carabidae*). *Syst Entomol* 24: 103–138.

Maddison, D.R., Ruvolo, M. and Swofford, D.R. (1992) Geographic origins of human mitochondrial DNA: Phylogenetic evidence from control region sequences. *Syst Biol* 41: 111–124.

Maddison, W.P. (1990) A method for testing the correlated evolution of two binary characters: Are gains or losses concentrated on certain branches of a phylogenetic tree? *Evolution* 44: 539–557.

Maddison, W.P. (1995) Phylogenetic histories within and among species. *In*: P.C. Hoch and A.G. Stephenson (eds): *Experimental and molecular approaches to plant biosystematics*. Monographs in systematics, Volume 53. Missouri Botanical Garden, St. Louis, pp. 273–287.

Maddison, W.P. (1997) Gene trees in species trees. *Syst Biol* 46: 523–536.

Maddison, W.P., Donoghue, M.J. and Maddison, D.R. (1984) Outgroup analysis and parsimony. *Syst Zool* 33: 83–103.

Maddison, W.P. and Maddison, D.R. (1992) MacClade. Sinauer Associates, Sunderland, MA.

Maddison, W.P. and Maddison, D.R. (1997) MacClade 3.07 (software and documentation). Sinauer Associates, Sunderland, MA.

Maddison, W.P. and Slatkin, M. (1991) Null models for the number of evolutionary steps in a character on a phylogenetic tree. *Evolution* 45: 1184–1197.

Majewski, J., Zawadzki, P., Pickerill, P., Cohan, F.M. and Dowson, C.G. (2000) Barriers to genetic exchange between bacterial species: *Streptococcus pneumoniae* transformation. *J Bacteriol* 182: 1016–1023.

Mallet, J. (1995) A species definition for the modern synthesis. *Trends Ecol Evol* 10: 294–298.

Mallet, J. (1996) Reply from J. Mallet. *Trends Ecol Evol* 11: 174–175.

Mallet, J. and Barton, N.H. (1989) Strong natural selection in a warning-color hybrid zone. *Evolution* 43: 421–431.

Mallet, J., McMillan, W.O. and Jiggins, C.D. (1998) Mimicry and warning color at the boundary between races and subspecies. *In*: D.J. Howard and S.H. Berlocher (eds): *Endless Forms*. Oxford University Press, Oxford, pp. 390–403.

Martin, W., Brinkmann, H., Savonna, C. and Cerff, R. (1993) Evidence for a chimeric nature of nuclear genomes: Eubacterial origin of eukaryotic glyceraldehyde-3-phosphate dehydrogenase genes. *Proc Natl Acad Sci USA* 90: 8692–8696.

Martin, W., Lagrange, T., Li YF, Bisanz-Seyer, C. and Mache, R. (1990) Hypothesis for the evolutionary origin of the chloroplast ribosomal protein L21 of spinach. *Curr Genetics* 18: 553–6.

Martin, W., Mustafa, A.Z., Henze, K. and Schnarrenberger, C. (1996) Higher-plant chloroplast and cytosolic fructose-1,6-bisphosphatase isoenzymes: Origins via duplication rather than prokaryote-eukaryote divergence. *Plant Mol Biol* 32: 485–491.

Martin, W. and Schnarrenberger, C. (1997) The evolution of the Calvin cycle from prokaryotic to eukaryotic chromosomes: a case study of functional redundancy in ancient pathways through endosymbiosis. *Curr Genetics* 32: 1–18.

Martin, W. (1999) Mosaic bacterial chromosomes: a challenge en route to a tree of genomes. *Bioessays* 21: 99–104.

Martin, W.F. (1998) Endosymbiosis and the origins of chloroplast-cytosol isoenzymes: a revision of the gene transfer corollary. *In*: Syvanen M. and C.I. Kado (eds): *Horizontal gene transfer*. Chapman and Hall, London-New York, pp. 363–376.

Martins, E.P. (1997) Phylogenies and the comparative method: A general approach to incorporating phylogenetic information into the analysis of interspecific data. *Amer Naturalist* 149: 646–667.

Martins, E.P. (1996) *Phylogenies and the Comparative Method in Animal Behavior*. Oxford University Press, Oxford.

Mason-Gamer, R.J. and Kellogg, E.A. (1996) Testing for phylogenetic conflict among molecular data sets in the tribe Triticeae (Gramineae). *Syst Biol* 45: 524–545.

Matic, I., Rayssiguier, C. and Radman, M. (1995) Interspecies gene exchange in bacteria: The role of SOS and mismatch repair systems in evolution of species. *Cell* 80: 507–515.

Mau, B., Newton, M.A. and Larget, B. (1999) Bayesian phylogenetic inference via Markov chain Monte Carlo methods. *Biometrics* 55: 1–12.

Maynard-Smith, J. (1996) Population genetics: an introduction. *In*: E.C. Neidhardt (ed.): *Escherichia coli and salmonella.* ASM Press, USA, pp. 2685–2690.

Mayr, E. (1940) Speciation phenomena in birds. *Amer Naturalist* 74: 249–278.

Mayr, E. (1942) *Systematics and the origin of species, from the viewpoint of a zoologist.* Columbia University Press, New York.

Mayr, E. (1963) *Animal Species and Evolution.* Belknap Press, Cambridge.

Mayr, E. (1969) *Principles of systematic zoology.* McGraw-Hill, New York.

Mayr, E. (1992) *The Growth of Biological Thought.* Belknap Press, Cambridge.

Mazodier, P. and Davies, J. (1991) Gene transfer between distantly related bacteria. *Annu Rev Genet* 25: 147–171.

McCracken, K.G., Harshman, J., McCleelan, D.A. and Afton, A.D. (1999) Data set incongruence and correlated character evolution: An example of functional convergence in the hind-limbs of stifftail diving ducks. *Syst Biol* 48: 683–714.

McFall-Ngai, M.J. and Ruby, E.G. (1991) Symbiont recognition and subsequent morphogenesis as early events in an animal-bacterial symbiosis. *Science* 254: 1491–1494.

McHugh, D. (1997) Molecular evidence that echiurans and pogonophorans are derived annelids. *Proc Natl Acad Sci USA* 94: 8006–8009.

Medawar, P.B. and Medawar, J.S. (1983) *Aristotle to zoos.* Harvard University Press, Cambridge.

Medigue, C., Rouxel, T., Vigier, P., Henaut, A. and Danchin, A. (1991) Evidence for horizontal gene transfer in *Escherichia coli* speciation. *J Mol Biol* 222: 851–856.

Meffe, G.K. and Carroll, R. (1997) *Principles of conservation biology.* Sinauer, Sunderland.

Messier, W. and Stewart, C. and-B. (1997) Episodic adaptive evolution of primate lysozymes. *Nature* 385: 151–154.

Meyer, A. (1998) Hox gene variation and evolution. *Nature* 391: 225–227.

Mickevich, M.F. and Farris, J.S. (1981) The implications of congruence in *Menidia. Syst Zool* 27: 143–158.

Mickoleit, G. (1973) Über den ovipositor der Neuropteroidea und Coleoptera und seine phylogenetische Bedeutung (Insecta, Holometabola). *Z Morphol Tiere* 74: 37–64.

Milinkovitch, M.C., LeDuc, R.G., Adachi, J., Farnir, F., Georges, M. and Hasegawa, M. (1996) Effects of character weighting and species sampling on phylogeny reconstruction: A case study based on DNA sequence data in Cetaceans. *Genetics* 144: 1817–1833.

Milinkovitch, M., Bérubé, M. and Palsbøll, P. (1998) Cetaceans are highly derived artiodactyls. *In*: J. Thewissen (ed.): *The emergence of whales, advances in vertebrate paleobiology.* Plenum, New York, pp. 113–131.

Milkman, R. and Bridges, M.M. (1990) Molecular evolution of the *Escherichia coli* chromosome. III. Clonal frames. *Genetics* 126: 505–517.

Mindell, D. (1991) Aligning DNA sequences: homology and phylogenetic weighting. *In*: M.J. Miyamoto and Cracraft J. (eds): *Phylogenetic Analysis of DNA Sequences.* Oxford University Press, New York, pp. 73–89.

Mindell, D.P., Sorenson, M.D., Dimcheff, D.E., Hasegawa, M., Ast, J.C. and Yuri, T. (1999) Interordinal relationships of birds and other reptiles based on whole mitochondrial genomes. *Syst Biol* 48: 138–152.

Mindell, D.P., Sorenson, M.D., Huddleston, C.J., Miranda, H.C., Jr., Knight, A., Sawchuck, S.J. and Yuri, T. (1997) Phylogenetic relationships among and within select avian orders based on mitochondrial DNA. *In*: M.J. Miyamoto and J. Cracraft (eds): *Phylogenetic Analysis of DNA Sequences.* Oxford University Press, New York, pp. 214–247.

Mishler, B.D. (1994) Cladistic analysis of molecular and morphological data. *Am J Physiol Anthropol* 94: 143–156.

Mishler, B.D. and Brandon, R.N. (1987) Individuality, pluralism and the phylogenetic species concept. *Biol Philosoph* 2: 397–414.

Mishler, B.D., Bremer, K., Humphries, C. and Churchill, S. (1988) The use of nucleic acid sequence data in phylogenetic reconstruction. *Taxon* 27: 391–395.

Mishler, B.D. and Donoghue, M.J. (1982) Species concepts: A case for pluralism. *Syst Zool* 31: 491–503.

Mita, K., Ichimura, S. and James, T. (1994) Highly repetitive structure and its organization of the silk fibroin gene. *J Mol Evol* 38: 583–592.

Miyamoto, M. (1985) Consensus cladograms and general classifications. *Cladistics* 1: 186–189.

Miyamoto, M.M. and Fitch, W.M. (1995) Testing species phylogenies and phylogenetic methods with congruence. *Syst Biol* 44: 64–76.

Montgomery, S.L. (1975) Comparative breeding site ecology and the adaptive radiation of picture-winged *Drosophila* (Diptera: Drosophilidae) in Hawaii. *Proc Hawaii Entomol Soc* 22: 65–103.

Mooers, A.O. and Schluter, D. (1999) Reconstructing ancestor states with Maximum Likelihood: Support of one- and two-rate models. *Syst Biol* 1999: 623–632.

Moon, J.W. and Steel, M.A. (1993) A limiting distribution for parsimoniously bicoloured trees. *Appl Math Lett* 6: 5–8.

Moran, N.A. and Telang, A. (1998) Bacteriocyte-associated symbionts of insects. *Bioscience* 48: 295–304.

Moran, N.A., Munson, M.A., Baumann, P. and Ishikawa, H. (1993) A molecular clock in endosymbiotic bacteria is calibrated using the insect hosts. *Proc R Soc Lond B* 253: 167–171.

Moreira, D. (2000) Multiple independent horizontal transfers of informational genes from bacteria to plasmids and phages: implications for the origin of bacterial replication machinery. *Mol Microbiol* 35: 1–5.

Moritz, C., Dowling, T.E. and Brown, W.M. (1987) Evolution of animal mitochondrial DNA: relevance for population genetics and systematics. *Ann Rev Ecol Syst* 18: 489–522.

Moritz, C. and Birmingham, P. (1998) Comparative phylogeography: concepts and applications. *Molec Ecol* 7(4): 367–369.

Morrison, D.A. and Ellis, J.T. (1997) Effects of nucleotide sequence alignment on phylogeny estimation: a case study of 18S rDNAs of Apicomplexa. *Mol Biol Evol* 14: 428–441.

Müller, W.E., Kruse, M., Koziol, C., Müller, J.M. and Leys, S.P. (1998) Evolution of early Metazoa: Phylogenetic status of the Hexactinellida within the phylum of Porifera (sponges). *Prog Mol Subcell Biol* 21: 141–156.

Munson, M.A., Baumann, P., Clark, M.A., Baumann, L., Moran, N.A. and Voegtlin, D.J. (1991) Evidence for the establishment of aphid-eubacterial endosymbiosis in an ancestor of four aphid families. *J Bacteriol* 173: 6321–6324.

Mushegian, A.R., Garey, J.R., Martin, J. and Liu, L.X. (1998) Large-scale taxonomic profiling of eukaryotic model organisms: a comparison of orthologous proteins encoded by the human, fly, nematode and yeast genomes. *Genome Res* 8: 590–598.

Muto, A. and Osawa, S. (1987) The guanine and cytosine content of genomic DNA and bacterial evolution. *Proc Natl Acad Sci USA* 84: 166–169.

Nardon, P. and Grenier, A. and-M. (1989) Endocytobiosis in Coleoptera: Biological, biochemical and genetic aspects. *In*: W. Schwemmler and G. Gassner (eds): *Insect endocytobiosis: Morphology, ecophysiology, isolation, identification, applications.* CRC Press, Boca Raton, pp. 175–216.

Naylor, G. and Brown, W. (1998) *Amphioxus* mt DNA, chordate phylogeny, and the limits of inference based on comparisons of sequences. *Syst Biol* 47: 61–76.

Nealson, K.H. and Hastings, J.W. (1991) The luminous bacteria. *In*: A. Balows, H.G. Truper, M. Dworkin, W. Harder and K.H. Schleifer (eds): *The Prokaryotes, a handbook on the biology of bacteria: ecophysiology, isolation, identification, applications*, 2nd edition. Springer-Verlag, New York, pp. 625–639.

Nee, S., Read, A.F. and Harvey, P.H. (1996) Why phylogenies are necessary for comparative analysis. *In*: E.P. Martins (ed.): *Phylogenies and the comparative method in animal behaviour.* Oxford University Press, New York.

Needleman, S.B. and Wunsch, C.D. (1970) A general method applicable to the search for similarities in the amino acid sequence of two proteins. *J Mol Biol* 48: 443–453.

Nei, M. and Kumar, S. (2000) *Molecular Evolution and Phylogenetics.* Oxford University Press, Oxford.

Nelson, G.J. (1978) Ontogeny, phylogeny, paleontology and the biogenetic law. *Syst Zool* 27: 324–345.

Nelson, G.J. (1994) Homology and systematics. *In*: B.K. Hall (ed.): *Homology: The hierarchical basis of comparative biology.* Academic Press, San Diego, pp. 101–149.

Nelson, G.J. and Platnick, N.I. (1981) *Systematics and biogeography. Cladistics and vicariance.* Columbia University Press, New York.

Nelson, K.E., Clayton, R.A., Gill, S.R., Gwinn, M.L., Dodson, R.J., Haft, D.H., Hickey, E.K., Peterson, J.D., Nelson, W.C., Ketchum, K.A. et al. (1999) Evidence for lateral gene transfer between Archaea and bacteria from genome sequence of *Thermotoga maritima. Nature* 399: 323–329.

Nikaido, M., Rooney, A. and Okada, N. (1999) Phylogenetic relationships among cetartiodactyls based on insertions of short and long interspersed elements: Hippopotamuses are the closest extant relatives of whales. *Proc Natl Acad Sci USA* 96: 10,261–10,266.

Nikoh, N., Iwabe, N., Kuma, K., Ohno, M., Sugiyama, T., Watanabe, Y., Yasui, K., Shi-cui, Z., Hori, K., Shimura, Y. et al. (1997) An estimate of divergence time of Parazoa and Eumetazoa and that of Cephalochordata and Vertebrata by aldolase and triose phosphate isomerase clocks. *J Mol Evol* 45: 97–106.

Nishiguchi, M.K., Ruby, E.G. and McFall-Ngai, M.J. (1997) Phenotypic bioluminescence as an indicator of competitive dominance in the *Euprymna-Vibrio* symbiosis. *In*: J.W. Hastings, J.W. Kricka and P.E. Stanley (eds): *Bioluminescence and chemiluminescence: molecular reporting with photons*. J. Wiley and Sons, New York, pp. 123–126.

Nishiguchi, M.K., Ruby, E.G. and McFall-Ngai, M.J. (1998) Competitive dominance among strains of luminous bacteria provides an unusual form of evidence for parallel evolution in Sepiolid squid-*Vibrio* symbioses. *Appl Environ Microbiol* 64: 3209–3213.

Nishiguchi, M.K. (2000) Temperature affects species distribution in symbiotic populations of *Vibrio* spp. *Appl Environ Microbiol* 66: 3550–3555.

Nishiguchi, M.K. (2001) Co-evolution of symbionts and hosts: The sepiolid-*Vibrio* model. *In*: J. Seckbach (ed.): *Cellular origin and life in extreme habitats*. Cole-Kluwer Academic Publishers, Dordrecht, The Netherlands; *in press*.

Nishiguchi, M.K. Host recognition is responsible for symbiont composition in environmentally ansmitted symbiosis. *Microb Ecol*; *in press*.

Nixon, K.C. and Carpenter, J.M. (1996) On simultaneous analysis. *Cladistics* 12: 221–41.

Nixon, K.C. and Wheeler, Q.D. (1990) An amplification of the phylogenetic species concept. *Cladistics* 6: 211–23.

Nixon, K.C. (1999) The Parsimony Ratchet, a new method for rapid parsimony analysis. *Cladistics* 15: 407–414.

Normark, B.B. (1999) Evolution in a putatively ancient asexual aphid lineage: Recombination and rapid karyotype change. *Evolution* 53: 1458–1469.

Novacek, M. and, Q. Wheeler (1992) Extinct taxa: Accounting for 99.999…% of the earth's biota. *In*: M. Novacek and Q.D. Wheeler (eds): *Extinction and phylogeny*. Columbia University Press, New York, pp. 1–16.

Nunn, G.B. and Stanley, S.E. (1998) Body size effects and rates of cytochrome *b* evolution in tube-nosed seabirds. *Mol Biol Evol* 15: 1360–1371.

O'Grady, P.M., Clark, J.B. and Kidwell, M.G. (1998) Phylogeny of the *Drosophila saltans* species group based on combined analysis of nuclear and mitochondrial DNA sequences. *Mol Biol Evol* 15: 656–664.

O'Grady, P.M. (1999) Reevaluation of phylogeny in the *Drosophila obscura* species group based on combined analysis of nucleotide sequences. *Mol Phylogenet Evol* 12: 124–139.

Oakley, T.H. and Cunningham, C.W. (2000) Independent contrasts succeed where ancestor reconstruction fails in a known bacteriophage phylogeny. *Evolution* 54: 397–405.

O'Brien, A.D., Newland, J.W., Miller, S.F., Holmes, R.K., Smith, H.W. and Formal, S.B. (1984) Shiga-like toxin-converting phages from *Escherichia coli* strains that cause hemorrhagic colitis or infantile diarrhea. *Science* 226: 694–696.

Ochman, H. and Groisman, E.A. (1994) The origin and evolution of species differences in *Escherichia coli* and *Salmonella typhimurium*. *Exs.* 69: 479–493.

Ochman, H. and Lawrence, J. (1996) Phylogenetics and the amelioration of bacterial genomes. *In*: F.C. Neidhardt (ed.): *Escherichia coli and Salmonella: Cellular and molecular biology*, ASM Press, Washington DC, pp.

Ochman, H., Lawrence, J. and Groisman, E.A. (2000) Lateral gene transfer and the nature of bacterial innovation. *Nature* 405: 299–304.

Ochman, H., Whittam, T.S., Caugant, D.A. and Selander, R.K. (1983) Enzyme polymorphism and genetic population structure in *Escherichia coli* and *Shigella*. *J Gen Microbiol* 129: 2715–2726.

Ohno, S. (1970) *Evolution by gene duplication*. Springer-Verlag, Berlin.

O'Leary, M.A. (1999) Parsimony analysis of total evidence from extinct and extant taxa and the cetacean-artiodactyl question (Mammalia, Ungulata). *Cladistics* 15: 315–330.

O'Leary, M. and Geisler, J. (1999) The position of Cetacea within Mammalia: Phylogenetic analysis of morphological data from extinct and extant taxa. *Syst Biol* 48: 455–490.

Olmstead, R.G. and Palmer, J.D. (1994) Chloroplast DNA systematics: a review of methods and data analysis. *Amer J Bot* 81: 1205–1224.

Olmstead, R.G., Reeves, P.A. and Yen, A.C. (1998) Patterns of sequence evolution and implications for parsimony analysis of chloroplast DNA. *In*: D.E. Soltis, P.S. Soltis and J.J. Doyle (eds): *Molecular systematics of plants II: DNA sequencing*. Kluwer Press, Boston, pp. 164–187.

Olmstead, R.G. and Sweere, J.A. (1994) Combining data in phylogenetic systematics: An empirical approach using three molecular data sets in the Solanaceae. *Syst Biol* 43: 467–481.

Olsen, G.J., Woese, C.R. and Overbeek, R. (1994) The winds of (evolutionary) change: Breathing new life into microbiology. *J Bacteriol* 176: 1–6.

Omland, K.E. (1999) The assumptions and challenges of ancestral state reconstructions. *Syst Biol* 48: 604–611.

Page, R.D.M. (1987) Graphs and generalized tracks: Quantifying Croizat's panbiogeography. *Syst Zool* 36: 1–17.

Page, R.D.M. (1993a) *Component 2.0*, software and documentation. Natural History Museum, London.

Page, R.D.M. (1993b) Genes, organisms, and areas: The problem of multiple lineages. *Syst Biol* 42: 77–84.

Page, R.D.M. (1994a) Maps between trees and cladistic analysis of historical associations among genes, organisms, and areas. *Syst Biol* 43: 58–77.

Page, R.D.M. (1994b) Parallel phylogenies: Reconstructing the history of host-parasite assemblages. *Cladistics* 10: 155–173.

Page, R.D.M. (1995) Treemap v. 1.0. Program and documentation (http://taxonomy.zoology.gla.ac.uk/rod/treemap.html).

Page, R.D.M. (1996) TreeMap. Program and documentation Div. of Env. and Evol. Bio., Institute of Biomedical and Life Sciences, University of Glasgow, Glasgow.

Page, R.D.M. (1998a) Genetree 1.0, software and documentation. University of Glasgow, Glasgow.

Page, R.D.M. (1998b) GeneTree: Comparing gene and species phylogenies using reconciled trees. *Bioinformatics* 14: 819–820.

Page, R.D.M. and Charleston, M.A. (1997) From gene to organismal phylogeny: Reconciled trees and the gene tree/species tree problem. *Mol Phylogenet Evol* 7: 231–240.

Page, R.D.M. and Holmes, E.C. (1998) *Molecular Evolution: A phylogenetic approach*. Blackwell Scientific, Oxford.

Pagel, M. (1994) Detecting correlated evolution on phylogenies: a general model for the comparative analysis of discrete characters. *Proc R Soc Lond B* 255: 37–45.

Pagel, M. (1999) Inferring the historical patterns of biological evolution. *Nature* 401: 877–884.

Palmer, J. (1985) The silk and silk production system of the funnel-web mygalomorph spider *Euagrus* (Araneae, Dipluridae). *J Morphol* 186: 195–207.

Panchen, A.L. (1992) *Classification, Evolution and the Nature of Biology*. Cambridge University Press, London.

Paquin, B., Heinfling, A. and Shub, D.A. (1999) Sporadic distribution of tRNA(Arg)CCU introns among alpha-purple bacteria: Evidence for horizontal transmission and transposition of a group I intron. *J Bacteriol* 181: 1049–1053.

Parker, A. (1997) Combining molecular and morphological data in fish systematics: Examples from Cyprinodontiformes. *In*: C.A. Stepien and T.D. Kocher (eds): *Molecular Evolution of Fishes*. Academic Press, San Diego, pp. 163–183.

Parkhe, A., Seeley, S., Gardener, K., Thompson, L. and Lewis, R. (1997) Structural studies of spider silk proteins in the fiber. *J Mol Recogn* 10: 1–6.

Parkhill, J., Achtman, M., James, K.D., Bentley, S.D., Churcher, C., Klee, S.R., Morelli, G., Basham, D., Brown, D., Chillingworth, T. et al. (2000a) Complete DNA sequence of a serogroup A strain of *Neisseria meningitidis* Z2491. *Nature* 404: 502–506.

Parkhill, J., Wren, B.W., Mungall, K., Ketley, J.M., Churcher, C., Basham, D., Chillingworth, T., Davies, R.M., Feltwell, T., Holroyd, S. et al. (2000b) The genome sequence of the food-borne pathogen *Campylobacter jejuni* reveals hypervariable sequences. *Nature* 403: 665–668.

Pashley, D.P., McPheron, B.A. and Zimmer, E.A. (1993) Systematics of holometabolous insect orders based on 18S ribosomal RNA. *Mol Phylogenet Evol* 2: 132–142.

Patterson, C. (1982) Morphological characters and homology. *In*: Joysey KA and Friday AE (eds): *Problems of phylogenetic reconstruction*. Academic Press, London, pp. 21–74.

Patterson, C. (1988) *Molecules and morphology in evolution: Conflict or compromise?* Cambridge University Press, Cambridge.

Patterson, C., Williams, D.M. and Humphries, C.J. (1993) Congruence between molecular and morphological phylogenies. *Annu Rev Ecol Syst* 24: 153–188.

Pearson, D.L., Blum, M.S., Jones, T.H., Fales, H.M., Gonda, E. and White, B.R. (1988) Historical perspective and the interpretation of ecological patterns: defensive compounds of tiger beetles
(Coleoptera: Cicindelidae). *Amer Naturalist* 132: 404–416.

Pecon Slattery, J. and Murphy, W. and, O'Brien, S. (2000) Patterns of diversity among SINE elements
isolated from three Y-chromosome genes in carnivores. *Mol Biol Evol* 17: 825–829.

Penalva, M.A., Moya, A., Dopazo, J. and Ramon, D. (1990) Sequences of isopenicillin N synthetase
genes suggest horizontal gene transfer from prokaryotes to eukaryotes. *Proc R Soc Lond B* 241:
164–169.

Penny, D. and Hasegawa, M. (1997) *Platypus* put in its place. *Nature* 387: 549–550.

Penny, D., Hendy, M.D., Lockhart, P.J. and Steel, M.A. (1996) Corrected parsimony, minimum evolution and Hadamard conjugations. *Syst Biol* 45: 593–603.

Penny, D., Hendy, M.D. and Steel, M.A. (1991) Testing the theory of descent. *In*: M. Miyamoto and
J. Cracraft (eds): *Phylogenetic Analysis of DNA sequences*. Oxford University Press. Oxford, pp.
155–183.

Penny, D., Steel, M.A., Lockhart, P.J. and Hendy, M.D. (1994) The role of models in reconstructing
evolutionary trees. *In*: R.W. Scotland, D.J. Siebert and D.M. Williams (eds): *Models in phylogeny
reconstruction*. Oxford University Press, Oxford, pp. 211–230.

Philippe, H. and Forterre, P. (1999) The rooting of the universal tree of life is not reliable. *J Mol Evol*
49: 509–523.

Philippe, H., Chenuil, A. and Adoutte, A. (1994) Can the Cambrian explosion be inferred through
molecular phylogeny? *Development* Suppl: 15–25.

Philippe, H., Lecointre, G., Le, H.L.V. and Le Guyander, H. (1996) A critical study of homology in
molecular data with the use of a morphologically based cladogram, and its consequences for character weighting. *Mol Biol Evol* 13: 1174–1186.

Phillips, A., Janies, D. and Wheeler, W.C. (2000) Multiple sequence alignment in phylogenetic analysis. *Mol Phylogenet Evol* 16: 317–330.

Piatigorsky, J. and Wistow, G.J. (1989) Enzyme/crystallins: gene sharing as an evolutionary strategy.
Cell 57: 197–199.

Platnick, N.I. (1979) Philosophy and the transformation of cladistics. *Syst Zool* 28: 537–546.

Poe, S. (1996) Data set incongruence and the phylogeny of crocodilians. *Syst Biol* 45: 393–414.

Poe, S. (1998) Sensitivity of phylogeny estimation to taxonomic sampling. *Syst Biol* 47: 18–31.

Poe, S. and Swofford, D.L. (1999) Taxon sampling revisited. *Nature* 398: 299–300.

Porterfield, J.C., Page, L.M. and Near, T.J. (1999) Phylogenetic relationships among fantail darters
(Percidae: *Etheostoma*: *Catonotus*): Total evidence analysis of morphological and molecular data.
Copeia 3: 551–564.

Powell, J.R. (1975) Protein variation in natural populations of animals. *Evol Biol* 8: 79–118.

Powell, J.R. (1997) Progress and prospects in evolutionary biology: The *Drosophila* model. Oxford
University Press, New York.

Powell, J.R. and DeSalle, R. (1995) *Drosophila* molecular phylogenies and their uses. *Evol Biol* 28:
87–138.

Purvis, A., Gittleman, J.L. and Luh, H.K. (1994) Truth or consequences—effects of phylogenetic
accuracy on two comparative methods. *J Theor Biol* 167: 293–300.

Purvis, A. and Rambaut, A. (1995) Comparative analysis by independent contrasts (CAIC): An Apple
Macintosh application for analysing comparative data. *Comput Appl Biosci* 11: 247–251.

Purvis, A. and Webster, A.J. (1999) Phylogenetically independent comparisons and primate phylogeny. *In*: P.C. Lee (ed.): *Comparative primate socioecology*. Cambridge University Press,
Cambridge, pp. 44–70.

Queralt, R., Adroer, R., Oliva, R., Winkfein, R. and Retief, J. and, G. Dixon (1995) Evolution of protamine P1 genes in mammals. *J Mol Evol* 40: 601–607.

Quicke, D.L.J. and Belshaw, R. (1999) Incongruence between morphological data sets: An example
from the evolution of endoparasitism among parasitic wasps (Hymenoptera: Braconidae). *Syst
Biol* 48: 436–454.

Raff, R.A., Field, K.G., Olsen, G.J., Giovannoni, S.J., Lane, D.J., Ghiselin, M.T., Pace, N.R. and Raff,
E.C. (1989) Metazoan phylogeny based on analysis of 18S ribosomal RNA. *In*: B. Fernhölm, K.
Bremer and H. Jurnvall (eds): *The hierarchy of life. Molecules and morphology in phylogenetic
analysis*. Excerpta Medica, Amsterdam-New York-Oxford, pp. 247–261.

Raff, R.A. and Kaufman, T.C. (1983) *Embryos, Genes, and Evolution*. MacMillan, New York.

Raymond, C.S., Shamu, C.E., Shen, M.M., Seifert, K.J., Hirsch, B., Hodgkin, J. and Zarkower, D.

(1998) Evidence for evolutionary conservation of sex-determining genes. *Nature* 391: 691–695.

Reeck, G.R., de Haen, C., Teller, D.C., Doolittle, R.F., Fitch, W.M., Dickerson, R.E., Chambon, P., McLachlen, A.D., Margoliash, E. and Jukes, T.H. (1987) "Homology" in proteins and nucleic acids: a terminology muddle and a way out of it. *Cell* 50: 667.

Regier, J.C. and Shultz, J.W. (1997) Molecular phylogeny of the major arthropod groups indicates polyphyly of crustaceans and a new hypothesis for the origin of hexapods. *Mol Biol Evol* 14: 902–913.

Regier, J.C. and Shultz, J.W. (1998) Molecular phylogeny of arthropods and the significance of the Cambrian "explosion" for molecular systematics. *Amer Zool* 38: 918–928.

Remsen, J. and DeSalle, R. (1998) Character congruence of multiple data partitions and the origin of the Hawaiian Drosophilidae. *Mol Phylogenet Evol* 9: 225–235.

Rhymer, J.M., Williams, M.J. and Braun, M.J. (1994) Mitochondrial analysis of gene flow between New Zealand mallards (*Anas platyrhynchos*) and grey ducks (*A. superciliosa*). *Auk* 111: 970–978.

Rice, K.A., Donoghue, M.J. and Olmstead, R.G. (1997) Analyzing large data sets: rbcL 500 revisited. *Syst Biol* 46: 554–562.

Ridley, M. (1983) *The explanation of organic diversity: The comparative method and adaptations for mating*. Oxford University Press, Oxford.

Ridley, M. (1989) The cladistic solution to the species problem. *Biol Philosoph* 4: 1–16.

Ridley, M. and Grafen, A. (1996) How to study discrete comparative methods. *In*: E.P. Martins (ed.): *Phylogenies and the comparative method in animal behaviour*. Oxford University Press, New York, pp. 76–103.

Rieppel, O. (1988) *Fundamentals of Comparative Biology*. Birkhäuser Verlag, Basel.

Rieppel, O. (1992) Homology and logical fallacy. *J Evol Biol* 5: 701–715.

Rieppel, O. (1994a) Homology, topology, and typology: The history of modern debates. *In*: Homology: B.K. Hall (ed.): *The Hierarchical Basis of Comparative Biology*. Academic Press, San Diego, pp. 63–100.

Rieppel, O. (1994b) Species and history. *In*: R.W. Scotland, D.J. Siebert and D.M. Williams (eds): *Models in phylogeny reconsruction*. Systematics Association special vol. 52. Clarendon Press, Oxford, pp. 31–50.

Rivera, M.C., Jain, R., Moore, J.E. and Lake, J.A. (1998) Genomic evidence for two functionally distinct gene classes. *Proc Natl Acad Sci USA* 95: 6239–6244.

Rivera, M.C. and Lake, J.A. (1992) Evidence that eukaryotes and eocyte prokaryotes are immediate relatives. *Science* 257: 74–76.

Roberts, M.S. and Cohan, F.M. (1993) The effect of DNA sequence divergence on sexual isolation in *Bacillus*. *Genetics* 134: 401–408.

Robinson, D. and Foulds, L.R. (1979) Comparison of weighted labeled trees. *In*: A. Dold and B. Eckmann (eds): *Lecture Notes in mathematics, Vol 748*. Springer-Verlag Berlin, pp. 119–126.

Rodrigo, A.G., Kelly-Borges, M., Bergquist, P.R. and Bergquist, P.L. (1993) A randomisation test of the null hypothesis that two cladograms are sample estimates of a parametric phylogenetic tree. *N Z J Bot* 31: 257–268.

Roger, A.J., Smith, M.W., Doolittle, R.F. and Doolittle, W.F. (1996) Evidence for the Heterolobosea from phylogenetic analysis of genes encoding glyceraldehyde-3-phosphate dehydrogenase. *J Eukaryot Microbiol* 43: 475–485.

Rokas, A., Kathirithamby, J. and Holland, P.W.H. (1999) Intron insertion as a phylogenetic character: the *engrailed* homeobox of Strepsiptera does not indicate affinity with Diptera. *Insect Mol Biol* 8: 527–530.

Rolfe, R. and Meselson, M. (1959) The relative homogeneity of microbial DNA. *Proc Natl Acad Sci USA* 45: 1039–1042.

Rosen, D.E. (1978) Vicariant patterns and historical explanation in biogeography. *Syst Zool* 27: 159–188.

Roth, V.L. (1984) On homology. *Biol J Linn Soc* 22: 13–29.

Roth, V.L. (1988) The biological basis of homology. *In*: C.J. Humphries (ed.): *Ontogeny and Systematics*. Columbia University Press, New York, pp. 35–49.

Roth, V.L. (1991) Homology and hierarchies: Problems solved and unresolved. *J Evol Biol* 4: 167–194.

Roux, K.H., Greenberg, A.S., Greene, L., Strelets, L., Avila, D., McKinney, E.C. and Flajnik, M.F. (1998) Structural analysis of the nurse shark (new) antigen receptor (NAR): Molecular convergence of NAR and unusual mammalian immunoglobulins. *Proc Natl Acad Sci USA* 95: 11,804–11,809.

Ruby, E.G. (1996) Lessons from a cooperative, bacterial-animal association: The *Vibrio fischeri-Euprymna scolopes* light organ symbiosis. *Annu Rev Microbiol* 50: 591–624.

Ruby, E.G. and Asato, L.M. (1993) Growth and flagellation of *Vibrio fischeri* during initiation of the sepiolid squid light organ symbiosis. *Arch Microbiol* 159: 160–167.

Ruiz-Trillo, I., Riutort, M., Littlewood, D.T., Herniou, E.A. and Baguñà, J. (1999) Acoel flatworms: earliest extant bilaterian Metazoans, not members of Platyhelminthes. *Science* 283: 1919–1923.

Russo, C.M.A., Takezaki, N. and Nei, M. (1995) Molecular phylogeny and divergence times of drosophilid species. *Mol Biol Evol* 12: 391–404.

Sander, K. (1983) The evolution of patterning mechanisms: gleanings from insect embryogenesis and spermatogenesis. *In*: B.C. Goodwin, N. Holder and C.C. Wylie (eds): *The Sixth Symposium of the British Society for Developmental Biology*. Cambridge University Press, Cambridge, pp. 137–159.

Sanderson, M.J. (1994) Reconstructing the history of evolutionary processes using maximum likelihood. *In*: D.M. Fambrough (ed.): *Molecular Evolution of Physiological Processes*. Rockefeller University Press, New York, pp. 13–26.

Sanderson, M.J. (1995) Objections to bootstrapping phylogenies: a critique. *Syst Biol* 44: 299–320.

Sanderson, M.J. and Donoghue, M.J. (1989) Patterns of variation in levels of homoplasy. *Evolution* 43: 1781–1795.

Sanderson, M.J. and Doyle, J.J. (1992) Reconstruction of organismal and gene phylogenies from data on multigene families: Concerted evolution, homoplasy and confidence. *Syst Biol* 41: 4–17.

Sanderson, M.J., Purvis, A. and Henze, C. (1998) Phylogenetic supertrees: assembling the trees of life. *TREE* 13: 105–109.

Sankoff, D. and Cedergren, R.J. (1983) Simultaneous comparison of three or more sequences related by a tree. *In*: D. Sankoff and J.B. Kruskal (eds): *Time warps, string edits, and macromolecules: The theory and practice of sequence comparison*. Addison-Wesley, Reading, pp. 253–263.

Sankoff, D., Morel, C. and Cedergren, R.J. (1973) Evolution of 5S RNA and the non-randomness of base replacement. *Nature New Biol* 245: 232–234.

Saunders, N.C. (1987) Intraspecific phylogeography: The mitochondrial DNA bridge between population genetics and systematics. *Annu Rev Ecol Syst* 18: 489–522.

Sawyer, S. (1989) Statistical tests for detecting gene conversion. *Mol Biol Evol* 6: 526–538.

Schlein, Y. (1980) Morphological similarities between the skeletal structures of siphonaptera and mecoptera. *Proc. International Conference on Fleas*: 359–367.

Schöniger, M. and von Haeseler, A. (1994) A stochastic model for the evolution of autocorrelated sequences. *Mol Phylogenet Evol* 3: 240–247.

Schuh, R.T. (2000) *Biological Systematics, principles and applications*. Cornell University Press, Ithaca.

Schuh, R.T. and Polhemus, J. (1980) Analysis of taxonomic congruence among morphological, ecological, and biogeographic data sets for the Leptopodomorpha (Hemiptera). *Syst Zool* 30: 309–325.

Schwartz, R.M. and Dayhoff, M.O. (1978) Origins of prokaryotes, eukaryotes, mitochondria, and chloroplasts. *Science* 199: 395–403.

Selden, P. (1989) Orb-web weaving spiders in the early Cretaceous. *Nature* 340: 711–713.

Semple, C. and Wolfe, K.H. (1999) Gene duplication and gene conversion in the *Caenorhabditis elegans* genome. *J Mol Evol* 48: 555–564.

Sharp, P.M., Cowe, E., Higgins, D.G., Shields, D.C., Wolfe, K.H. and Wright, F. (1988) Codon usage patterns in *Escherichia coli*, *Bacillus subtilis*, *Saccharomyces cerevisiae*, *Schizosaccharomyces pombe*, *Drosophila melanogaster* and *Homo sapiens*; A review of the considerable within- species diversity. *Nucl Acid Res* 16: 8207–8211.

Sharp, P.M. and Li WH (1987) The codon Adaptation Index—a measure of directional synonymous codon usage bias, and its potential applications. *Nucl Acid Res* 15: 1281–1295.

Sharp, P.M. and Matassi, G. (1994) Codon usage and genome evolution. *Curr Opin Genet Develop* 4: 851–860.

Shatters, R.G. and Kahn, M.L. (1989) Glutamine synthetase II in *Rhizobium*: reexamination of the proposed horizontal transfer of DNA from eukaryotes to prokaryotes. *J Mol Evol* 29: 422–428.

Shaw, K.L. (1996) What are 'good' species? *Trends Ecol Evol* 11: 174.

Shear, W. (1986) *Spiders: Webs, Behavior, and Evolution*. Stanford University Press, Stanford.

Sheldon, F.H. and Bledsoe, A.H. (1993) Avian. *Molec Syst* 1970s to 1990s. *Annu Rev Ecol Syst* 24: 243–278.

Sheppard, P.M., Turner, J.R.G., Brown, K.S., Benson, W.W. and Singer, M.C. (1985) Genetics and the evolution of Muellerian mimicry in *Heliconius* butterflies. *Phil Trans R Soc Lond B* 308: 433–613.

Shimamura, M., Yasue, H., Ohahima, K., Abe, H., Kato, H., Kishiro, T., Goto, M., Munechika, I. and Okada, N. (1997) Molecular evidence from retroposons that whales form a clade within even-toed ungulates. *Nature* 388: 666–670.

Shubin, N.H. (1994) The phylogeny of development and the origin of homology. *In*: L. Grande and O. Rieppel (eds): *Interpreting the Hierarchy of Nature*. Academic Press, San Diego, pp. 201–225.

Shultz, J. (1987) The origin of the spinning apparatus in spiders. *Biol Rev* 62: 89–113.

Shultz, J.W. and Regier, J.C. (2000) Phylogenetic analysis of arthropods using two nuclear protein-encoding genes supports a crustacean + hexapod clade. *Proc R Soc Lond B* 267: 1011–1019.

Sibley, C.G. (1960) The electrophoretic patterns of egg-white proteins as taxonomic characters. *Ibis* 102: 215–284.

Sibley, C.G. and Ahlquist, J.E. (1988) A classification of the living birds of the world based on DNA-DNA hybridization studies. *Auk* 105: 409–423.

Sibley, C.G. and Ahlquist, J.E. (1990) *Phylogeny and Classification of Birds. A Study in Molecular Evolution*. Yale University Press, New Haven.

Siddall, M.E. (1997) Prior agreement: Arbitration or arbitrary. *Syst Biol* 46: 765–769.

Siddall, M.E. (1998) Success of parsimony in the four-taxon case: Long branch repulsion by likelihood in the Farris zone. *Cladistics* 14: 209–220.

Siddall, M.E. and Kluge, A.G. (1997) Probabilism and phylogenetic inference. *Cladistics* 13: 313–336.

Siddall, M.E. and Whiting, M. (1999) Long-branch abstractions. *Cladistics* 15: 9–24.

Sidow, A. (1996) Genome duplications in the evolution of early vertebrates. *Curr Opin Genet Develop* 6: 715–722.

Siegel-Causey, D. (1997) Phylogeny of the Pelecaniformes: molecular systematics of a privative group. *In*: D.P. Mindell (ed.): *Avian Molecular Evolution and Systematics*. Academic Press, New York, pp. 159–171.

Simmons, A., Ray, E. and Jelinski, L. (1994) Solid state ^{13}C NMR of *N. clavipes* dragline silk establishes structure and identity of crystalline regions. *Macromolecules* 27: 5235–5237.

Simon, C., Frati, F., Beckenbach, A., Crespi, B., Liu, H. and Flook, P. (1994) Evolution, weighting, and phylogenetic utility of mitochondrial gene sequences and a compilation of conserved polymerase chain reaction primers. *Annu Rev Entomol* 87: 651–701.

Simonet, M., Riot, B., Fortineau, N. and Berche, P. (1996) Invasin production by *Yersinia pestis* is abolished by insertion of an IS200-like element within the inv gene. *Infect Immunity* 64: 375–379.

Simpson, G.G. (1961) *Principles of animal taxonomy*. Columbia University Press, New York.

Slowinski, J.B. (1998) The number of multiple alignments. *Mol Phylogenet Evol* 10: 264–266.

Slowinski, J.B. and Page, R. (1999) How should species phylogenies be inferred from sequence data? *Syst Biol* 48: 814–825.

Smith, D.C. and Douglass, A.E. (1987) *Biology of symbiosis*. Edward Arnold, London.

Smith, D.J., Burnham, M.K., Bull, J.H., Hodgson, J.E., Ward, J.M., Browne, P., Brown, J., Barton, B., Earl, A.J. and Turner, G. (1990) *Beta*-lactam antibiotic biosynthetic genes have been conserved in clusters in prokaryotes and eukaryotes. *EMBO J* 9: 741–747.

Smith, J.M., Dowson, C.G. and Spratt, B.G. (1991) Localized sex in bacteria. *Nature* 349: 29–31.

Smith, J.M., Smith, N.H., O'Rourke, M. and Spratt, B.G. (1993) How clonal are bacteria?. *Proc Natl Acad Sci USA* 90: 4384–4388.

Smith, M.W. and Doolittle, R.F. (1992) A comparison of evolutionary rates of the two major kinds of superoxide dismutase. *J Mol Evol* 34: 175–184.

Smith, M.W., Feng, D.F. and Doolittle, R.F. (1992) Evolution by acquisition: the case for horizontal gene transfers. *Trends Biochem Sci* 17: 489–493.

Sneath, P.H.A. and Sokal, R.R. (1973) *Numerical taxonomy: the principles and practice of numerical classification*. Freeman, San Francisco.

Snel, B., Bork, P. and Huynen, M.A. (1999) Genome phylogeny based on gene content. *Nat Genet* 21: 108–110.

Sober, E. (1985) A likelihood justification of parsimony. *Cladistics* 1: 209–233.

Sober, E. (1988) *Reconstructing the past: Parsimony, evolution and inference*. MIT Press, Cambridge.

Sokal, R.R. and Rohlf, F.J. (1981) *Biometry*. W.H. Freeman & Co., New York, N.Y.

Song, Y. and Fambrough, D. (1994) Molecular evolution of the calcium-transporting ATPases analyzed by the maximum parsimony method. *In*: D.M. Fambrough (ed.): *Molecular Evolution of*

Physiological Processes. Rockefeller University Press, New York, pp. 271–283.

Soullier, S., Jay, P., Poulat, F., Vanacker, J.,-M., Berta, P. and Laudet, V. (1999) Diversification of the HMG and SOX family members during evolution. *J Mol Evol* 48: 517–527.

Souza, V., Nguyen, T.T., Hudson, R.R., Pinero, D. and Lenski, R.E. (1992) Hierarchical analysis of linkage disequilibrium in *Rhizobium* populations: Evidence for sex? *Proc Natl Acad Sci USA* 89: 8389–8393.

Sowers, K.R. and Schreier, H.J. (1999) Gene transfer systems for the Archaea. *Trends Microbiol* 7: 212–219.

Spicer, G.S. (1988) Molecular evolution among some *Drosophila* species groups as indicated by two-dimensional electrophoresis. *J Mol Evol* 27: 250–260.

Spieth, H.T. (1982) Behavioral biology and evolution of the Hawaiian picture-winged species group of *Drosophila*. *Evol Biol* 14: 351–437.

Spring, J. (1997) Vertebrate evolution by interspecific hybridisation—are we polyploid? *FEBS Lett* 400: 2–8.

Stalker, H.T. (1972) Intergroup phylogenies in *Drosophila* as determined by comparisons of salivary banding patterns. *Genetics* 70: 457–474.

Stanhope, M., Smith, M., Waddell, V., Porter, C., Shivji, M. and Goodman, M. (1996) Mammalian evolution and the interphotoreceptor retinoid binding protein (IRBP) gene: Convincing evidence for several superordinal clades. *J Mol Evol* 43: 83–92.

Stauffer, S., Coguill, S. and Lewis, R. (1994) Comparison of the physical properties of three silks from *Nephila clavipes* and *Araneus gemmoides*. *J Arachnol* 22: 5–11.

Steel, M.A. (1993) Distributions on bicoloured binary trees arising from the principle of parsimony. *Discrete Appl Math* 41: 245–261.

Steel, M.A., Goldstein, L. and Waterman, M. (1996) A central limit theorem for parsimony length of trees. *Adv Appl Probab* 28: 1051–1071.

Steel, M., Hendy, M.D. and Penny, D. (1992) Significance of the length of the shortest tree. *J Classif* 9: 71–90.

Steel, M.A., Hendy, M.D. and Penny, D. (1998) Reconstructing phylogenies from nucleotide pattern probabilities: A survey and some new results. *Discrete Appl Math* 88: 367–396.

Steel, M.A., Lockhart, P.J. and Penny, D. (1993b) Confidence in evolutionary trees from biological sequence data. *Nature* 364: 440–442.

Steel, M.A., Lockhart, P.J. and Penny, D. (1995) A frequency-dependent significance test for parsimony. *Mol Phylogenet Evol* 4: 64–71.

Steel, M.A. and Penny, D. (2000) Parsimony, likelihood and the role of models in molecular phylogenetics. *Mol Biol Evol* 17: 839–850.

Steel, M.A., Penny, D. and Hendy, M.D. (1993a) Parsimony can be consistent! *Syst Biol* 42: 581–587.

Steel, M.A., Székely, L.A. and Hendy, M.D. (1994) Reconstructing trees from sequences whose sites evolve at variable rates. *J Comp Biol* 1: 153–163.

Sterling, T., Salmon, J., Becker, D. and Savarese, D. (1999) *How to Build a Beowulf. A guide to the implementation and application of PC Clusters*. MIT Press, Cambridge.

Stiller, J.W.H. and Hall, B.D. (1999) Comment. *Science* 286: 1443.

Strong, E.E. and Lipscomb, D. (1999) Character coding and inapplicable data. *Cladistics* 15: 363–372.

Sueoka, N. (1959) A statistical analysis of deoxyribonucleic acid distributionin density gradient centrifugation. *Proc Natl Acad Sci USA* 45: 1480–1490.

Sueoka, N. (1962) On the genetic basis of variation and heterogeneityof DNA base composition. *Proc Natl Acad Sci USA* 48: 582–592.

Sueoka, N. (1988) Directional mutation pressure and neutral molecular evolution. *Proc Natl Acad Sci USA* 85: 2653–2657.

Sueoka, N. (1992) Directional mutation pressure, selective constraints, and genetic equilibria. *J Mol Evol* 34: 95–114.

Sueoka, N. (1993) Directional mutation pressure, mutator mutations, and dynamics of molecular evolution. *J Mol Evol* 37: 137–153.

Sullivan, J. (1996) Combining data with different distributions of among-site rate variation. *Syst Biol* 35: 470–488.

Sullivan, J. and Swofford, D. (1997) Are guinea pigs rodents? The importance of adequate models in molecular phylogenetics. *J Mammal Evol* 4: 77–86.

Sumi, M., Sato, M.H., Denda, K., Date, T. and Yoshida, M. (1992) A DNA fragment homologous to

F1-ATPase beta subunit was amplified from genomic DNA of *Methanosarcina barkeri*. Indication of an archaebacterial F-type ATPase. *FEBS Lett* 314: 207–210.

Swanson, W. and Vacquier, V. (1998) Concerted evolution in an egg receptor for a rapidly evolving abalone sperm. *Protein Sci* 281: 710–712.

Swofford, D.L., Olsen, G.J., Waddell, P.J. and Hillis, D.M. (1996) Phylogenetic inference. *In*: D.M. Hillis, C. Moritz and B.K. Mable (eds): *Molecular systematics*. Sinauer Associates, Sunderland, pp. 407–514.

Swofford, D.L. (1999) *PAUP*: Phylogenetic Analysis Using Parsimony (* and Other Methods)*, software and documentation. Sinauer, Sunderland.

Syvanen, M. (1985) Cross-species gene transfer; Implications for a new theory of evolution. *J Theor Biol* 112: 333–343.

Syvanen, M. (1994) Horizontal gene transfer: evidence and possible consequences. *Annu Rev Genet* 28: 237–261.

Székely, L.A. and Steel, M.A. (1999) Inverting random functions. *Ann Combin* 3: 103–113.

Takaya, N.K., Kobayashi, M. and Shoun, H. (1998) Fungal denitrification, a respiratory system possibly acquired by horizontal gene transfer from prokaryotes. *In*: M. Syvanen and C.I. Kado (ed.) *Horizontal gene transfer*. Chapman and Hall, London-New York, pp. 321–337.

Tanaka, H., Ren, F., Okayama, T. and Gojobori, T. (1999) Topology selection in unrooted molecular phylogenetic tree by mimimum model-based complexity. *Pac Symp Biocomput* 4: 326–337.

Tatusov, R.L., Koonin, E.V. and Lipman, D.J. (1997) A genomic perspective on protein families. *Science* 278: 631–637.

Tautz, D. (1999) Evolutionary biology; Debatable homologies. *Nature* 395: 17–19.

Teichmann, S.A. and Mitchison, G. (1999) Is there a phylogenetic signal in prokaryote proteins? *J Mol Evol* 49: 98–107.

Templeton, A.R. (1985) The phylogeny of the hominoid primates: A statistical analysis of the DNA-DNA hybridization data. *Mol Biol Evol* 2: 420–433.

Templeton, A.R. (1998) Nested clade analyses of phylogeographic data: Testing hypotheses about gene flow and population history. *Mol Ecol* 7: 381–397.

Tettelin, H., Saunders, N.J., Heidelberg, J., Jeffries, A.C., Nelson, K.E., Eisen, J.A., Ketchum, K.A., Hood, D.W., Peden, J.F., Dodson, R.J. et al. (2000) Complete genome sequence of *Neisseria meningitidis* serogroup B strain MC58. *Science* 287: 1809–1815.

Thomas, H. (1994) Anatomie cranienne et relations phylogenetiques du nouveau bovide (*Pseudoryx nghetinhensis*) decouvert dans la cordillere annamitique au Vietnam. *Mammalia* 58: 453–481.

Thomas, R.H. and Hunt, J.A. (1991) The molecular evolution of the alcohol dehydrogenase locus and the phylogeny of Hawaiian *Drosophila*. *Mol Biol Evol* 8: 687–702.

Thompson, J.D., Gibson, T.J., Plewniak, F., Jeanmougin, F. and Higgins, D.G. (1997) The CLUSTAL_X windows interface: flexible strategies for multiple sequence alignment aided by quality analysis tools. *Nucl Acid Res* 25: 4876–4882.

Thompson, J.D., Higgins, D.G. and Gibson, T.J. (1994) CLUSTAL W: Improving the sensitivity of progressive multiple sequence alignment through sequence weighting, position specific gap penalties and weight matrix choice. *Nucl Acid Res* 22: 4673–4680.

Thompson, J.D., Higgins, D.G. and Gibson, T.J. (1996) Clustal W, version 1.6. Program and documentation.

Thorne, J.L., Kishino, H. and Felsenstein, J. (1991) An evolutionary model for maimum likelihood alignment of D sequences. *J Mol Evol* 33: 114–124.

Thorne, J.L. and Churchill, G.A. (1995) Estimation and reliability of molecular sequence alignments. *Biometrics* 51: 100–113.

Thorne, J.L., Goldman, N. and Jones, D.T. (1996) Combining protein evolution and secondary structure. *Mol Biol Evol* 13: 666–673.

Thorne, J.L. and Kishino, H. (1992) Freeing phylogenies from artifacts of alignment. *Mol Biol Evol* 9: 1148–1162.

Thorne, J.L., Kishino, H. and Felsenstein, J. (1992) Inching toward reality: an improved likelihood model of sequence evolution. *J Mol Evol* 34: 3–16.

Thornton, J.W. and DeSalle, R. (2000) A new method to localize and test the significance of incongruence: detecting domain-shuffling in the nuclear receptor superfamily. *Syst Biol* 49: 183–201.

Thornton, J.W. and Kelley, D.B. (1998) Evolution of the androgen receptor: structure-function implications. *Bioessays* 20: 860–860.

Throckmorton, L.H. (1966) The relationships of the endemic Hawaiian Drosophilidae. *Univ TX Publ*

6615: 335–396.

Throckmorton, L.H. (1975) The phylogeny, ecology and geography of *Drosophila. In*: R.C. King (ed.): *Handbook of Genetics: Invertebrates of Genetic Interest.* Plenum Publishing Co., New York, pp. 421–469.

Tiboni, O., Cammarano, P. and Sanangelantoni, A.M. (1993) Cloning and sequencing of the gene encoding glutamine synthetase I from the archaeum *Pyrococcus woesei*: Anomalous phylogenies inferred from analysis of archaeal and bacterial glutamine synthetase I sequences. *J Bacteriol* 175: 2961–2969.

Titus, T.A. and Frost, D.R. (1996) Molecular homology assessment and phylogeny in the lizard family Opluridae (Squamata: Iguania). *Mol Phylogenet Evol* 6: 49–62.

Tsuji, S., Qureshi, M.A., Hou, E.W., Fitch, W.M. and Li SS (1994) Evolutionary relationships of lactate dehydrogenases (LDHs) from mammals, birds, an amphibian, fish, barley, and bacteria: LDH cDNA sequences from *Xenopus*, pig, and rat. *Proc Natl Acad Sci USA* 91: 9392–9396.

Tuffley, C. and Steel, M. (1997a) Modeling the covarion hypothesis of nucleotide substitution. *Math Biosci* 147: 63–91.

Tuffley, C. and Steel, M. (1997b) Links between maximum likelihood and maximum parsimony under a simple model of site substitution. *Bull Math Biol* 59: 581–607.

Turbeville, J.M., Field, K.G. and Raff, R.A. (1992) Phylogenetic position of phylum Nemertini, inferred from 18S rRNA sequences: Molecular data as a test of morphological character homology. *Mol Biol Evol* 9: 235–249.

Turbeville, J.M., Pfeifer, D.M., Field, K.G. and Raff, R.A. (1991) The phylogenetic status of arthropods, as inferred from 18S rRNA sequences. *Mol Biol Evol* 8: 669–686.

Turbeville, J.M., Schulz, J.R. and Raff, R.A. (1994) Deuterostome phylogeny and the sister group of the chordates: evidence from molecules and morphology. *Mol Biol Evol* 11: 648–655.

Turner, J.R.G., Johnson, M.S. and Eanes, W.F. (1979) Contrasted modes of evolution in the same genome: allozymes and adaptive change in *Heliconius. Proc Natl Acad Sci USA* 76: 1924–1928.

Urry, D., Luan, C.,-H. and Peng, S. (1995) Molecular biophysics of elastin structure, function, and pathology. *Ciba Found Symp* 192: 4–30.

Ursing, B. and Arnason, U. (1998a) Analyses of mitochondrial genomes strongly support a hippopotamus-whale clade. *Proc R Soc Lond B* 265: 2251–2255.

Ursing, B. and Arnason, U. (1998b) The complete mitochondrial DNA sequence of the pig (*Sus scrofa*). *J Mol Evol* 47: 302–306.

Van de Peer, Y., Rensing SA, Maier, U.,-G. and De Wachter, R. (1996) Substitution rate calibration of small subunit ribosomal subunit RNA identifies Chlorarachnida nucleomorphs as remnants of green algae. *Proc Natl Acad Sci USA* 93: 7732–7736.

Van Dijk, A., De Boef, E., Bekkers, A., Van Wijk, L., Van Swieten, E., Hamer, R. and Robillard, G. (1997) Structure characterization of the central repetitive domain of high molecular weight gluten proteins. II. Characterization in solution and the dry state. *Protein Sci* 6: 649–656.

van Tuinen, M., Sibley, C.G. and Hedges, S.B. (2000) The early history of modern birds from DNA sequences of nuclear and mitochondrial ribosomal genes. *Mol Biol Evol* 17: 451–457.

van Valen, L.V. (1982) Homology and causes. *J Morphol* 173: 305–312.

Vellai, T., Kovács, A.L., Kovács, G., Ortutay, C. and Vida, G. (1999) Genome economization and a new approach to the species concept in bacteria. *Proc. Biol. Sci.* 266: 1953–1958.

Vogler, A.P. and Kelley, K.C. (1996) At the interface of phylogenetics and ecology: the case of chemical defenses in *Cicindela. Ann Zool Fenn* 33: 39–47.

Vogler, A.P. and Kelley, C.K. (1998) Covariation of defensive traits in *Cicindela* tiger beetles: A phylogenetic approach using mtDNA. *Evolution* 52: 357–366.

Vogler, A.P. and Goldstein, P.Z. (1997) Adaptation, cladogenesis, and the evolution of habitat association in North American tiger beetles: A phylogenetic perspective. *In*: T. Givnish and K. Systma (eds): *Molecular evolution and adaptive radiation.* Cambridge University Press, Cambridge, pp. 353–373.

Vollrath, F. and Edmonds, D. (1989) Modulation of the mechanical properties of spider silk by coating with water. *Nature* 340: 305–307.

Von Dohlen, C.D. and Moran, N.A. (1995) Molecular phylogeny of the Homoptera: A paraphyletic taxon. *J Mol Evol* 41: 211–223.

Vulic, M., Dionisio, F., Taddei, F. and Radman, M. (1997) Molecular keys to speciation: DNA polymorphism and the control of genetic exchange in enterobacteria. *Proc Natl Acad Sci USA* 94: 9763–9767.

Vulic, M., Lenski, R.E. and Radman, M. (1999) Mutation, recombination, and incipient speciation of bacteria in the laboratory. *Proc Natl Acad Sci USA* 96: 7348–7351.

Waddell, P.J. (1996) *Statistical methods of phylogenetic analysis*. PhD thesis. Massey University, Palmerston North, New Zealand.

Waddington, C.H. (1940) *Organizers and Genes*. Cambridge University Press, Cambridge.

Wägele, J.W. and Stanjek, G. (1995) Arthropod phylogeny inferred from partial 12S rRNA revisited: Monophyly of the Tracheata depends on sequence alignment. *J Zool Syst Evol Res* 33: 75–80.

Wagner, G.P. (1989a) The biological homology concept. *Annu Rev Ecol Syst* 20: 51–69.

Wagner, G.P. (1989b) The origin of morphological characters and the biological basis of homology. *Evolution* 43: 1157–1171.

Wainright, P.O., Hinkle, G., Sogin, M.L. and Stickel, S.K. (1993) Monophyletic origins of the Metazoa: An evolutionary link with fungi. *Science* 260: 340–342.

Waldor, M.K. (1998) Bacteriophage biology and bacterial virulence. *Trends Microbiol* 6: 295–297.

Waldor, M.K. and Mekalanos, J.J. (1996) Lysogenic conversion by a filamentous phage encoding cholera toxin. *Science* 272: 1910–1914.

Waterman, M.S., Eggert, M. and Lander, E. (1992) Parametric sequence comparisons. *Proc Natl Acad Sci USA* 89: 6090–6093.

Watrous, L.E. and Wheeler, Q.D. (1981) The out-group comparison method of character analysis. *Syst Zool* 30: 1–11.

Wenzel, J.W. and Carpenter, J.M. (1994) Comparing methods: Adaptive traits and tests of adaptation. *In*: P. Eggleton and R. Vane-Wright (eds): *Phylogenetics in ecology*. Harcourt Brace, London, pp. 79–101.

Wetterer, J.K., Schultz, T.R. and Meier, R. (1998) Phylogeny of fungus-growing ants (Tribe Attini) based on mtDNA sequence and morphology. *Mol Phylogenet Evol* 9: 42–47.

Wheeler, Q.D. (1990) Ontogeny and character phylogeny. *Cladistics* 6: 225–268.

Wheeler, Q.D. (1995) The "Old Systematics": Classification and phylogeny. *In*: J. Pakaluk and S.A. Slipinski (eds): *Biology, phylogeny and classification of Coleoptera: papers celebrating the 80th birthday of Roy A. Crowson*. Muzeum i Instytut Zoologii PAN, Warszawa, pp. 31–62.

Wheeler, W.C. (1990) Nucleic acid sequence phylogenies and random outgroups. *Cladistics* 6: 363–367.

Wheeler, W.C. (1992) Extinction, sampling, and molecular phylogenetics. *In*: M.J. Novacek and Q.D. Wheeler (eds): *Extinction and phylogeny*. Columbia University Press, New York, pp. 205–215.

Wheeler, W.C. (1993) The triangle inequality and character analysis. *Mol Biol Evol* 10: 707–712.

Wheeler, W.C. (1994) Sources of ambiguity in nucleic acid sequence alignment. *In*: B. Schierwater, B. Streit, G.P. Wagner and R. DeSalle (eds): *Molecular Ecology and Evolution: Approaches and Applications*. Birkhäuser, Basel, pp. 323–352.

Wheeler, W.C. (1995) Sequence alignment, parameter sensitivity, and the phylogenetic analysis of molecular data. *Syst Biol* 44: 321–331.

Wheeler, W.C. (1996) Optimization alignment: the end of multiple sequence alignment in phylogenetics? *Cladistics* 12: 1–9.

Wheeler, W.C. (1998) Alignment characters, dynamic programming and heuristic solutions. *In*: R. DeSalle and B. Schierwarter (eds): *Molecular Approaches to Ecology and Evolution*. Birkhäuser Verlag, Basel, pp. 243–251.

Wheeler, W.C. (1999a) Measuring topological congruence by extending character techniques. *Cladistics* 15: 131–135.

Wheeler, W.C. (1999b) Fixed character states and the optimization of molecular sequence data. *Cladistics* 15: 379–386.

Wheeler, W.C. (2000a) Heuristic reconstruction of hypothetical-ancestral DNA sequences: sequence alignment vs direct optimization. *In*: R. Scotland and R.T. Pennington (eds): *Homology and Systematics*, Taylor and Francis, London, pp. 106–113.

Wheeler, W.C. (2000b) Homology and DNA sequence data. *In*: G.P. Wagner (ed.): *The character concept in evolutionary biology*, Academic Press, San Diego, pp. 303–317.

Wheeler, W.C., Cartwright, P. and Hayashi, C. (1993) Arthropod phylogeny: A combined approach. *Cladistics* 9: 1–39.

Wheeler, W.C., Gatesy, J. and DeSalle, R. (1995) Elision: A method for accommodating multiple molecular sequence alignments with alignment-ambiguous sites. *Mol Phylogenet Evol* 4: 1–9.

Wheeler, W.C. (2002) Optimization alignment: down, up, error and improvements. *In*: R. DeSalle et al. (eds): *Techniques in molecular systematics and evolution*. Birkhäuser Verlag, Basel,

Switzerland, pp. 55–69.

Wheeler, W.C. and Gladstein, D.S. (1994) MALIGN: a multiple sequence alignment program. *J Hered* 85: 417–418.

Wheeler, W.C. and Gladstein, D.S. (1995) Malign 2.7 (software and documentation). American Museum of Natural History, New York.

Wheeler, W.C. and Gladstein, D.S. (2000) MALIGN 2.7, Parallel version 1.5. American Museum of Natural History, New York. ftp://ftp.amnh.org/pub/molecular/malign.

Wheeler, W.C. and Hayashi, C. (1998) The phylogeny of extant chelicerate orders. *Cladistics* 14: 173–192.

Whiting, M.F. (1998a) Long-branch distraction and the Strepsiptera. *Syst Biol* 47: 134–138.

Whiting, M.F. (1998b) Phylogenetic position of the Strepsiptera: Rev. molecular and morphological evidence. *Int J Morphol Embryol* 27: 53–60.

Whiting, M.F., Carpenter, J., Wheeler, Q. and Wheeler, W. (1997) The Strepsiptera problem: Phylogeny of the holometabolous insect orders inferred from 18S and 28S ribosomal DNA sequences and morphology. *Syst Biol* 46: 1–68.

Whiting, M.F. and Wheeler, W.C. (1994) Insect homeotic transformation. *Nature* 368: 696.

Whittam, T.S. and Ake, S. (1992) Genetic polymorphisms and recombination in natural populations of *Escherichia coli. In*: N. Takahata and A.G. Clark (eds): *Mechanisms of molecular evolution: introduction to molecular paleopopulation biology*. Japan Scientific Society Press, Tokyo, pp. 223–246.

Whittam, T.S., Ochman, H. and Selander, R.K. (1983a) Geographic components of linkage disequilibrium in natural populations of *Escherichia coli. Mol Biol Evol* 1: 67–83.

Whittam, T.S., Ochman, H. and Selander, R.K. (1983b) Multilocus genetic structure in natural populations of *Escherichia coli. Proc Natl Acad Sci USA* 80: 1751–1755.

Wiley, E.O. (1978) The evolutionary species concept reconsidered. *Syst Zool* 27: 17–26.

Wilkie, T.M., Gilbert, D.J., Olsen, A.S., Chen, X.N., Amatruda, T.T., Korenberg, J.R., Trask, B.J., deJong, P., Reed, R., Simon, M.I. et al. (1992) Evolution of the mammalian G protein alpha-subunit multigene family. *Nat Genet* 1: 85–92.

Wilkinson, H.H., Spoerke, J.M. and Parker, M.A. (1996) Divergence in symbiotic compatibility in a legume-*Bradyrhizobium* mutualism. *Evolution* 50: 1470–1477.

Williams, P.L. and Fitch, W.M. (1989) Finding the minimal change in a given tree. *In*: B. Fernhölm, K. Bremer, H. Jurnvall (eds): *The hierarchy of life. Molecules and morphology in phylogenetic analysis*. Excerpta Medica, Amsterdam-New York-Oxford, pp. 453–470.

Willmann, R. (1987) The phylogenetic system of the Mecoptera. *Syst Entomol* 12: 519–524.

Wilson, A.C., Carlson, S.S. and White, T.J. (1977) Biochemical evolution. *Annu Rev Biochem* 46: 573–639.

Wilson, E.O. (1988) The current state of biological diversity. *In*: E.O. Wilson (ed.): *Biodiversity*. National Academy Press, Washington DC, pp. 3–18.

Winnepenninckx, B. and Backeljau, T. (1996) 18S rRNA alignments derived from different secondary structure models can produce alternative phylogenies. *J Zool Syst Evol Res* 34: 135–143.

Winnepenninckx, B., Backeljau, T. and De Wachter, R. (1995) Phylogeny of protostome worms derived from 18S rRNA sequences. *Mol Biol Evol* 12: 641–649.

Winnepenninckx, B., Backeljau, T. and Kristensen, R.M. (1998) Relations of the new phylum Cycliophora. *Nature* 393: 636–638.

Woese, C.R. (1987) Bacterial evolution. *Microbiol Rev* 51: 221–271.

Woese, C.R. (1998) The universal ancestor. *Proc Natl Acad Sci USA* 95: 6854–6859.

Woese, C.R. and Fox, G.E. (1977) Phylogenetic structure of the prokaryotic doma*In*: The primary kingdoms. *Proc Natl Acad Sci USA* 74: 5088–5090.

Woese, C.R., Kandler, O. and Wheelis, M.L. (1990) Towards a natural system of organisms: Proposal for the domains Archaea, Bacteria, and Eucarya. *Proc Natl Acad Sci USA* 87: 4576–4579.

Woodger, J.H. (1937) *The Axiomatic Method in Biology*. Cambridge University Press, London.

Work, R. and Young, C. (1987) The amino acid compositions of major and minor ampullate silks of certain orb-web weaving spiders (Araneae, Araneidae). *J Arachnol* 15: 65–80.

Wray, G.A., Levinton, J.S. and Shapiro, L.H. (1996) Molecular evidence for deep Precambrian divergences among metazoan phyla. *Science* 274: 568–573.

Xu, M. and Lewis, R. (1990) Structure of a protein superfiber: spider dragline silk. *Proc Natl Acad Sci USA* 87: 7120–7124.

Xu, X. and Arnason, U. (1994) The complete mitochondrial DNA sequence of the horse, *Equus cabal-*

lus: Extensive heteroplasmy of the control region. *Gene* 148: 357–362.

Yang, Z. (1994) Statistical properties of the maximum likelihood method of phylogenetic estimation and comparison with distance matrix methods. *Syst Biol* 43: 329–342.

Yang, Z. (1996) Phylogenetic analysis using parsimony and likelihood methods. *J Mol Evol* 42: 294–307.

Yang, Z. and Rannala, B. (1997) Bayesian phylogenetic inference using DNA sequences: A Markov chain Monte Carlo method. *Mol Biol Evol* 14: 717–724.

Yanofsky, C. and Le, S. (1959) Transduction and recombination study of linkage relationshipsamong the genes controlling tryptophan synthesisin *Escherichia coli*. *Virology* 8: 425–447.

Yap, W.H., Zhang, Z. and Wang, Y. (1999) Distinct types of rRNA operons exist in the genome of the actinomycete *Thermomonospora chromogena* and evidence for horizontal transfer of an entire rRNA operon. *J Bacteriol* 181: 5201–5209.

Yeates, D.K. and Wiegmann, B.M. (1999) Congruence and controversy: Toward a higher-level phylogeny of the Diptera. *Annu Rev Entomol* 44: 397–428.

Yeates, D.K. (1995) Groundplans and exemplars: paths to the tree of life. *Cladistics* 11: 343–357.

Yokoyama, S. and Radlwimmer, F.B. (1999) The molecular genetics of red and green color vision in mammals. *Genetics* 153: 919–932.

Zar, J. (1974) *Biostatistical Analysis*. Prentice-Hall, Englewood Cliffs.

Zhang, D. and Hewitt, G.M. (1996) Nuclear integrations: challenges for mitochondrial DNA markers. *TREE* 11: 247–251.

Ziebuhr, W., Ohlsen, K., Karch, H., Korhonen, T. and Hacker, J. (1999) Evolution of bacterial pathogenesis. *Cell Mol Life Sci* 56: 719–728.

Zinder, N.D. and Lederberg, J. (1952) Genetic exchange in *Salmonella*. *J Bacteriol* 679–699.

Zrzavy, J., Mihulka, S., Kepka, P., Bezdek, A. and Tietz, D. (1998) Phylogeny of the Metazoa based on morphological and 18S ribosomal DNA evidence. *Cladistics* 14: 249–285.

Subject index